T0256037

The Standard Model in a Nutshell

The Standard Model in a Nutshell

Dave Goldberg

PRINCETON UNIVERSITY PRESS · PRINCETON AND OXFORD

Names: Goldberg, Dave, 1974– author.
Title: The standard model in a nutshell / Dave Goldberg.
Other titles: In a nutshell (Princeton, N.J.)
Description: Princeton, New Jersey ; Oxford : Princeton University Press,
    [2017] | Series: In a nutshell | Includes bibliographical references and
    index.
Identifiers: LCCN 2016040024 | ISBN 9780691167596 (hardcover ; alk. paper) |
    ISBN 0691167591 (hardcover ; alk. paper)
Subjects: LCSH: Standard model (Nuclear physics) | Particles (Nuclear
    physics) | Symmetry (Physics)
Classification: LCC QC794.6.S75 G65 2017 | DDC 539.7/2—dc23 LC record available at
    https://lccn.loc.gov/2016040024

British Library Cataloging-in-Publication Data is available

This book has been composed in Scala

Printed on acid-free paper. ∞

Typeset by Nova Techset Pvt Ltd, Bangalore, India
Printed in the United States of America

1 3 5 7 9 10 8 6 4 2

Contents

Preface for Instructors

For us physicists, the Standard Model is part of our everyday vocabulary. We might make casual reference to fundamental particles and forces with the expectation that our students should have picked them up at some point in their undergraduate education. However, it is rare for a curriculum to spend a full term addressing the question of what the Standard Model *is* and why, given the complexity of the particle zoo, it's supposed to be so elegant.

This work is a study in just-in-time instruction. It was developed in response to a very real need to give students context for the rest of their education. Thus, while many key concepts are derived rigorously, others are motivated by simple examples and appeals to reasonableness. This book is, in the most literal sense, intended to serve as the Standard Model in a nutshell. Students are expected to come away with not only an appreciation of the beauty of the Model but a recognition of the many remaining problems therein.

I first started writing this book because I was teaching a course to an advanced, but general, physics audience, and neither the minutiae of quantum field theory nor a focus on particle phenomenology seemed right for the audience. Those tended to be the approaches of the extant textbooks. My hope is that you might design your course similarly—as an advanced survey for cosmology or astrophysics students, or uncommitted theorists of any stripe.

This book is intended to serve a stand-alone, one-term course for advanced undergraduates and first- and second-year graduate students in physics who have already seen the following:

1. Classical electromagnetism
2. Classical mechanics, including Lagrangians
3. Nonrelativistic quantum mechanics

That's it.

I don't assume any knowledge of particle physics phenomenology, special relativity, relativistic quantum mechanics, group theory, or quantum field theory. If your curriculum differs, I anticipate a couple of other paths, including the following.

The "Classical Only" Sequence

Virtually all discussion of quantum field theory can be saved until a later course. This involves excising §5.7 as well as §7.3 and 7.4 (the initial sections, on Fermi's golden rule, remain to motivate the importance of a scattering amplitude in general). For the weak and strong interactions, instructors may skip §8.5, 9.3.3, 9.4, 10.1.4, and 11.1.4 through the end of Chapter 11.

The "Advanced Background" Sequence

While many courses will be aimed at a joint undergraduate and graduate student audience, some instructors may focus their courses on a more advanced audience. In that case, so long as students are comfortable with tensor and 4-vector notation, Chapter 1 may be skipped entirely, with the exception of §1.3.2 on natural units; §2.1 and §3.1 may also be skipped for students with a very strong background in classical mechanics. While most physics curricula do not require group theory at either the undergraduate or graduate level, for those that include a discussion of Lie groups and generators, Chapter 4 can be skipped with few consequences, though §4.4 and 4.5 on SU(2) and SU(3) are still likely to be useful.

While some graduate students may be comfortable with the Dirac equation, some care should be taken with the decision to skip Chapter 5, as the chiral representation, the symmetry properties of a Dirac field, and the quantization of the field are all likely to be new even to students who have seen relativistic quantum mechanics.

This work is not the end of the story, especially for students who want to go on to particle physics research. While I think it forms a strong foundation, departments are encouraged to develop this as the first course in a sequence which might include experimental particle physics or advanced quantum field theory—topics which students will likely take to more easily with a solid background in their motivation.

Acknowledgments

This book has been a labor of love. While I first became interested in the deep question of symmetry and classical fields as a cosmologist, I didn't appreciate their true strength until I wrote about them for a popular audience in my last book. I "road-tested" this material with the wonderful graduate and undergraduate students in the Drexel physics program, and I am deeply indebted to my students: Matthias Agne, Eric Carchidi, Jeremy Gaison, Bao Huynh, Mike Jewell, Cindy Lin, David Lioi, Kat Netherton, Sean Robinson, Mike Schlenker, Courtney Slocum, Tori Tielebein, Lise Wills, Megan Wolfe, and Jacob Zettlemoyer. My current class, Kelley Commeford, Dan Douglas, Mark Giovinazzi, T. J. McSorley, Alex Morrese, Tyler Rehak, Tyler Reisinger, Ben Relethford, Jim Streuli, Joe Tomlinson, Charles Unruh, and Joe Wraga, along with my colleague Jim McCray, had the dubious pleasure of learning from the semifinal manuscript. I appreciate their flyspecking and recommendations for the many points that could use additional clarification.

I am also thankful to Sean Carroll, Michelle Dolinski, Bob Gilmore, John Peacock, Naoko Kurahashi Neilson, and Mark Trodden for immensely useful comments and discussions on early drafts. I am grateful to the developers of the feynmf LaTeX package, without whom I would have been lost in a morass of hand-drawn Feynman diagrams, and likewise to Howard Georgi, who was kind enough to contribute his portrait. The manuscript and my sanity have been greatly improved by the ministrations of my agent, Andrew Stuart; my editorial team, Ingrid Gnerlich, Barbara Liguori, Nathan Carr, and Eric Henney; and the thoughtful comments of the anonymous referees.

I am always grateful for my wonderful wife, Emily Joy, whose unfailing love and support have made this possible.

Introduction

If you've made it this far in your physics education, you may have been struck by the realization that as elegant as you may find Lagrangian mechanics or Maxwell's equations or the Schrödinger wave equation, there *must* be something deeper underneath.

Along the way, you may well have heard of something called the *Standard Model* of particle physics. It is normally spoken of, quite rightly in my opinion, in a tone of hushed reverence. If you've encountered the Standard Model only in passing, you may be underwhelmed. It's usually represented as a ranked list of fundamental interactions: *strong, electromagnetic, weak*, and (if it must be mentioned at all in this context) *gravity*.[1] The Standard Model is also a collection of particles and how they respond (or don't) to those fundamental forces (Figure 1).

For a theory that is meant to be elegant and to do away with so much of the rote memorization that characterizes early courses in physics, the Standard Model can seem to the uninitiated to be just a laundry list of things that happen.

It is anything but.

At its heart, the Standard Model is the theory of the symmetry of empty space, and the rules by which **classical fields** can occupy and interact within that space. You've likely already been exposed to at least one classical field: electromagnetism, the properties of which can described by Maxwell's equations and the Lorentz force law.

We will explore the symmetries of classical fields. Indeed, they will be the central focus of our attention. But we will ultimately need to deal with the quantum nature of the universe—which will in turn give rise to particles.

There are important differences between quantum mechanics and classical fields. Classical systems are deterministic, while quantum systems by necessity contain

---

[1] Gravity is not, in fact, part of the Standard Model at all—an omission that we as a physics community will need to deal with at some point.

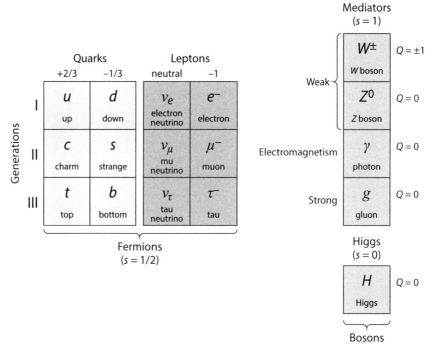

Figure 1. The Standard Model particle zoo. For the moment, "Quarks," "Leptons," "Mediators," and so on, are simply labels. Throughout this course, we'll delve into where this structure comes from.

uncertainty and randomness. But quantum mechanics and classical fields can be unified. For electromagnetism (and the other forces of the Standard Model) we have a *quantum field* version of the theory (QFT), wherein the field is broken down into indivisible chunks: the photons. While our main focus in this book is on the classical side of things, to produce any useful results, we'll need to do a few direct QFT calculations.

Don't fret.

We'll develop just-in-time plausibility relationships to indicate how these calculations should work. Should you wish to do the calculations in greater detail, you can find the *Feynman rules* for doing QFT calculations in Appendix C. Better yet, if you are planning on becoming a particle physicist, you can and should do this course in sequence with a formal QFT course.

We will focus our attention on the Standard Model fields: electromagnetism, the weak interaction, and strong force, as well as unifications among these. We'll see how they are derived from simple statements of symmetry, and along the way, we'll develop an understanding of group theory, Lagrangian mechanics, and symmetry breaking. By the end, we'll be prepared to talk meaningfully about electroweak unification and the Higgs boson, color confinement in the strong force, and what questions remain to be answered.

Symbols

We will use a number of mathematical conventions and symbols in this work. In an effort to maintain consistency, we'll use most symbols (especially Greek symbols) in only one context or, alternatively, in such widely different contexts that the meaning will be clear. Here we present a table of symbols used throughout the text, along with the numbered equation in which each is introduced.

Symbol	Description	Equation First Used
$\mathcal{A}$	The scattering amplitude of a QFT process	7.14
d^4x	The 4-space volume element	2.7
E_{Pl}	The Planck energy	1.24
$F^{\mu\nu}$	The Faraday tensor in electromagnetism	6.13
$\vec{F}^{\mu\nu}$	The SU(2) Faraday tensor	8.23
g_i	Element i of a group	4.1
g_{ij}	The components of a metric tensor	1.8
G_F	The Fermi constant	8.3
g_W	The weak coupling constant	8.7
$\mathbf{I}$	The identity matrix or element	4.2
$J^{\pm}$	The charged weak current	8.25
$\mathcal{L}$	The Lagrangian density	2.7
$\mathbf{M}(\theta)$	The matrix representation of a group element	4.4
p	The 4-momentum of a particle	1.34

Symbol	Description	Equation First Used
$\{q_i\}$	A set of independent degrees of freedom for a dynamic system	2.4
S	The dynamical action	2.1
S_{ij}	The amplitude of transition over infinite time	7.6
$T^{\mu\nu}$	The stress-energy tensor	3.11
T_3	Weak isospin	9.14
u	The 4-velocity of a particle	1.31
$u_s(p)$	The normalized electron spinor basis	5.18
$\hat{U}(t, t_0)$	The unitary evolution operator	7.4
$v_s(p)$	The normalized positron spinor basis	5.19
w	The equation of state (P/ρ) of a fluid	3.26
$W^{\pm}$	The 4-vector describing a W-boson	8.27
x	A 4-vector spacetime coordinate	1.32
$\vec{x}$	A 3-vector spacetime coordinate	1.1
x^i	Component of a 3-vector (italic $i = \{1, 2, 3\}$).	1.4
$\mathbf{X}$	The generator of a symmetry transform	4.5
$\hat{\mathcal{X}}$	The particle exchange operator	5.59
Y_W	The weak hypercharge	9.2
α_e	The fine-structure constant	7.34
δ^i_j	The Kronecker delta function	1.11
ϵ	A continuous parameter for a transformation	3.1
ϵ_{ijk}	The Levi-Civita cyclic tensor	4.15
γ	The relativistic time dilation factor	1.27
γ^{μ}	The gamma matrices in the Dirac equation	5.5
γ^5	The chirality matrix	5.33
$\Lambda^{\bar{i}}_{\phantom{\bar{i}}i}$	The transformation matrix between two frames	1.12
λ_f	The Yukawa coupling constant	9.42
σ_i	The Pauli spin matrices	4.13
τ	The proper time coordinate	1.29
ϕ	The amplitude of a scalar field	2.6
Φ	A multiplet (written as a column vector) of scalar fields	4.10
ψ	A bispinor field	5.12

Symbol	Description	Equation First Used
$\overline{\psi}$	The adjoint spinor	5.17
Ψ	A multiplet (written as a column vector) of bispinor fields	4.21
$[A, B]$	The commutation operator	4.3
$\{A, B\}$	The anticommutation operator	5.6
∂_i	The partial derivative with respect to x^i	1.18
$\circ$	The general "multiplication" operator of group elements	4.1
$\hat{X}$	An operator on a field or wavefunction	2.16
$\Box$	The d'Alembertian, $\partial_\mu \partial^\mu$	2.11
$\displaystyle{\not{p}}$	The contraction of a 4-vector with γ-matrices	5.10

The Standard Model in a Nutshell

1 | Special Relativity

Figure 1.1. Albert Einstein (1879–1955), c. 1947. Einstein developed the principles of special relativity and much else that will be useful in this text.

The Standard Model is a study in symmetry. Throughout this volume, we'll explore different extrinsic and intrinsic symmetry relations, introduce notation for handling them economically, and delve into the physical manifestations of these symmetries. But before we do any of that, it might help if we describe what a symmetry actually is. The mathematician Hermann Weyl [159] had a pithy definition:

> A thing is symmetrical if there is something you can do to it so that after you have finished doing it, it looks the same as before.

The "thing," in the case of the Standard Model is "the laws of physics themselves."

As for what "you can do to it," the list of possible manipulations is almost without limit. These might include shifting every atom in the universe by some fixed displacement, or rotating all creation by some angle around a fixed point. In practice, we can't do either of these things, but they bring to mind the more general questions, Are the laws of physics the same everywhere? and Is there a preferred direction in the universe? respectively.

It's natural, therefore, to begin with a simple question that we *can* explore: Can an observer tell if he or she is moving at a constant rate or standing still? This question forms the basis of relativity, which, in turn, provides the set of ground rules for our development of physical law.

1.1 Galileo

In 1632, Galileo Galilei speculated about the the nature of motion in his *Dialogue Concerning the Two Chief World Systems* [71]:

> Shut yourself up with some friend in the largest room below decks of some large ship. ...
> And casting anything toward your friend, you need not throw it with more force one way than
> another, provided the distances be equal; and leaping with your legs together, you will reach
> as far one way as another. Having observed all these particulars, though no man doubts that,
> so long as the vessel stands still, they ought to take place in this manner, make the ship move
> with what velocity you please, so long as the motion is uniform and not fluctuating this way
> and that. You will not be able to discern the least alteration in all the forenamed effects, nor
> can you gather by any of them whether the ship moves or stands still.

Galileo's main argument was in favor of a heliocentric model of the universe,[1] since one of the chief counterarguments was that if the earth were to travel around the sun, then surely, the argument goes, we'd feel the sense of the motion.

Galileo's insight—and it still informs our understanding of physical space today—is that there is no experiment you can perform that will establish whether you are at rest or whether you are traveling at constant speed and direction, what we know call an inertial **frame of reference**. A frame is a hypothetical construct wherein there are an arbitrarily large number of observers who appear stationary to one another and have calibrated their metersticks and timepieces. A frame, in other words, defines an origin and a set of coordinate axes.

1.1.1 Galilean Relativity

Galileo argued that a coordinate transformation of the form

$$\vec{x}' = \vec{x} + \vec{v}t \tag{1.1}$$

would leave all the equations of physics equally valid. Coordinate transformations of this sort are known as **boosts** and are illustrated in Figure 1.2. In the transformation, the "primed" coordinate represents the measurements determined by an observer boosted by a fixed velocity $\vec{v}$ with respect to the "unprimed" observer, whom we conveniently label as "at rest."

There is *no such thing* as an absolute rest frame, which is rather the point of relativity (Galilean and special). Two different observers can each assert, with equal legitimacy, that he or she is at rest and the other is moving, and nothing in the laws of physics can resolve the dispute one way or another. The two observers each make their measurements in different inertial reference frames, each secure in the consistency of their measurements.

---

[1] An argument that, for the purposes of the current work, we'll consider settled.

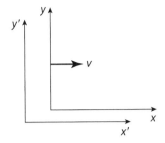

Figure 1.2. A boost transform between two frames.

Within any given frame, we can define the velocity of a particle traveling between two events:[2]

$$\vec{u} \equiv \frac{\Delta \vec{x}}{\Delta t}. \tag{1.2}$$

An event is nothing more than a label corresponding to a particular point in space and time. Everything we're going to do in Galilean and special relativity will revolve around how the coordinates for each event change from one frame to another.

We can ask how fast that particle might be seen to be traveling in another frame:

$$\vec{u}' = \frac{(\Delta \vec{x} + \vec{v}\Delta t)}{\Delta t}$$
$$= \vec{u} + \vec{v}. \tag{1.3}$$

This is exactly what intuition would tell you. An arrow fired at 100 m/s from the back of a plane traveling at 300 m/s (apart from being staggeringly dangerous) will have a net speed of 200 m/s relative to the ground. Further, provided $\vec{v}$ is constant:

$$\frac{d\vec{u}'}{dt} = \frac{d\vec{u}}{dt}.$$

In Galilean relativity, acceleration is manifestly frame independent, which is why it's so central to Newton's second law.

At this point in history, it's hard to feel the shock of this result anymore. It feels natural and intuitive that inertial frames are all equivalent. But the implications are incredibly far-reaching. Whatever fundamental laws govern the universe, they appear to be structured in such a way as to be invariant under a boost.

1.2 **Vectors and Tensors**

Recognizing these symmetries will be incredibly helpful. If we were to try to formulate every possible theory of the universe, we'd be here forever, but anticipating that the final result has to conform to a particular set of guidelines is going to speed things up considerably.

---

[2] In relativity, v is typically reserved to refer to the relative speed between two frames, and u is generally used for velocities within a particular frame.

1.2.1 3-Vector Notation

Our work is slowly but surely taking us away from three-dimensional space and toward a four-dimensional spacetime. Before bringing in time, it will help to clean up our vector notation. Vectors may be written as the sum of coefficients and unit vectors:

$$\vec{v} = \sum_{i=1}^{3} v^i \vec{e}_i. \tag{1.4}$$

We've numbered our various dimensions: v^1, v^2, v^3, where v^2 (for instance) isn't the square of a number but rather the value of the y-component of a vector. Likewise, v^i represents some (any) component of the vector, from one through three. The index label i is totally arbitrary. Any Roman letter will do, and the expression will mean the same thing: in this instance, that v^i (or v^j or what have you) represents the components of a vector. The choice of index matters only within an equation, in that the notation on the left of the equality and the right must match. Unit vectors are also written in a general way, as $\vec{e}_i$. They also have a subscripted index (stay tuned for the significance of upstairs versus downstairs indices).

The summation form of equation (1.4) is still a bit clunky but can be dealt with using a space-saving notation. When there is a matching dummy index on the top and bottom, we may sum terms explicitly using the **Einstein summation convention**

$$\vec{v} = v^i \vec{e}_i.$$

which is identical in content to equation (1.4). As a matter of shorthand, we will use the term v^i to refer to a vector rather than, more properly, to the *components* of the vector. This distinction is much more important in curved coordinate systems than it is here and won't cause too many complications.

Our study of fields will introduce objects more complicated than vectors, including those with a downstairs index, called **one-forms**. Fortunately, the Einstein summation convention works for any combination of terms. We can sum over matching indices upstairs and downstairs. For instance,

$$A_i B^i = A_1 B^1 + A_2 B^2 + A_3 B^3 \tag{1.5}$$

regardless of what A_i represents.

This result looks very much like a dot product, and indeed it is. But before we get into how dot products work in general (and answer the nagging questions about what a downstairs index really means), we need to delve into the world of **tensors.**

Example 1.1: Consider a vector $A^i = \begin{pmatrix} 2 \\ 3 \\ -1 \end{pmatrix}$ and a one-form $B_j = (0\ \ 2\ \ 1)$. Compute $A^k B_k$.

Solution: The specific choice of index label doesn't matter. Relabeling the indices A^k and B_k in the sum is arbitrary. However, it *is* important that the contracted vectors have the same dummy index. This computation yields a scalar:

$$A^k B_k = (2 \cdot 0) + (3 \cdot 2) + (-1 \cdot 1) = 5$$

1.2.2 *A Few Rules about Tensors*

You likely have a pretty good sense of what a vector is: it's an object with both a magnitude and a direction. Given some coordinate system, we can specify a vector by simply giving a list of numbers. That is, in fact, what we're doing when we talk about v^i.

Tensors are a generalization of vectors with more than one index. As a simple example, we can generate a tensor by taking the outer product of two vectors:

$$M^{ij} = A^i B^j$$

where, in Euclidean space, i and j can each take on three different values. M^{ij} represents a table of nine numbers, each indexed by an ordered pair. The number of distinct indices is known as the **order** of a tensor, so M^{ij} is second order, while an ordinary vector is a first order tensor.

In principle, we can imagine a tensor of just about any order (including zero—a scalar), but there are a few bookkeeping rules that will keep you out of trouble.

1. The positions and order of indices matter.

   In addition to vectors, we are going to encounter tensorial objects with indices of every number and position. For instance,

   $$u_i; \quad g_{ij}; \quad M^i{}_j; \quad \Gamma^i{}_{jk}.$$

   Every one of these objects can be specified by the total number of indices (the order) and whether each index is upstairs (formally known as **contravariant**) or downstairs (**covariant**).

   Some of these tensor have special names. For example, as we've seen, an object with one index downstairs is known as a one-form, while an object with two downstairs indices is a **two-form**, and so forth.

   You *cannot* simply interchange an upstairs and a downstairs index; that is, an equation of the form

   $$\cancel{A_i = B^i}$$

   is not allowed. We've put a line through invalid equations throughout to prevent anyone from flipping through the book in search of easy answers and inadvertently writing down a mathematical abomination. A satisfying explanation of *why* upstairs and downstairs indices matter will have to wait until we've explored coordinate transformations, but the rule will have to suffice for now.

   Likewise, the sequence of the indices matters. For a simple but illuminating case, suppose M^{ij} is an asymmetric tensor. In that case,

   $$M^{ij} = -M^{ji}.$$

   A careless swap of the order of indices will, in this case, introduce an erroneous minus sign.

2. To be valid an equation must match indices.

   that which we call a rose
   By any other name would smell as sweet;

   (*Romeo and Juliet*, Act II, Scene II)

It does not, obviously, matter whether a tensor is labeled T^{ij} or T^{kl}. Those are simply labels, and it is understood that in Euclidean space, i or j or k or l can take the values 1, 2, or 3. Dummy indices, especially, can be labeled as desired:

$$A^i B_i = A^j B_j.$$

But to be valid, an equation must match the same nondummy indices. That is, an expression like

$$M^{ij} A_j = B^i$$

is mathematically valid (whether it's physically correct is another matter), and represents three linearly independent equations.

However, the expression

$$\cancel{M^{ij} = A^k}$$

is complete gibberish.

Likewise, the same dummy index can't appear twice, either upstairs or downstairs. While

$$\cancel{M_{ii}}$$

may make a sort of intuitive sense, it is not meaningful in tensor algebra.

3. Tensors are not matrices.

Throughout our study of fields, we're going to encounter a lot of second-order tensors. As these have two indices, your natural inclination will be to treat them like matrices. Don't. Or, at least, be aware that tensors don't multiply in the same way as matrices.

The closest approximation to what we'd normally call a matrix is a tensor of the form

$$A^i = M^i{}_j B^j, \tag{1.6}$$

which multiplies a tensor and a vector, producing another vector. But such clean results are the exception rather than the rule.

Consider the two different tensor contractions

$$A_i = M_{ij} B^j$$

and

$$A_i = M_{ji} B^j.$$

Depending on which index gets contracted, the products will be a totally different, and both results will be one-forms rather than vectors. The point is simply that while you're undoubtedly quite adept at multiplying matrices times themselves or vectors, you should be extremely cautious before doing so.

With those rules in mind, we're prepared to manipulate tensors and relate them to the physical world.

1.2.3 The Metric

A meterstick has the very useful property that it is a meter no matter which direction it's oriented, and Euclidean geometry accounts for this quite nicely. By the Pythagorean theorem,

$$\text{length}^2 = \Delta \vec{x} \cdot \Delta \vec{x} = \Delta x^2 + \Delta y^2 + \Delta z^2. \tag{1.7}$$

Though Δx or Δy will vary as we rotate the meterstick, the total length will stay the same. The **metric tensor** is a geometric tool that allows to take a dot product no matter how complicated the geometry. Think of it as a function in which the arguments are two vectors, and out pops a scalar.

As normally written, the metric tensor is a two-form—two downstairs indices—and almost universally given the letter g. In Cartesian coordinates, the form of the metric is especially simple:

$$g_{ij} = \begin{pmatrix} 1 & 0 & 0 \\ 0 & 1 & 0 \\ 0 & 0 & 1 \end{pmatrix}, \tag{1.8}$$

where we've written it as a matrix because its symmetry makes the ordering of indices irrelevent. If you are underwhelmed, don't be. The metric will not be so simple in all coordinate systems, and it certainly won't be in special relativity. The metric performs two main functions. First, it can be used to pull indices downstairs. For instance, a vector

$$A^i = \begin{pmatrix} 1 \\ 2 \\ 0 \end{pmatrix}$$

can be converted into the downstairs version by contracting:

$$A_i = g_{ij} A^j. \tag{1.9}$$

In this particularly simple case,

$$A_i = g_{ij} A^j = \begin{pmatrix} 1 & 2 & 0 \end{pmatrix}.$$

The metric can also be used to lower an index of a tensor of any rank, but must be done with great care. For instance

$$g_{ik} M^{ij} = M_k{}^j$$

works only if the first index of M is lowered by the operation, leaving j as second index in the raised position.

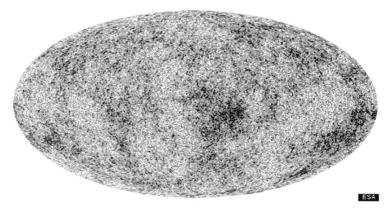

Figure 1.3. A temperature map of the whole sky in microwaves, as imaged by the *Planck* satellite. The image is a Mollweide projection in which the *x*-axis corresponds to the celestial equator. Grayscale variations indicate fractional temperature differences of about 10^{-5}, and there is little variance on smoothing scales larger than $\sim 1°$. Credit: ESA and the Planck Collaboration.

The metric is primarily an engine to turn two vectors into a scalar via the dot product; that is,

$$\vec{A} \cdot \vec{B} = g_{ij} A^i B^j. \tag{1.10}$$

The metric must be symmetric, since the dot product is commutative. Additionally, since the metric is itself a tensorial object, there are upstairs and downstairs versions which serve as inverses of each other:

$$g^{ij} g_{jk} = \delta^i_{\ k}, \tag{1.11}$$

where $\delta^i_{\ k}$ is the Kronecker-delta function (defined to be 1 if $i = k$, but 0 otherwise). In other words, the upstairs version of the metric is simply the inverse of the downstairs version. This relation will come in handy when *raising* tensor indices.

1.2.4 Coordinate Transformations

Invariances are at the heart of the Standard Model, which means that we are particularly interested in exploring quantities that are unchanged under various transformations. For example, the universe seems not to have any preferred direction. This is one of the *assumptions* underlying the **cosmological principle.** The other, that there is no preferred location in the universe, provides another important symmetry. These are assumptions, to be sure, but large-scale surveys of both galaxies [15] and the cosmic microwave background [56, 130] suggest that on scales well below the cosmic horizon, the universe is largely homogeneous and isotropic (Figure 1.3). By extension, the homogeneity and isotropy of the universe reflect a homogeneity and isotropy of physical laws.

Under the cosmological principle, *all* the laws of the universe remain unchanged under a rotation of coordinate axes or with a shift of origin. This seems like a minor point, but it

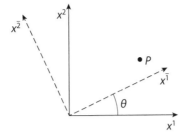

Figure 1.4. Rotated coordinate axes, with the z-axis (out of the page) suppressed.

implies that, for instance, only dot products, rather than individual components of vectors, will be found in fundamental physical laws.

To make the concept concrete, consider two different reference frames, which we'll label "barred" and "unbarred." The coordinates as measured in one frame are related to another via some sort of yet-to-be-determined coordinate transformation:

$$x^i \rightarrow x^{\bar{i}}.$$

To transform between the two, we introduce a coordinate transformation tensor $\Lambda^{\bar{i}}{}_i$, such that

$$x^{\bar{i}} = \Lambda^{\bar{i}}{}_i x^i, \tag{1.12}$$

where

$$\Lambda^{\bar{i}}{}_i = \frac{\partial x^{\bar{i}}}{\partial x^i}. \tag{1.13}$$

In general, determining the Λ tensor is the hard part of the process. Once you've done it, transformation of coordinates is a breeze.

Example 1.2: How do the coordinates of a vector change upon rotation of the coordinate axes by an amount θ around the z-axis (Figure 1.4)?

Solution: We can express the transformation of coordinates (and thus all vector components) as

$$x^{\bar{1}} = x^1 \cos\theta + x^2 \sin\theta$$
$$x^{\bar{2}} = -x^1 \sin\theta + x^2 \cos\theta$$
$$x^{\bar{3}} = x^3,$$

from which we can use the relation in equation (1.13) to compute the elements of Λ directly. Computing one of these terms explicitly, we get

$$\Lambda^{\bar{2}}{}_1 = \frac{\partial x^{\bar{2}}}{\partial x^1} = -\sin\theta,$$

and similarly for the other terms.

Writing all these out, we find a transformation matrix (and, yes, it's a matrix):

$$\Lambda^{\bar{i}}_{\;i} = \begin{pmatrix} \cos\theta & \sin\theta & 0 \\ -\sin\theta & \cos\theta & 0 \\ 0 & 0 & 1 \end{pmatrix}, \tag{1.14}$$

where we need to make sure we've identified the rows with the upper (barred) index and the columns with the lower (unbarred) index. This matrix, in turn, allows rotation of any arbitrary vector from the old frame to the new.

For any coordinate transformation, there is an inverse such that

$$\Lambda^{i}_{\;\bar{i}} = \frac{\partial x^i}{\partial x^{\bar{i}}},$$

since the choice of frame to call barred and the one to call unbarred is completely arbitrary. Applying the coordinate transformation and then the inverse must necessarily lead back to the original state of affairs:

$$\Lambda^{i}_{\;\bar{i}}\Lambda^{\bar{i}}_{\;j} = \delta^{i}_{\;j}. \tag{1.15}$$

In example 1.2 we computed the coordinate transformation matrix for a rotation around the z-axis. The inverse is simple enough. Instead of rotating through an angle θ, we simply rotate back through an angle $-\theta$:

$$\Lambda^{i}_{\;\bar{i}} = \begin{pmatrix} \cos\theta & -\sin\theta & 0 \\ \sin\theta & \cos\theta & 0 \\ 0 & 0 & 1 \end{pmatrix}. \tag{1.16}$$

It can readily be verified that the inverse relation (equation 1.15) is satisfied by this transform.

Transformation matrices can be used on *any* type of tensorial object, not only vectors. For instance,

$$A_{\bar{i}} = \Lambda^{i}_{\;\bar{i}} A_i$$

transforms the components of a one-form.

To transform tensors with more than one index, we simply need to sum over all of them. For instance, a metric can be represented in a new frame by

$$g_{\bar{i}\bar{j}} = \Lambda^{i}_{\;\bar{i}}\Lambda^{j}_{\;\bar{j}} g_{ij}. \tag{1.17}$$

Note that to transform two indices, we required the product of two Λ terms. By examining the matching indices, we note that the first transformation matrix lowers the first index of g, and the second lowers the second index. If we were to write the sum explicitly, each element of $g_{\bar{i}\bar{j}}$ would require summing over $3 \times 3 = 9$ elements in a three-dimensional space. However, most of those elements would be zero.

Example 1.3: What happens to the metric if we transform the coordinate frame by a rotation (equation 1.16)?

Solution:

$$g_{\bar{i}\bar{j}} = \Lambda^i_{\bar{i}}\Lambda^j_{\bar{j}}g_{ij}$$

$$= \begin{pmatrix} \cos^2\theta + \sin^2\theta & \cos\theta\sin\theta - \cos\theta\sin\theta & 0 \\ \cos\theta\sin\theta - \cos\theta\sin\theta & \cos^2\theta + (-\sin\theta)^2 & 0 \\ 0 & 0 & 1 \end{pmatrix}$$

$$= \begin{pmatrix} 1 & 0 & 0 \\ 0 & 1 & 0 \\ 0 & 0 & 1 \end{pmatrix}.$$

Rotations (around the x- and y-axes as well as around the z) leave the metric unchanged. *That's* incredibly powerful! We'll learn later that this means that rotations are elements of a **symmetry group** of Cartesian space known as SO(3). In systems with SO(3) symmetry, all measurably quantities remain invariant under arbitrary rotations.

Example 1.4: Using the coordinate transformation matrix, compute the metric of a two-dimensional flat space in polar coordinates.

Solution: For convenience, we'll label Cartesian coordinates as unbarred, and polar (r, θ) as barred. The coordinate transformation is

$$x = r\cos\theta$$

$$y = r\sin\theta,$$

and so the transformation matrix is

$$\Lambda^i_{\bar{i}} = \begin{pmatrix} \cos\theta & -r\sin\theta \\ \sin\theta & r\cos\theta \end{pmatrix}.$$

We can readily get the metric for the new frame:

$$g_{\bar{i}\bar{j}} = \begin{pmatrix} g_{xx}\Lambda^x_r\Lambda^x_r + g_{yy}\Lambda^y_r\Lambda^y_r & g_{xx}\Lambda^x_r\Lambda^x_\theta + g_{yy}\Lambda^y_r\Lambda^y_\theta \\ g_{xx}\Lambda^x_\theta\Lambda^x_r + g_{yy}\Lambda^y_\theta\Lambda^y_r & g_{xx}\Lambda^x_\theta\Lambda^x_\theta + g_{yy}\Lambda^y_\theta\Lambda^y_\theta \end{pmatrix} = \begin{pmatrix} 1 & 0 \\ 0 & r^2 \end{pmatrix}.$$

So, for example, given a particle moving in polar coordinates, we can compute the components of a velocity by taking a simple time derivative:

$$v^{\bar{i}} = \dot{x}^{\bar{i}} = \begin{pmatrix} \dot{r} \\ \dot{\theta} \end{pmatrix}.$$

You may notice that the two terms in the velocity do not have the same units. That's okay! We can compute the overall speed via

$$|\vec{v}|^2 = g_{\bar{i}\bar{j}} v^{\bar{i}} v^{\bar{j}}$$
$$= \dot{r}^2 + r^2 \dot{\theta}^2,$$

which you may recall from orbital dynamics problems.

1.2.5 *The Real Difference between Vectors and One-Forms*

We developed a tool for lowering and raising indices of tensors (the metric) and for contracting indices (and thus reducing the order of the tensor by two). As our study of classical fields will also be a study in dynamics, it's important that we have the necessary tools to *increase* the order of a tensor. We thus need one more operation—coordinate derivatives.

Consider a scalar function $f(\vec{x})$ defined and continuously differentiable everywhere in space. This, incidentally, is basically the definition of a scalar field.

Spatial derivatives can be expressed via

$$\partial_i f(\vec{x}) \equiv \frac{\partial f}{\partial x^i}, \tag{1.18}$$

which are the components of a gradient. They also leave the expression with a downstairs component. A gradient, in other words, is a one-form as opposed to a vector, which finally gives us an opportunity to understand the real difference between upstairs and downstairs indices.

A gradient of a scalar field will have units of inverse length, while a position vector will have units of length. The same coordinate transformation that *decreases* the number of standard length units between two points will *increase* the rate at which a scalar field changes per unit length. Put another way, the *product* of a vector and a one-form produces a scalar invariant.

More generally, derivatives add an additional downstairs index to a tensor; that is,

$$\partial_i v^j = \frac{\partial v^j}{\partial x^i}$$

is a rank 2, $\binom{1}{1}$, tensor and represents nine different numbers. However, if the two indices are the same, we get

$$\partial_i v^i = \nabla \cdot \vec{v},$$

which contracts the upper and lower indices via a *divergence* and yields a scalar.

What works well in space will turn out to work equally well in spacetime, which is good, because our intuition will need the support of mathematical structure.

1.3 Foundations of Relativity

1.3.1 Einstein's Postulates

Galilean relativity produced a remarkably intuitive statement about how vectors—and especially how velocity vectors—transform from one inertial frame to another. A bullet fired toward the front of a moving train will appear to travel faster when observed from outside the train than from within. By that logic, the same should be true for a beam of light. If a laser is fired from a moving source, it's reasonable to expect that the photons will travel faster than if they were shot from a source at rest (e.g., in a boosted frame, as in Figure 1.2).

This hypothesis was put to the test by a number of researchers, including Hippolyte Fizeau and Leon Foucault at the end of the nineteenth century, but the compelling experimental evidence came from the interferometer designed by Albert Michelson and Edward Morley.

In 1887, Michelson and Morley [109] published their famous result demonstrating that the speed of light from the sun (and consequently from any other source) is constant throughout the year. Or, more bluntly, light moves at a constant speed regardless of the relative state of motion of the observer and source.

The constancy of the speed of light directly contradicts Galilean relativity. The only way to resolve the tension is through the possibility that *time* as well as space, transforms in different inertial frames. This was exactly the possibility explored by Albert Einstein in his seminal 1905 paper [53].

While there's some debate about how much direct inspiration Einstein drew from the Michelson-Morley experiment, there can be little doubt that their work provided substantial support for Einstein's fundamental postulates of **special relativity**:

1. The laws by which the states of physical systems undergo change are not affected, whether these changes of state be referred to the one or the other of two systems of coordinates in uniform translationary motion.
2. As measured in any inertial frame of reference, light is always propagated in empty space with a definite velocity c that is independent of the state of motion of the emitting body.

These postulates seem to accurately describe the laws of nature. Any equation that purports to adequately reflect physical law needs to satisfy these symmetry constraints. We already know that one of the classics,

$$\vec{F} = m\vec{a},$$

doesn't fit the bill, as it allows for the possibility of superluminal motion.[3]

Our notational goal will be to develop a formalism that doesn't *allow us* to write physical laws that violate the constant speed of light and nonpreeminence of any given inertial frame.

---

[3] Newton never actually formulated his second law as $F = ma$, but rather used $F = dp/dt$, which remains true in relativistic systems. However, his formulation of momentum was decidedly nonrelativistic.

1.3.2 Natural Units

The speed of light is central to special-relativistic arguments—so special, in fact, that we generally want to remove it from our notation entirely. Astronomers occasionally talk about distances to stars in terms of "light-years," the distance that light can travel in a year:

light-year $= c \times 1\,\mathrm{yr} \simeq 9.5 \times 10^{15}$ m.

We could equally well talk about a light-second, $300,000$ km, or any other unit of light-time.

There is a fundamental, almost intuitive sense in which units of length and units of time can be said to be equivalent, with the speed of light used as the tool of currency conversion. We'll break down the distinction between space and time entirely, by setting

$$c = \hbar = 1, \tag{1.19}$$

a convention known as **natural units.** Using natural units, we can express all quantities as energy to some power:

$$[m] = [E]^1, \tag{1.20}$$

for example. This one should be obvious, since $E = mc^2$ is probably the most famous equation in physics. In natural units, the equivalence between mass and energy can be seen almost immediately on dimensional grounds. For example, the mass of a proton is approximately 935 MeV, while that of an electron is only 0.511 MeV.

Things are a little less intuitive when we refer to length, but the conversion can be made clear by computing the **Compton wavelength** of a particle:

$$\lambda_C = \frac{\hbar}{mc}.$$

The physical interpretation of the Compton wavelength is that it is the smallest scale on which a single particle can be identified. On smaller scales, the energy goes up, and particles can be created out of the vacuum. Thus, in natural units:

$$[L] = [E]^{-1}. \tag{1.21}$$

Large scales are low energies and vice-versa. Angstrom scales, for instance have energies in the inverse kilo-electron-volt range, while energies corresponding to femtometer scales are in the giga-electron-volt range.

Finally, since distance and time have the same units,

$$[T] = [E]^{-1}. \tag{1.22}$$

These units can be combined in all sorts of ways. For example, energy density is expressed in units of $[E]^4$, speeds are dimensionless (fractions of c), and so on. In many cases, we won't even find it necessary to compute physical quantities in real units, but we *will* find it useful to make sure that if $A = B$, then both A and B have the *same* units.

Table 1.1. Converson of MKS Units to Natural Units.

Unit	Natural Units
$1\,\text{kg}$	$5.63 \times 10^{26}\,\text{GeV}$
$1\,\text{m}$	$\left(1.97 \times 10^{-16}\,\text{GeV}\right)^{-1}$
$1\,\text{s}$	$\left(6.58 \times 10^{-25}\,\text{GeV}\right)^{-1}$
E_{Pl}	$1.22 \times 10^{19}\,\text{GeV}$

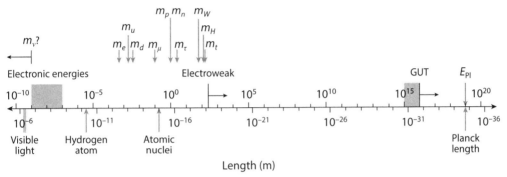

Figure 1.5. Characteristic energy/length scales in particle physics.

One particularly interesting case involves Newton's constant, G. Though we won't be dealing with any gravity in this book, in natural units, G has a value of

$$G = \frac{1}{E_{Pl}^2},$$ (1.23)

where E_{Pl} is known as the **Planck energy** and represents the energy scale on which quantum mechanics and gravity are both important. Expanding the terms, we get

$$E_{Pl} = \sqrt{\frac{\hbar c}{G}} \simeq 1.22 \times 10^{19}\,\text{GeV}.$$ (1.24)

This is another way of saying that in particle calculations, gravity is staggeringly weak. We write out a few more useful conversions to natural units in Table 1.1, and plot a range of energy scales in Figure 1.5.

1.4 Spacetime

1.4.1 4-Vectors

Having manipulated units and tensor conventions, we're *finally* prepared to delve head-first into relativity. Because we have the advantage of history, we present results slightly out of order of their discovery, beginning with the fundamental interconnectedness of space and time. As Hermann Minkowski put it in 1908 [112]:

> Henceforth space by itself, and time by itself, are doomed to fade away into mere shadows, and only a kind of union of the two [what we now call **spacetime**] will preserve an independent reality.

Indeed, we now define positions and other vectorial quantities in terms of a **4-vector** [111, 131]:

$$x^{\mu} = \begin{pmatrix} t \\ x \\ y \\ z \end{pmatrix}, \tag{1.25}$$

where μ (and other Greek-letter indices) may take the values 0, 1, 2, 3, and by convention, Roman indices may take the values $i = 1, 2, 3$. While all vectors in Euclidean space have three components (and the vector itself is labeled $\vec{x}$), all well-defined vectors in Minkowski spacetime have four (with the vector simply labeled as x).

1.4.2 Lorentz Transforms

Almost immediately following the Michelson-Morley result, George FitzGerald (1889) [66] and Hendrik Antoon Lorentz (1892) [104] attempted to explain away the constant speed of light by proposing that measurement equipment was deformed in a particular way when traveling through a hypothetical luminiferous aether. This aether was to be the medium through which electromagnetic radiation propagated.

While the concept of an aether was ultimately abandoned, the mathematical grounding developed by FitzGerald, Lorentz, and others (though now almost exclusively referred to as **Lorentz transforms**) gives the relationship between the 4-vector coordinates of two frames in relative motion (Figure 1.2). For the simplest case of a relative speed v in the x-direction,

$$\Lambda^{\bar{\mu}}_{\mu}(v) = \begin{pmatrix} \gamma & v\gamma & 0 & 0 \\ v\gamma & \gamma & 0 & 0 \\ 0 & 0 & 1 & 0 \\ 0 & 0 & 0 & 1 \end{pmatrix}. \tag{1.26}$$

For the inverse, the sign of the velocity is simply switched. Boosts in other directions can be developed essentially by inspection. The "gamma factor" is defined as

$$\gamma \equiv \frac{1}{\sqrt{1 - v^2}}. \tag{1.27}$$

The Lorentz transforms are at the center of relativistic physics. Equations are said to be **Lorentz invariant** if they are identical in any rotated or boosted frame, and it is our aim throughout to derive only Lorentz-invariant quantities and expressions. As a helpful hint, these relations will involve only free 4-vector indices on both sides of an equation or, even better, well-defined scalars in which all indices are contracted.

The Lorentz transforms have some nice features. In the nonrelativistic limit, $\gamma \simeq 1$, which means that the Lorentz transforms approach the Galilean transform relation (equation 1.1).

A simple reading of the Lorentz tranform matrix demonstrates an important physical manifestation of relativity, in that clocks run slow by a factor of γ compared with their stationary counterparts. This is *not* simply an optical illusion as *any* measurement of time will display the same time dilation: the ticking of a clock, the beating of a heart, or even the decay of particles.

In 1941, Bruno Rossi and David Hall [140] found that charged elementary particles known as **muons** created in the upper atmosphere survive to the surface of the earth instead of being destroyed, despite their short mean lifetime of $2.4\,\mu s$ [119]. The explanation was that muons travel at relativistic speeds, their internal "clocks" are slowed relative to the earth, and thus their effective decay time is effectively increased.

Example 1.5: Muons are created approximately 10 km above the surface of the earth. What is the typical speed of atmospheric muons if 5% of them reach the surface of the earth?

Solution: Even traveling at the speed of light, it should take muons $33\,\mu s$ to reach the surface from the upper atmosphere, many their mean lifetime. If 5% of muons survive, then in the frame of the muons, the trip to the earth takes only

$$e^{-t/\tau} = 0.05; \quad t = 3\tau = 7.2\,\mu s.$$

Time must be dilated by a factor of $\gamma = 33\,\mu s / 7.2\,\mu s = 4.58$.

Inverting the γ relation, we get

$$v = \sqrt{1 - \frac{1}{\gamma^2}} = 0.976.$$

Among other things, the Lorentz transforms guarantee that massive particles move at sublight speeds in all frames. Likewise, if a photon travels at the speed of light in one frame, then it quickly follows that it moves at the speed of light in all boosted frames.

1.4.3 The Minkowski Metric

We introduced a metric as a way of computing the lengths of vectors. In special relativity, we know that time and space have *very* similar behavior, but they are not identical.

The Minkowski metric plays a role in spacetime to equivalent the Pythagorean theorem:[4]

$$g_{\mu\nu} = \begin{pmatrix} 1 & 0 & 0 & 0 \\ 0 & -1 & 0 & 0 \\ 0 & 0 & -1 & 0 \\ 0 & 0 & 0 & -1 \end{pmatrix}. \tag{1.28}$$

The metric is especially useful because (as with the metric in Euclidean space) it allows us to tell the distance between two nearby event separated by a 4-vector dx^μ:

$$d\tau^2 = g_{\mu\nu} dx^\mu dx^\nu.$$

This distance is known as the **interval**.[5] Supposing the interval is positive, τ is known as the **proper time** between two events. While the flow of time is a function of the relative velocity between an observer and coordinate axes, the proper time, by definition, is the flow as measured by an observer in his/her own frame.

Note also that the interval is a scalar. There are no remaining indices, which means that if we've set our notation adequately (and we have), the entire expression should be Lorentz invariant. In other words, regardless of the relative state of motion,

$$dt^2 - dl^2 = d\bar{t}^2 - d\bar{l}^2,$$

which can be shown algebraically using the Lorentz boost (equation 1.26).

This is going to be true of dot products in general. And, indeed, the interval is nothing more than the dot product of two displacement vectors. We can express it even more succinctly in familiar notation:

$$d\tau^2 = dx^\mu dx_\mu = dx \cdot dx, \tag{1.29}$$

where, as a reminder, the lack of a vector arrow above the dx terms indicates that they are 4-vectors rather than 3-vectors. Because the indices are contracted, it does not matter which is upstairs and which is downstairs The sign of the interval immediately yields some very useful information:

- $d\tau^2 = 0$: **Lightlike** or **null** separation.
  A photon can—must—travel exactly at the speed of light, which means that the emission and observation of a photon will always be separated by an interval of zero (which is another way of saying that time does not pass for a photon or any other massless particle).
- $d\tau^2 < 0$: **Spacelike** separation.
  The time order of the two events cannot be determined unambiguously, and neither can have a causal connection to the other.

---

[4] In many texts, the sign convention is reversed, with a "−" on the time component and a "+" on the others. If you ever want to be sure, people will usually refer to the trace of the metric. Ours is negative.

[5] Like the metric itself, it's sometimes given with the opposite sign convention.

- $d\tau^2 > 0$: **Timelike** separation.

  Likewise, a positive interval means that two events are separated by **timelike** separation.

The beauty of the Minkowski metric is that the metric *itself* doesn't change upon a Lorentz boost, just as the Euclidean metric didn't change upon coordinate rotation. That is,

$$g_{\bar{\mu}\bar{\nu}} = \Lambda^{\mu}_{\ \bar{\mu}}(v)\Lambda^{\nu}_{\ \bar{\nu}}(v)g_{\mu\nu} \tag{1.30}$$

will again produce an identical copy of the Minkowski metric. This result is nothing more than a consequence of Einstein's first postulate of special relativity. In problem 1.9 you will have the opportunity to show this for boosts as well. That the metric is invariant under Lorentz boosts demonstrates that our initial choice of transform was correct.

1.5 Relativistic Dynamics

1.5.1 *The 4-Velocity*

Having developed the rules for vectors in general, we may construct dynamical quantities that will be useful in understanding particles and their interactions with fields. For instance, in Newtonian mechanics, we have the 3-velocity

$$v^i = \frac{dx^i}{dt},$$

but this is clearly not Lorentz invariant, because time and space aren't being treated on the same footing. Indeed, the t coordinate is no longer fixed in special relativity, so derivatives with respect to time are no longer well determined.

In special relativity, as with Galilean, we can define a velocity. The definition of the **4-velocity** is quite simple:

$$u^{\mu} \equiv \frac{dx^{\mu}}{d\tau}, \tag{1.31}$$

where τ is the proper time. The 4-velocity allows us to take derivatives of an arbitrary function f with respect to the proper time in a convenient way:

$$\frac{df}{d\tau} = \partial_{\mu}f\frac{dx^{\mu}}{d\tau} = \frac{\partial f}{\partial x^{\mu}}u^{\mu}$$

where, exactly as with 3-vectors, derivatives generate a downstairs index.

For a particle at rest, the velocity is simply

$$u^{\mu}_{(rest)} = \begin{pmatrix} 1 \\ 0 \\ 0 \\ 0 \end{pmatrix},$$

which can quickly be shown to have the normalization

$$u \cdot u = g_{\mu\nu} u^{\mu} u^{\nu} = 1. \tag{1.32}$$

As we showed earlier, dot products of vectors are Lorentz invariant. If this is true in one frame, it's true in all.

But the individual *components* of the 4-velocity will change, even if the magnitude of the 4-velocity doesn't. Boosted by an arbitrary velocity, the 4-velocity can be computed as

$$u^{\mu} = \begin{pmatrix} \gamma \\ \gamma v^1 \\ \gamma v^2 \\ \gamma v^3 \end{pmatrix}. \tag{1.33}$$

We leave it as an exercise for you in problem 1.7 to show algebraically that $u \cdot u = 1$ in the boosted frame.

Of greatest immediate interest is generally the component $u^0 = dt/d\tau$, how quickly the particle is moving through time. This is simply γ, which we previously identified as the classic time-dilation factor.

Photons (and all massless particles) travel at the speed of light and thus have an infinite time-dilation factor. As a consequence, there is no such thing as "the 4-velocity of a photon" for the simple reason that there is no such thing as "the rest frame of a photon." Momentum, however, is another matter.

1.5.2 The 4-Momentum

As a final physical quantity, we define a **4-momentum** for a particle of mass m:

$$p^{\mu} = mu^{\mu}, \tag{1.34}$$

which has the pleasing property

$$p \cdot p = m^2, \tag{1.35}$$

derivable directly from the normalization condition on u. The timelike (zeroth) component of the 4-momentum is just the energy. And thus

$$p^{\mu} = \begin{pmatrix} E \\ p^1 \\ p^2 \\ p^3 \end{pmatrix} = \begin{pmatrix} m\gamma \\ mv^1\gamma \\ mv^2\gamma \\ mv^3\gamma \end{pmatrix}. \tag{1.36}$$

We can consider the normalization condition on the momentum in both natural units

$$E^2 - |\vec{p}|^2 = m^2$$

(where $\vec{p}$ represents the spacelike terms) and in MKS units, wherein we insert c to get the correct dimensionality:

$$E^2 = (mc^2)^2 + (|\vec{p}|c)^2. \tag{1.37}$$

Light has no mass, but it still carries momentum, as was suggested as early as 1619 by Johannes Kepler [91], who noted that the tail of a comet points away from the sun. It is also a natural consequence of Maxwell's equations and can be seen by even a tabletop radiometer. The 4-momentum gives us an immediate relation between momentum and energy for a massless particle.

$$E = \pm|\vec{p}|,$$

where logic dictates that only the positive energy case is valid. In our discussion of the Dirac equation, we'll call this assumption into question.

1.5.3 A Final Note on Notation and Symmetry

Our discussion of tensor notation and special relativity is ultimately both an insurance policy and a shortcut. Special relativity has survived every challenge thrown at it for more than a century, which means that any accurate description of the physical world must be Lorentz invariant. It must be expressed in terms of 4-vectors, tensors in four-dimensional spacetime[6] or, better yet, scalars.

Newton's law of gravity, with its dependence on the inverse square of distance and no dependence on time, is manifestly *not* Lorentz invariant, and thus there must be a deeper physical law describing gravity. Indeed there is: general relativity. We can immediately rule out huge swaths of physical theories by inspection alone without proving that each one isn't Lorentz invariant. Given that the space of possible physical theories is virtually infinite, this narrows things down enormously.

Problems

1.1 Consider a vector field in three-dimensional Cartesian space:

$$u^i = \begin{pmatrix} xy \\ x^2 + 2 \\ 3 \end{pmatrix}.$$

    (a) Compute the components of $\partial_j u^i$.
    (b) Compute $\partial_i u^i$.
    (c) Compute $\partial_j \partial^j u^i$.

---

[6] In principle, our 3+1 dimensional spacetime might be embedded in a higher-dimensional space, so strictly we might say that a theory must be expressed in $d+1$ dimensional spacetime. However, in this text, we'll confine ourself to the known macroscopic dimensions.

1.2 Consider the two-dimensional Cartesian plane with coordinates x^i. Apply the following coordinate transformation:

$$x^{\bar{1}} = 2x^1$$
$$x^{\bar{2}} = x^2$$

Consider a vector $\vec{v}$ with components in the unbarred frame:

$$v^i = \begin{pmatrix} 4 \\ -3 \end{pmatrix}.$$

(a) What is the metric in the barred frame?
(b) What are the components of $v^{\bar{i}}$ in the barred frame?
(c) What are the components of $v_{\bar{i}}$ in the barred frame?
(d) What is the magnitude of $v^{\bar{i}} v_{\bar{i}}$ in the barred frame? Compare this result with that from the unbarred frame.

1.3 Consider a transformation from three-dimensional Cartesian coordinates (x, y, z) to spherical coordinates (r,θ,ϕ):

$$x = r \sin\theta \cos\phi$$

$$y = r \sin\theta \sin\phi$$

$$z = r \cos\theta$$

(a) Labeling the spherical coordinates as the "barred" frame, compute the transformation matrix $\Lambda^{\bar{i}}_{\ i}$.
(b) Compute the metric for spherical coordinates.

1.4 Express the following quantities in natural units, in the form $(\# \text{ GeV})^n$.
(a) The current energy density of the universe: $\sim 10^{-26} \text{ kg/m}^3$
(b) 1 angstrom
(c) 1 nanosecond
(d) 1 gigaparsec $\simeq 3 \times 10^{25}$ m
(e) The luminosity of the sun $\simeq 4 \times 10^{26}$ W

1.5 The universe has a horizon size of a few gigaparsecs. Using the results of the previous problem, estimate the total energy within the observable universe.

1.6 Consider a 4-vector:

$$A^\mu = \begin{pmatrix} 2 \\ 3 \\ 0 \\ 0 \end{pmatrix}.$$

(a) Compute $A \cdot A = A^\mu A_\mu$.
(b) What are the components $A^{\bar{\mu}}$ if you rotate the coordinate frame around the z-axis through an angle $\theta = \pi/3$?
(c) For your answer in part (b), verify that $A^{\bar{\mu}} A_{\bar{\mu}}$ is the same as in part (a).
(d) What are the components $A^{\bar{\mu}}$ if you boost the frame (from part a) a speed $v = 0.6$ in the x-direction?
(e) For your answer in part (d), verify that $A^{\bar{\mu}} A_{\bar{\mu}}$ is the same as in part (a).

1.7 Show that the 4-velocity, as defined in equation (1.33), satisfies the relation $u \cdot u = 1$ from the definition of γ.

1.8 Consider two events in a 1+1-dimensional spacetime:

$$x_1 = 2\,\text{s} \quad t_1 = 4\,\text{s}$$
$$x_2 = -2\,\text{s} \quad t_2 = 9\,\text{s}$$

(a) What is the interval between the two events?
(b) Is this a spacelike or a timelike separation?
(c) If the interval is timelike, what is the 3-velocity required to get from event 1 to event 2? What is the corresponding γ factor?

1.9 Show that a Lorentz boost (equation 1.26) leaves the Minkowski metric unchanged.

1.10 Consider a scalar field

$$\phi(x) = 2t^2 - 3x^2.$$

(a) Compute the components of $\partial_\mu \phi$.
(b) Compute the components of $\partial^\mu \phi$.
(c) Compute $\partial_\mu \partial^\mu \phi$. This operation is the **d'Alembertian operator** on the field and is vital for wave propagation.

1.11 Consider a tensor field in a 1+1-dimensional Minkowski space:

$$W^\alpha = \begin{pmatrix} t^2 - x^2 \\ 2x \end{pmatrix}.$$

(a) Compute W_α.
(b) Compute $\partial_\beta W_\alpha$.
(c) Compute $\partial_\alpha W^\alpha$.

1.12 In April, 2015, proton beams in the Large Hadron Collider were brought up to an energy of 6.5 TeV.
(a) What is the γ factor of an LHC proton?
(b) What is the approximate speed of a proton in the beam? Express your answer in terms of $1 - \delta v$, where δv is the amount by which a proton lags light.

1.13 An excited hydrogen atom emits a 10.2 eV Lyman-α photon.
(a) What is the momentum of the photon? Be sure to express your answer in natural units.
(b) As Newton's third law remains in force, what is the kinetic energy of the recoiling ground-state hydrogen atom?
(c) What is the recoil speed of the proton?

Further Readings

Different texts will use different (sometimes wildly different) conventions. With that in mind, you may find the following helpful supplements to the material in this chapter.

- Einstein, Albert. *The Meaning of Relativity*, Princeton, NJ: Princeton University Press, 2005. Einstein wrote a number of semipopular works on relativity, though all contain mathematics. While only the first section of *Meaning* focuses on special relativity, interested readers will also find the last two sections (on general relativity) written at a great introductory level.
- Helliwell, T. M. *Special Relativity*, Sausalito, CA: University Science Books, 2009. The whole of Helliwell's book is well worth a read.
- Schutz, Bernard. *A First Course in General Relativity*, 2nd ed., Cambridge: Cambridge University Press, 2009. I would especially recommend chapters 1 through 3, which focus on tensor notation and special relativity. With the exception of the signature of the metric, I generally follow Schutz's notation.
- Synge, J. L., & Schild, A. *Tensor Calculus*, Toronto: Dover, 1949. While much of the book focuses on curved spaces and differential geometry, the interested reader will find the first two chapters, which provide a general introduction to tensors, especially useful.

2 | Scalar Fields

Figure 2.1. Sir William Rowan Hamilton (1805–1865). Hamilton developed many of the fundamentals of variational mechanics, including many of the equations that now bear Lagrange's and Euler's names.

We have thus far focused primarily on the dynamics of individual particles, but particles are nothing more than a shadow of deeper physics—physics dominated by **fields.** Fields are a reflection of the fact that physics extends over the entire universe and can be quantified and measured in all places and times.

The tensor notation we used in the last chapter will come in extremely handy, except that instead of defining the properties (4-momentum, for instance) of a particle, a tensor field represents a continuous array of values which may be differentiated with respect to space and time. Space and time serve as a background of fixed frame-dependent coordinates, and the fields change and interact upon that background.

The electromagnetic field—defined by the vector potential $\vec{A}(x)$ and the scalar potential $\Phi(x)$—is probably the one familiar to most readers. It took the genius of James Clerk Maxwell to unify their interactions into a single theory, a unification that played a key role in Einstein's development of the theory of special relativity.

Electromagnetism will be an excellent, albeit somewhat overworked, model for us. It will serve as both an outstanding example of a well-known classical theory, as well as a jumping-off point to start talking about the photon and quantization of the field—the existence of which Einstein helpfully demonstrated through his interpretation of the photoelectric effect [52] (and for which he was subsequently awarded the 1921 Nobel Prize in Physics).

Other fields will follow in the same mold, and we will find it convenient to move fluidly (as it were) between discussions of continuous fields and the particles they represent. Eventually. Field quantization is going to require a bit of machinery first, and even our best-understood **vector field**, the electromagnetic potential, is a bit complicated to start with.

It's far simpler to start with a **scalar field**, defined by only a single number everywhere in spacetime. Fortunately for us, this isn't entirely a toy model. Scalar fields describe both the physics of the **Higgs boson** [55, 84, 88], as well as the mechanics of the inflationary early universe [85, 103]. But first, we need some intuition about the physical interpretation of field mechanics.

2.1 The Principle of Least Action

2.1.1 Historical Motivation

Newton's laws work extremely well for describing the unconstrained motions of particles. However, once we introduce constraints or, even worse, *fields*, all bets are off.

Take light, for instance. In the first century CE, Hero of Alexandria wrote the following in his work *Catoptrics* [126]:

> [W]hat moves with constant velocity follows a straight line.... The moving body strives to follow the shortest path since it cannot afford the time for a slower motion, that is a longer path.... The shortest of all lines is the straight line.

This principle, he argued, can explain the reflection of light off a mirror, wherein the angle of incidence equals the angle of reflection.

This same idea was expanded by Pierre de Fermat a millennium and a half later, in 1662 [106]. Fermat's principle, as it's come to be known, supposes that the actual path taken by a beam light between two points is the one that minimizes the total light travel time, thus deriving Snell's law (Figure 2.2) from first principles:

$$n_1 \sin \theta_1 = n_2 \sin \theta_2.$$

Fermat's principle is remarkable in many ways. Starting from global boundary conditions, it becomes possible to derive the path of a beam of light—almost as though the light itself had some sort of purpose in trying to traverse the distance as efficiently as possible.

For the next several centuries (which also saw the development of Newtonian mechanics), applications of minimization principles were further extended to trying to solve dynamical problems. Among the most famous of these was the brachistochrone

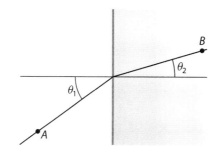

Figure 2.2. Refraction of light in glass.

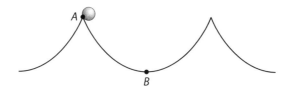

Figure 2.3. The inverted cycloid solves the brachistochrone problem.

problem—to find the curve that would most quickly move a particle, acted on only by gravity, between two designated endpoints.

In rather brash language, in 1696 Johann Bernoulli—one of most eminent of a famous family of mathematical geniuses—offered the brachistochrone problem as a challenge to the mathematics and physics communities:

> If someone communicates to me the solution of the proposed problem, I shall publicly declare him worthy of praise.

Newton, having written the *Principia Mathematica* a decade earlier, was able to solve it in a long evening [115]. He found that the solution was an inverted cycloid (Figure 2.3).

The brachistochrone problem demonstrates that dynamics and minimization problems are intimately related. By the 1750s, Leonhard Euler and his student Joseph Louis Lagrange had developed, in essence, a generalized approach to generating Fermat's theorem.

Their methodology was, at least in part, a consequence of the introduction of the concept of **action** as suggested by Pierre-Louis Moreau de Maupertuis in 1747 (in language reminiscent of the principle introduced by Hero of Alexandria):

> This is the principle of least Action, a principle so wise and so worthy of the supreme Being, and intrinsic to all natural phenomena; one observes it at work not only in every change, but also in every constancy that Nature exhibits. In the collision of bodies, motion is distributed such that the quantity of Action is as small as possible, given that the collision occurs. At equilibrium, the bodies arrange such that, if they were to undergo a small movement, the quantity of Action would be smallest.

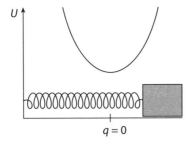

Figure 2.4. A simple harmonic oscillator, along with its corresponding potential.

The action itself wasn't well defined in the modern sense until 1834, when William Hamilton offered the relation

$$S \equiv \int dt\, L(\dot{q}, q, t),$$ (2.1)

where S is the action, q is the coordinate of a system with one degree of freedom, and L is the **Lagrangian** (named in Lagrange's honor by Hamilton). A particle will traverse the path that locally minimizes the action. For particles moving under a conservative force, the Lagrangian is simply the difference of the kinetic and potential energies:

$$L = K - U.$$ (2.2)

Historically, particle dynamics began with the development of the equations of motion, and the Lagrangian formalism was created as a more elegant approach. However, for our analysis of fields, we'll find it far easier to start with a Lagrangian and work our way backward to equations of motion, rather than the other way around. Doing so will lead inexorably to Lorentz-invariant field equations.

2.1.2 The Euler-Lagrange Equations

Consider an example that puts the "classic" in classical mechanics: the mass on a spring. Lessons learned from the simple harmonic oscillator will serve us well.

The mass on a spring has a single degree of freedom, the extension or compression of the spring, q. Given some initial values of q and $\dot{q}$, it is possible to predict the complete future evolution of the oscillator. This is true not only for a mass on a spring but— extended to many particles—for the entire Newtonian framework. The corresponding Lagrangian for the harmonic oscillator is

$$L = \frac{1}{2}m\dot{q}^2 - \frac{1}{2}kq^2,$$ (2.3)

where k is the spring constant of the system, and m is the oscillator mass.

The Lagrangian allows us to compute the action for any potential particle path. In principle, a minimum action path $q_{min}(t)$ may be found through trial and error, but in practice, trial and error isn't necessary. Indeed, we never need to figure out the path of extremized action at all. Instead, we merely need to *suppose* that we have identified an extremum of the action. We then consider what would happen for small perturbations away from that extremum. That is, for any given coordinate

$$\delta L = \frac{\partial L}{\partial q_i}\delta q_i + \frac{\partial L}{\partial \dot{q}_i}\delta \dot{q}_i,$$ (2.4)

where $\delta q_i(t)$ represents the small perturbation from the "true" path. Note that the perturbation of the Lagrangian uses up to only first-order derivatives of q_i. That is because q_i and $\dot{q}_i$ are the only terms to enter the Lagrangian.

Near the minimum, the perturbation of the action should be zero:

$$\delta S = 0 = \int dt \left[\frac{\partial L}{\partial q_i}\delta q_i + \frac{\partial L}{\partial \dot{q}_i}\delta \dot{q}_i \right];$$

or, integrating by parts, we obtain

$$\delta S = \int dt \left[\frac{\partial L}{\partial q_i}\delta q_i - \frac{d}{dt}\left(\frac{\partial L}{\partial \dot{q}_i} \right)\delta q_i \right] + \delta q_i \frac{\partial L}{\partial \dot{q}_i}\bigg|_A^B.$$

The last term integrates zero from the boundary conditions on δq_i, and thus

$$\int dt\, \delta q_i \left[\frac{\partial L}{\partial q_i} - \frac{d}{dt}\left(\frac{\partial L}{\partial \dot{q}_i} \right) \right] = 0$$

for any arbitrary segment of time, and we find the familiar **Euler-Lagrange equations**:

$$\frac{d}{dt}\left(\frac{\partial L}{\partial \dot{q}_i} \right) = \frac{\partial L}{\partial q_i}.$$ (2.5)

Indeed, we could easily imagine deriving classical mechanics in the reverse order of that in which it was discovered: invent a Lagrangian assuming a variational principle and see how a particle evolves under those constraints. This is, incidentally, almost precisely what we *will do* when describing the dynamics of fields, because unlike with individual particles, there is no preexisting equivalent to Newton's second law of motion.

The one-dimensional harmonic oscillator Lagrangian has a corresponding Euler-Lagrange equation:

$$m\ddot{q} = -kq,$$

which you may recognize as simply Hooke's law. Simple physics, to be sure, but the universe of possibilities becomes much more interesting when we allow an *infinite* array of oscillators to serve as a model for continuous fields.

Example 2.1: What are the equations of motion for a particle moving in a central potential $U(r)$, in polar coordinates?

Solution: The Lagrangian of the system is

$$L = \frac{1}{2}m\left(\dot{r}^2 + r^2\dot{\theta}^2 \right) - U(r),$$

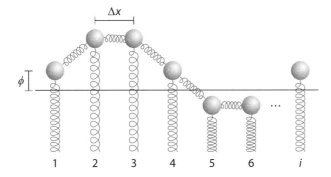

Figure 2.5. A string envisioned as an infinite array of oscillators.

where the kinetic energy term was derived from our work in example 1.4. We begin by noting

$$\frac{\partial L}{\partial \dot{r}} = m\dot{r},$$

which quickly yields two equations of motion:

$$m\ddot{r} = mr\dot{\theta}^2 - \frac{\partial U}{\partial r}$$

$$\frac{d}{dt}\left(mr^2\dot{\theta}\right) = 0,$$

the latter of which is a statement of conservation of angular momentum.

2.2 Continuous Fields

2.2.1 Strings

A scalar field, as we've already established, has a single value defined at all points in spacetime. And since it's a little tricky to visualize a scalar field in three-dimensions, we'll confine ourselves to a one-dimensional string for the time being. We start by laying out an infinite array of simple harmonic oscillators, each one coupled to its immediately adjacent neighbor, as illustrated in Figure 2.5. For each oscillator, the only free parameter is the displacement from equilibrium ϕ_i (and for which we constrain the particles to have only vertical motion). We can write a kinetic energy

$$K = \frac{1}{2}\sum_i m\dot{\phi}_i^2,$$

as well as a potential energy

$$U = \sum_i \left(\frac{1}{2}k_s\phi_i^2 + \frac{1}{2}k_c\left[(\phi_i - \phi_{i-1})^2 + \Delta x^2\right]\right),$$

where k_s is the spring constant of each "self-coupling" spring, and k_c is the coupling between adjacent oscillators. Ignoring the final term in the potential energy (it's a constant), we get for the Lagrangian

$$L = \sum_i \left(\frac{1}{2} m \dot{\phi}_i^2 - \frac{1}{2} k_s \phi_i^2 - \frac{1}{2} k_c (\phi_i - \phi_{i-1})^2 \right).$$

We deal with the spatial boundary conditions by assuming that the oscillators form either an infinite array or a closed loop. Under those conditions, the Euler-Lagrange equations are easy to solve:

$$m \ddot{\phi}_i = -k_c (2\phi_i - \phi_{i-1} - \phi_{i+1}) - k_s \phi_i,$$

where the unexpected terms arise from considering neighbors both to the left and right of oscillator i. If we space the oscillators ever closer together, the preceding solution takes a pleasing form. By redefining the spring parameters

$$\mu \equiv \frac{m}{\Delta x}; \quad T \equiv k_c \Delta x; \quad \sigma \equiv \frac{k_s}{\Delta x}$$

as the mass density, the tension, and the stiffness of the string, respectively and invoking the relation

$$\lim_{\Delta x \to 0} \frac{\phi_{i+1} - 2\phi_i + \phi_{i-1}}{\Delta x^2} = \frac{\partial^2 \phi}{\partial x^2}$$

we obtain a continuous equation of motion from the Euler-Lagrange equations:

$$\mu \ddot{\phi}(x) = T \frac{\partial^2 \phi}{\partial x^2} - \sigma \phi(x), \tag{2.6}$$

Which describes the motion of a massive string under tension.

By a similar substitutions, we can write the Lagrangian for a continuous string:

$$L = \int dx \left(\frac{1}{2} \mu \dot{\phi}^2 - \frac{1}{2} T \left(\frac{\partial \phi}{\partial x} \right)^2 - \frac{1}{2} \sigma \phi^2 \right).$$

We're almost there, but the final steps in going from oscillator to string to hypertrampoline to field involves exploiting a bit of relativity.

2.2.2 From String to Classical Field

The action for our one-dimensional string may be written as

$$S = \int dt \, dx \left(\frac{1}{2} \dot{\phi}^2 - \frac{1}{2} \frac{T}{\mu} \left(\frac{\partial \phi}{\partial x} \right)^2 - \frac{1}{2} \frac{\sigma}{\mu} \phi^2 \right),$$

where we have exploited the fact that minimizing a functional also works if we divide by a fixed constant.

While we derived all of this in 1+1 dimensions (one dimension of space, one of time), there's no reason we can't extend it to 3+1:

$$S = \int d^4 x \left(\frac{1}{2} \dot{\phi}^2 - \frac{1}{2} \frac{T}{\mu} (\nabla \phi)^2 - \frac{1}{2} m^2 \phi^2 \right).$$

You'll note that in addition to increasing the dimensionality of space, we've also made a couple of substitutions. For one, we've defined $m^2 = \sigma/\mu$, a suggestive variable definition that will prove to be well named in short order.

In addition, we've substituted

$$\int dt\, d^3x = \int d^4x.$$

The chief advantage to working in natural units is that space and time are dimensionally equivalent. Our action—that is, the action of all physics over all times and places in our entire universe—is designed to keep space and time on equal footing. With that in mind, we define a **Lagrangian density**:

$$S = \int d^4x \mathcal{L}. \tag{2.7}$$

The Lagrangian density has three major requirements:

1. Terms must be functions of only the field(s), ϕ, or first derivatives of the field(s), $\partial_\mu \phi$. This is the same sort of constraint placed on Lagrangians of particles.
2. The Lagrangian must be a real-valued scalar function with the dimensionality of an energy density: $[E/l^3] = [E]^4$.
3. The functional form of the Lagrangian must be invariant under Lorentz boosts and rotations.

The last requirement seems complicated but isn't. All our work developing special relativity led up to a denouement in which constructing Lorentz-invariant quantities became the easiest thing in the world. We simply need to make sure that all tensor relations are contracted over 3+1 dimensional space. Going from a string or trampoline to a classical Lorentz-invariant field requires us to recognize that we're using an analogy. We need to finally let go of the analogy. That is, if we set $T/\mu = 1$, we produce a Lorentz-invariant Lagrangian for a real-valued scalar field:

$$\mathcal{L} = \frac{1}{2}\partial^\mu \phi \partial_\mu \phi - \frac{1}{2}m^2\phi^2, \tag{2.8}$$

where the raised derivative will produce a negative sign for the spatial derivatives only. Further, from dimensionality considerations, the field itself has units of energy

$$[\phi] = [E],$$

as does the quantity m, which, yes, will turn out to represent the mass of the associated particle.

Throughout, we've said we're describing classical fields, and yet, here we are, talking about particle masses. Don't let the casual mention of particles throw you. We'll see how we get something somewhat like particles even before formally quantizing our fields.

2.2.3 Minimizing the Action

We've introduced a free-field Lagrangian, but now we need to figure out how to minimize the action in general. Though the number of degrees of freedom is (infinitely) larger, the derivation is identical to that for single-particle systems. Perturbing around a minimization of the action alters the action by

$$\delta S = \int d^4 x \left[\frac{\partial \mathcal{L}}{\partial \phi} \delta \phi + \frac{\partial \mathcal{L}}{\partial (\partial_\mu \phi)} \delta (\partial_\mu \phi) \right] = 0.$$

We next integrate by parts:

$$\delta S = \int d^4 x \left[\frac{\partial \mathcal{L}}{\partial \phi} \delta \phi - \partial_\mu \left(\frac{\partial \mathcal{L}}{\partial \partial_\mu \phi} \right) \right] \delta \phi + \frac{\partial \mathcal{L}}{\partial (\partial_\mu \phi)} \delta \phi \bigg|_A^B = 0,$$

where, again, the last term vanishes by the boundary value conditions ($\delta \phi X_A^\mu = \delta \phi X_B^\mu = 0$). Thus, the terms in the square brackets must cancel for any parameterization of the minimized action:

$$\boxed{\partial_\mu \left(\frac{\partial \mathcal{L}}{\partial (\partial_\mu \phi_a)} \right) = \frac{\partial \mathcal{L}}{\partial \phi_a}} \tag{2.9}$$

Hurrah! The expression is the relativistic form of the Euler-Lagrange equations! The subscript a just admits the possibility that we've got more than one field involved.

2.3 The Klein-Gordon Equation

Applying the Euler-Lagrange equations in 4-space can be a frustrating exercise in index accounting when you first encounter it. Consider the free-field scalar Lagrangian that we "derived" earlier (equation 2.8). The right-hand side of the Euler-Lagrange is simple:

$$\frac{\partial \mathcal{L}}{\partial \phi} = -m^2 \phi,$$

but the left-hand side,

$$\partial_\mu \left(\frac{\partial \left(\frac{1}{2} \partial^\nu \phi \partial_\nu \phi \right)}{\partial (\partial_\mu \phi)} \right), \tag{2.10}$$

is considerably tougher.[1] Simply consider the terms inside the parentheses in equation (2.10). There is only one free index, μ, and it is downstairs and also in the denominator. Just as with ordinary fractions, this means that the term will ultimately yield an upstairs index:

$$\left(\frac{\partial \left(\frac{1}{2} \partial^\nu \phi \partial_\nu \phi \right)}{\partial (\partial_\mu \phi)} \right) = \partial^\mu \phi,$$

---

[1] Note that we've changed the dummy index in the interior of the Lagrangian from μ to ν. As we're contracting, it can be anything we like, but it must be a different index from those attached to the derivatives *on* the Lagrangian.

which you can confirm by explicitly expanding the expression for $\mu = 0, 1, 2, 3$. This is very similar to the result you might expect from your experience taking derivatives with respect to quadratic terms generally.

Putting the entire Euler-Lagrange equation together, we get

$$\partial_\mu \partial^\mu \phi + m^2 \phi = 0.$$

The first term is an operator known as the **d'Alembertian**,[2] which can more simply be written as

$$\Box \phi + m^2 \phi = \left(\frac{\partial^2}{\partial t^2} - \nabla^2 + m^2 \right) \phi = 0. \tag{2.11}$$

The d'Alembertian is the central operator in wave mechanics.

Equation (2.11) is known as the **Klein-Gordon equation.** Starting with a second order Lagrangian, we ultimately produced a homogeneous, linear differential equation, which means that we can express solutions as a linear superposition of plane waves:

$$\phi(x) \propto e^{\pm i p \cdot x}. \tag{2.12}$$

The terms in the exponents are most easily understood by separating out the time- and spacelike components:

$$-i p \cdot x = i(\vec{p} \cdot \vec{x} - E_p t),$$

where the timelike component of the wavenumber is just the energy of the wave (presumed to be positive). Application of the Klein-Gordon equation to the field (equation 2.12) requires a dispersion relation such that

$$(-E_p^2 + |\vec{p}|^2 + m^2)\phi = 0,$$

which should look familiar from our derivation of the special-relativistic energy-momentum relation, equation (1.37).

We have *not* come up with a general rule for quantizing our fields, but nevertheless, we have a good sense of what our scalar field represents: the dynamics of a particle of mass m. Since the solutions are linear in ϕ, we superpose an arbitrary number of particles, each evolving independently. We will see, further, that these particles are also spinless, and chargeless, but proof of that will have to wait a bit.

2.4 Which Lagrangians Are Allowed?

The only Lagrangian we've seen so far was, in a sense, pulled out of thin air. Sure, we argued on plausibility grounds that it strongly resembled the physics of an oscillating string (or, in higher dimensions, a trampoline), but that sort of reasoning is only suggestive. While ultimately, we're going to generate the Lagrangians for observed

---

[2] Named after the eighteenth-century mathematician and physicist Jean-Baptiste le Rond d'Alembert.

particles in real physical systems, the space of possibilities is almost too complicated to consider.

However, we are helped by the fact that Lagrangians can't be any arbitrary combination of fields and their derivatives but must obey certain constraints in addition to those noted in § 2.2.2, namely:

1. The Lagrangian must be a function of the field and first derivatives of the field.
2. The Lagrangian must be real and have dimensions of $[E]^4$.
3. The Lagrangian must be Lorentz invariant.

We can further limit the possibilities with the following considerations:

4. All terms in a Lagrangian should be at least quadratic in the field(s) or their derivatives.

   As the dynamical information from a Lagrangian is obtained by taking derivatives with respect to the field, any constant terms will necessarily vanish. Moreover, even linear terms can readily be removed through a redefinition. For instance,

   $$\phi \rightarrow \phi + \Delta\phi$$

   will always allow us to swallow linear terms in such a way as to make a potential vanish around an equilibrium.

5. Higher-order terms may not be important at all energies.

   For scalar fields, there are remarkably few ways to produce dynamical expressions which are Lorentz-invariant real scalars. They typically take the form

   $$\partial_\mu \phi \partial^\mu \phi.$$

   But the same is true for the potential—the terms that include only the field itself. Using the particle Lagrangian formalism as our example ($L = K - V$), we see that the potential for our free-field particle is

   $$V = \frac{1}{2}m^2\phi^2.$$

   But this is not our only option. We might well imagine a potential of the form

   $$V = c_2\phi^2 + c_3\phi^3 + c_4\phi^4 + \ldots.$$

   Potentials of this form have the interesting property that their behavior changes depending on the amplitude of the field. Note, for instance, that c_2 has dimensions of $[E]^2$, while c_3 has dimensions of $[E]^1$, and c_4 is dimensionless.

   For small perturbations, only the c_2 term in the potential will matter. Or, rather, the quadratic term will dominate unless

   $$\phi \gtrsim \frac{c_2}{c_3},$$

   at which point, the cubic term kicks in.

   This is important to remember. In certain physical systems, the "true" Lagrangian may not be the one we see in a particular energy regime. This is the underlying motivation for exploring increasingly energy sectors in particle accelerators.

Given the constraints listed, the *only* real-valued scalar field with a quadratic (low-energy) potential is the one first given in equation (2.8). But nothing prevents us from imagining a universe made of many such fields.

2.5 Complex Scalar Fields

2.5.1 *The Lagrangian*

Quadratic Lagrangians dominate at low energies. They also produce linear dynamical equations, which makes them especially simple to solve. In general, we'll refer to them as **free-field** Lagrangians, and we can easily set up a toy model filled with more than one type of scalar field. For instance,

$$\mathcal{L} = \frac{1}{2}\partial_\mu\phi_1\partial^\mu\phi_1 - \frac{1}{2}m_1^2\phi_1^2 + \frac{1}{2}\partial_\mu\phi_2\partial^\mu\phi_2 - \frac{1}{2}m_2^2\phi_2^2. \tag{2.13}$$

The Euler-Lagrange equations are linearly independent. Particle species "1" has no impact on species "2" and vice-versa. They simply flow past one another through space.

We can continue adding species to our heart's content, but we're able to simplify our expression if the two species have the same mass, $m_1 = m_2 = m$. The potential term becomes

$$\frac{1}{2}m^2(\phi_1^2 + \phi_2^2).$$

Inspection of the Lagrangian (equation 2.13) yields a strange insight: what we call one particle or another is really a matter of mathematical convenience. For instance, suppose we define two new particles, A and B such that

$$\phi_A = \phi_1\cos\theta + \phi_2\sin\theta$$

$$\phi_B = -\phi_1\sin\theta + \phi_2\cos\theta,$$

where θ is an arbitrary phase. The new delineations of these particles are, in a sense, "rotations" of the old versions. Nevertheless, the two versions produce very similar Lagrangian mass terms:

$$\frac{1}{2}m^2(\phi_1^2 + \phi_2^2) = \frac{1}{2}m^2(\phi_A^2 + \phi_B^2).$$

It's a little harder (but still straightforward) to show that

$$\frac{1}{2}\partial_\mu\phi_1\partial^\mu\phi_1 + \frac{1}{2}\partial_\mu\phi_2\partial^\mu\phi_2 = \frac{1}{2}\partial_\mu\phi_A\partial^\mu\phi_A + \frac{1}{2}\partial_\mu\phi_B\partial^\mu\phi_B.$$

The original Lagrangian (equation 2.13) can be rewritten identically for particles A and B instead of 1 and 2. But remember, A and B are *different particles* from the original ones. They are superpositions of 1 and 2, which means, in a sense, that that which we call a particle of a particular type or another is entirely a matter of the basis in which we measure it. This language, incidentally, may be reminiscent of your work on spin in ordinary quantum mechanics.

Since linear combinations of fields can be rewritten to produce Lagrangians, we can consider the two species to be part of a single **complex scalar field**; that is,

$$\phi = \left(\frac{\phi_1 + \phi_2}{2}\right) + i\left(\frac{\phi_1 - \phi_2}{2}\right). \tag{2.14}$$

We leave it as an exercise (problem 2.7) to show that equation (2.13) can be rewritten as

$$\mathcal{L} = \partial_\mu \phi \partial^\mu \phi^* - m^2 \phi \phi^*. \tag{2.15}$$

This relation is necessarily real (satisfying the constraints on Lagrangian densities). In this case, we treat ϕ and ϕ^* like separate fields for the purpose of computing the Euler-Lagrange equations. We will make a lot of use of this toy model for fields, since it's the simplest one we've yet encountered that has more than one particle.

We'll give away some of the surprise and let you know now that complex phases correspond to electric charge, and fields that are complex conjugates of one another have equal and opposite charge. From that logic, purely real fields are necessarily neutral.

Treating ϕ and ϕ^* as independent fields quickly yields the Euler-Lagrange equations:

$$(\Box + m^2)\phi = 0$$
$$(\Box + m^2)\phi^* = 0.$$

Both the ϕ field and its conjugate evolve according to the Klein-Gordon equation, as expected. It's an oversimplification to say that ϕ and ϕ^* represent different particles; one field completely describes the other. Indeed, they represent **antiparticles** of one another.

We are still within the regime of toy models, and thus ϕ and ϕ^* were *designed* to be antiparticles of one another. But as we shall see, all particles have a corresponding antiparticle even if, as is is the case with the Higgs boson, the photon, and the Z^0 particle of the weak interaction, the antiparticle and particle are one and the same. Those three particles are necessarily electrically neutral, since a particle and antiparticle have opposite electric charge, as well as opposite weak charges (weak hypercharge and weak isospin, which we'll explore in Chapter 9) and opposite strong charges (conventionally known as **color**, which we'll discuss in Chapter 11).

This description of the ϕ field as a combination of particles and antiparticles has an interesting consequence. Exchanging ϕ_1 for ϕ_2 has the exact same effect as taking the complex conjugate of the field; that is,

$$\phi_1 \leftrightarrow \phi_2 \text{ is equivalent to } \phi \leftrightarrow \phi^*.$$

But taking the complex conjugate of the field doesn't affect the Lagrangian one iota. This suggests an important symmetry in our toy model.

Example 2.2: Compute the equations of motion for a complex scalar field with a potential

$$V(\phi) = \alpha(\phi\phi^*)^{1/2}.$$

Solution: From the symmetry of the calculation, we need only compute

$$\frac{\partial V}{\partial \phi^*} = \frac{\alpha}{2} \frac{\phi}{|\phi|},$$

where θ is the phase of the field. Thus,

$$\Box \phi = -\frac{\alpha}{2} \frac{\phi}{|\phi|},$$

which can create circular "orbits" for fixed $|\phi|$.

2.5.2 C, P, and T

Our complex scalar field suggests a very nice symmetry between particles and antiparticles. Indeed, from the perspective of our theory there is *no* objective reason to say that ϕ is the particle and ϕ^* is the antiparticle rather than the other way around. A number of **discrete symmetries** are introduced in this simple system, but they show up throughout our study of classical fields.

The complex scalar Lagrangian remains unchanged, both in form and for any particular solution provided we make the following transformations:

- **C (Charge) Transformation** Turn all particles into antiparticles. In this case, it means all ϕ_1 are exchanged with ϕ_2, or, as we've seen,

$$\hat{C}\phi = \phi^*, \tag{2.16}$$

  where the "hat" notation indicates an operator. Objects more complicated than scalar fields will tend to have more complicated charge conjugation operations.

- **P (Parity) Transformations** Invert the spatial coordinates of all the vectors in a system:

$$\hat{P}\vec{x} = -\vec{x}. \tag{2.17}$$

  Thus, a scalar field will simply be turned into its mirror image:

$$\hat{P}\phi(t, \vec{x}) = \phi(t, -\vec{x}).$$

  The parity transform can shed light on a number of familiar concepts. For instance, position (as we've seen) and momentum reflect to include a minus sign:

$$\hat{P}\vec{p} = -\vec{p},$$

  and thus, the cross product of the two is

$$\hat{P}(\vec{x} \times \vec{p}) = (-\vec{x}) \times (-\vec{p}) = \vec{x} \times \vec{p}.$$

Angular momentum, in other words, isn't an ordinary vector. Since the parity operator leaves angular momentum unchanged, angular momentum transforms as what is known as an **axial vector** (or **pseudovector**, but in this text, we prefer the former label). Spin, one of the most important quantities in quantum mechanics, is thus also an axial vector, as are magnetic fields.

You might object that spin and angular momentum are governed by the "right-hand rule," suggesting, naively, that angular momentum should swap under parity transformations. However, under a P symmetry transformation, *every* thing is reflected, including your left and right hands.

In describing the symmetry of theories, we're looking to see whether the theory itself will remain invariant under the transformation. For instance, parity transformation of the Lorentz force law,

$$\vec{F} = q(\vec{E} + \vec{v} \times \vec{B}),$$

produces a negative for the force, E-field, and velocity, leaving the entire equation unchanged.

Likewise, the Lorentz force equation is also unchanged upon C transformations. While individual charges will be inverted, we need to remember that the electromagnetic fields are ultimately generated by charges, and they'll be inverted as well. The net effect will be that a global C transformation will leave the interaction invariant.

- **T (Time) Transformations** Invert the time like coordinate of all vectors.

$$\hat{T}x^0 = -x^0, \tag{2.18}$$

which is most evident in time derivatives:

$$\hat{T}\dot{\phi} = -\dot{\phi} \;\; ; \;\; \hat{T}\ddot{\phi} = \ddot{\phi}.$$

The transformation properties of C, P, and T should make it clear why second derivatives and quadratic first derivatives are so ubiquitous in real fields: those are the combinations which will leave Lagrangians and equations of motions invariant under discrete transformations.

These operators may be combined, and their effects on scalar plane waves are particularly interesting. For instance,

$$\hat{C}e^{-ip\cdot x} = \hat{P}\hat{T}e^{-ip\cdot x} = e^{ip\cdot x}.$$

In this transformation, the elements of p should be treated as fixed numbers (unchanged by the transformations), while x inverts under PT. The combination of parity and time inversions produces the same effect on a field as taking a charge conjugation: both produce a complex conjugate. Applying all three leaves a scalar plane wave completely unaltered. Scalar fields, then, are invariant under CPT transformations.

Our discussion thus far has focused on toy models with scalar particles, but the interrelations among time, space, and charge were noted early on. As Richard Feynman put it in his 1965 Nobel Lecture [62]:

I received a telephone call one day at the graduate college at Princeton from Professor [John Archibald] Wheeler, in which he said, "Feynman, I know why all electrons have the same charge and the same mass." Why? "Because, they are all the same electron!"...But, Professor

I said, there aren't as many positrons[3] as electrons. Well, maybe they are hidden in the protons or something he said. I did not take the idea that all the electrons were the same one from him as seriously as I took the observation that positrons could simply be represented as electrons going from the future to the past in a back section of their world lines. That, I stole!

As we move forward, this convention of assigning antiparticles as the backward-traveling version of their ordinary-matter counterpart will prove to be extremely useful.

2.5.3 C, P, or T Violation

We're often less concerned with the symmetry of a particle or field as we are with the symmetry in the Lagrangian describing a theory. The Klein-Gordon Lagrangian is manifestly invariant under parity inversion simply because scalar field particles don't carry intrinsic angular momentum. Likewise, combinations of $\phi\phi^*$ required the Lagrangian to be invariant under charge conjugation.

Time inversion is a little trickier. A first-derivative velocity term will naturally produce a negative sign upon time inversion T. As only squares of velocities show up in the Lagrangian, the overall expression remains blissfully unaware of time inversion.

It's a simple theory, and thus the Klein-Gordon Theory is invariant under C, P, and T transformations individually, as well as under the combination of CPT. Indeed, *every* theory yet discovered remains unchanged under CPT. Indeed, in the 1950s, Gerhart Lüders [105] and Wolfgang Pauli [123] showed that CPT symmetry, along with an assumption of **locality**,[4] led inexorably to Lorentz invariance.

Our scalar Lagrangians are more special still (along with, we'll note, both the electromagnetic and strong forces), because the Lagrangian remains unchanged under C, P, and T individually. This will *not* be true in general and, in particular, in the weak interaction.

In 1957, Chien-Shiung Wu [166] and her collaborators studied the decay of cobalt-60 and showed that the weak interaction did not respect mirror symmetry (P) but still respected the symmetry of C and P combined, as we'll see in detail in Chapters 5 and 8. This was merely the first indication of symmetry breaking in the fundamental forces.

In 1964, James Cronin and Val Fitch [38] found CP violations in the decays of neutral composite particles known as kaons, which we'll explore further in Chapter 10. For now, we'll finish our discussion by simply noting the significance of CP-violating terms.

Partially in response to the Cronin-Fitch experiments, in 1967, the Soviet physicist Andrei Sakharov [144] argued that for there to be a net of protons over antiprotons (and neutrons over antineutrons), there *needed* to be violation of CP symmetry, in addition to processes that simply *allowed* the creation of a net number of protons or neutrons. The Cronin-Fitch experiments showed that the first criterion was satisfied. Unfortunately, there's been no experimental detection of **baryon number** violation.[5] Given that matter

---

[3] The antiparticle of the electron.

[4] Roughly, that systems are affected only by their immediate surroundings, and thus (as we're assuming in this book) all forces require a mediator to propagate.

[5] Baryons are, for this purpose, protons and neutrons.

seems to have dominated over antimatter in our universe, at some energies or timescales beyond current constraints, the remaining Sakharov condition needs to be met.

Problems

2.1 Snell's law is a classic minimization problem. Consider two points, A and B (as illustrated in Figure 2.2), at $(-d, 0)$ and (d, h), respectively. There is an interface at $x = 0$, to the left of which photons travel at $c = 1$, and to the right of which they travel at $v = 1/n$.

(a) If a photon takes two straight legs to travel from A to B, compute the travel time if it passes through the interface at position y.

(b) Find a relationship that extremizes the light-travel time as a function of y. You need not solve for y.

(c) Noting

$$\sin \theta_1 = \frac{y}{\sqrt{y^2 + d^2}},$$

compute a relationship between $\sin(\theta_2)$ for the extremized path.

2.2 Consider the motion of a particle of mass m on a spring with fundamental frequency $\omega = \sqrt{k/m}$ around the origin. The mass starts at rest at $t = 0$ and undergoes an oscillation of amplitude x_0:

$$x(t) = x_0 \cos(\omega t).$$

(a) Though we almost never do so in practice, compute the *action* of the system over one oscillation.

(b) Now imagine a perturbation

$$\delta x(t) = \epsilon t \left(\frac{2\pi}{\omega} - t \right)$$

which is constrained to be zero at the extremes of the oscillation. Compute the integrated action of

$$x(t) = x_0 \cos(\omega t) + \epsilon t \left(\frac{2\pi}{\omega} - t \right).$$

You should find that $\epsilon = 0$ minimizes the action.

2.3 Consider two particles of equal mass m connected by a spring of constant k and confined to move in one dimension. The entire system moves without friction. At equilibrium, the spring has a length L.

(a) Write down the Lagrangian of this system as a function of x_1 and x_2 and their time derivatives. Assume $x_2 > x_1$.

(b) Write the Euler-Lagrange equation for this system.

(c) Make a change of variables:

$$\Delta \equiv x_2 - x_1 - L$$

$$X \equiv \frac{1}{2}(x_1 + x_2).$$

Write the Euler-Lagrangian in these new variables.

2.4 As a first, simple model for the behavior of interacting fields, consider a mass m on a one-dimensional spring with constant k and an additional, coefficient α on a cubic term such that

$$V(x) = \frac{1}{2}kx^2 + \alpha x^3$$

under the general assumption that the system oscillates with amplitudes of $x \ll k/\alpha$.

(a) Write the Euler-Lagrange equation for the system.

(b) The solution to the Euler-Lagrange equation for a spring may be written as

$$x_0(t) = Xe^{i\omega_0 t},$$

where X is a complex number, and $\omega_0 = \sqrt{k/m}$ as usual. Compute a *perturbation* term $x_1(t)$ by assuming

$$x(t) = x_0(t) + x_1(t)$$

and that $|x_1| \ll |x_0|$.

2.5 Starting with the equation of motion for a string under tension (equation 2.6), compute the oscillation frequency as a general function of mass density μ, tension T, and stiffness σ, for the fundamental mode of a string of length L. Assume a solution of the form

$$\phi(x, t) = \phi_0(t) \sin(kx).$$

Does your answer in any way explain the relative sizes of a bass and a violin?

2.6 We assert that $\partial_\mu \phi$ transforms like a 1-form; that is,

$$\Lambda^\mu_{\ \bar{\mu}} \partial_\mu \phi = \partial_{\bar{\mu}} \phi.$$

(a) Relate the components of $\partial_\mu \phi$ to $\partial_{\bar{\mu}} \phi$ in (1+1)-dimensional spacetime with a boost in the x-direction of speed v.
(b) Show that the kinematic component of the Klein-Gordon Lagrangian (equation 2.8) remains invariant upon a boost.

2.7 Show that the complex Lagrangian

$$\mathcal{L} = \partial_\mu \phi \partial^\mu \phi^* - m^2 \phi \phi^*$$

is algebraically identical to

$$\mathcal{L} = \frac{1}{2} \partial_\mu \phi_1 \partial^\mu \phi_1 - \frac{1}{2} m_1^2 \phi_1^2 + \frac{1}{2} \partial_\mu \phi_2 \partial^\mu \phi_2 - \frac{1}{2} m_2^2 \phi_2^2$$

if $m_1 = m_2 = m$, and

$$\phi = \left(\frac{\phi_1 + \phi_2}{2} \right) + i \left(\frac{\phi_1 - \phi_2}{2} \right).$$

2.8 Consider a complex plane wave of the form

$$\phi = \phi_0 e^{-ik \cdot x},$$

where k and x are both 4-vectors, and $\omega = k^0$ should be interpreted as the frequency, while k^i are the components of the wavenumber. Recalling the Klein-Gordon equation for massive fields,

$$\Box \phi + m^2 \phi = 0,$$

compute the dispersion relation for the wave and from that compute the **phase velocity** ω/k and the **group velocity** $d\omega/dk$. The group velocity is the one that actually carries information and so, presumably, should be subluminal.

2.9 Consider a Lagrangian of a real-valued scalar field:

$$\mathcal{L} = \frac{1}{2} \partial_\mu \phi \partial^\mu \phi - \frac{1}{2} m^2 \phi^2 - \frac{1}{6} c_3 \phi^3.$$

(a) Is this Lagrangian Lorentz invariant? Is it invariant under C, P, and T transformations individually?
(b) What are the dimensions of c_3?
(c) What is the Euler-Lagrange equation for the field?

(d) Ignoring the c_3 contribution, a free-field solution may be written

$$\phi_0(x) = Ae^{-ip\cdot x} + A^* e^{ip\cdot x}$$

for a complex coefficient A. Consider a lowest-order contribution for $\phi_1 \ll A$ to a perturbation such that

$$\phi(x) = \phi_0 + \phi_1.$$

Derive a dynamical equation for ϕ_1.

2.10 Show that a complex scalar Lagrangian with $V = V(\phi\phi^*)$ is invariant under C, P, and T transformations.

Further Readings

- Cottingham, W. N., and D. A. Greenwood. *An Introduction to the Standard Model of Particle Physics.* Cambridge: Cambridge University Press, 1998. Chapter 3 has an especially nice discussion of the Klein-Gordon equation.
- Neuenschwander, Dwight E. *Emmy Noether's Wonderful Theorem.* Baltimore, MD: Johns Hopkins University Press, 2011. Neuenschwander's book is somewhat mathematical for a popular science book, but a bit general for students considering field dynamics at the advanced undergraduate level. However, it contains many interesting historical discussions.
- Taylor, John R. *Classical Mechanics.* Herndon, VA: University Science Books, 2005. Taylor is an outstanding general introduction to nonrelativistic particle mechanics. Chapters 6 and 7, which cover Lagrangian mechanics, will be especially useful.
- Tong, David. Lectures on Quantum Field Theory. http://www.damtp.cam.ac.uk/user/tong/qft.html. University of Cambridge, 2006. Readers may find sections 1, "Classical Field Theory," and 2, "Canonical Quantization," particularly useful.
- Zee, A. *Quantum Field Theory in a Nutshell.* Princeton, NJ: Princeton University Press, 2003. Our discussion of the derivation of field mechanics was largely inspired by Zee's approach.

3 | Noether's Theorem

Figure 3.1. Emmy Noether (1882–1935), c. 1910. Noether developed the general connection between symmetry and conservation laws.

Conservation laws are the bread and butter of physics. Quantities like angular or linear momentum or electric charge are useful because you can measure them at one moment, and so they will remain forever. For much of the history of physics, conservation laws were taken to be almost axiomatic facts of nature. Galileo argued for something very much like the conservation of momentum in his *Two New Sciences* [72] in 1638, which ultimately gave rise to Newton's first law of motion. Benjamin Franklin, in a letter to Peter Collinson in 1747 [67] noted a similar effect for what we now call electrical charges:

> the Equality is never destroyed, the Fire only circulating. Hence have arisen some new Terms among us. We say *B* (and other Bodies alike circumstanced) are electrised *positively*; *A negatively*: Or rather *B* is electrised *plus* and *A minus*.

Energy conservation, too, seems experimentally to be a fact of nature, a property noted by Galileo in the motion of pendulums, and culminating, in more modern language, in the first law of thermodynamics as elucidated by Rudolf Clausius in 1850 [39].

Conservation laws are useful to be sure, but by the turn of the twentieth century, no one had any real idea of why nature conserved some quantities and not others. The connection had to wait until 1918, when the mathematician Amalie "Emmy" Noether published her eponymous theorem [117, 115] relating conservation laws to symmetries. We've already seen how useful symmetries can be by construction of only Lorentz-invariant Lagrangians. However, Noether's theorem demonstrated that the seemingly "obvious" symmetries of nature—invariance over time and throughout space, for instance—can have profound implications and predict (or really, retrodict, given the order of scientific discovery) important conservation laws.

Noether's theorem can be paraphrased conceptually:

Symmetries give rise to conservation laws.

Noether's contributions tend to get a rather cursory treatment in classical mechanics texts, since many of the most important conserved quantities in particle systems can be derived or deduced almost by inspection. For fields, conservation laws are another matter entirely; we'd be completely lost without Noether's insights as to what's symmetric and what isn't.

3.1 Conserved Quantities for Particles

Noether symmetries are transformations that leave the underlying Lagrangian of a system unchanged. Suppose a system is observed in a particular frame. We may then switch to other reference frames using a coordinate transformation via a single continuous parameter. For instance, in one dimension

$$q \to q + \epsilon \tag{3.1}$$

is equivalent to moving the origin to the left a small amount ϵ: a small but arbitrary parameter describing the transformation. In the limit of $\epsilon \to 0$, the coordinates (and thus the Lagrangian) remain unchanged.

We are especially interested in systems such that

$$\delta L = \sum \left[\frac{\partial L}{\partial q_i} \delta q_i + \frac{\partial L}{\partial \dot{q}_i} \delta \dot{q}_i \right] = 0, \tag{3.2}$$

where δq_i and $\delta \dot{q}_i$ refer to the change in the particle coordinates due to the transformation. If the Lagrangian remains unchanged by a transform, then we're dealing with a symmetry of the system. This is, indeed, the very *definition* of a symmetry in the sense described by Weyl in Chapter 1.

The first term under the sum in equation (3.2) can be substituted by a judicious use of the Euler-Lagrange equations:

$$\delta L = \sum \left[\frac{d}{dt} \left(\frac{dL}{d\dot{q}_i} \right) \delta q_i + \frac{\partial L}{\partial \dot{q}_i} \delta \dot{q}_i \right]$$

$$= \frac{d}{dt} \sum \frac{dL}{d\dot{q}_i} \delta q_i.$$

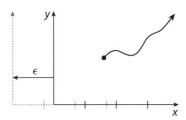

Figure 3.2. The Lagrangian remains unchanged under a linear translation of coordinates.

Assuming that all coordinates can be written as a function of the transformation parameter ϵ,

$$\frac{d}{dt}\left[\sum_i \frac{\partial L}{\partial \dot{q}_i}\frac{dq_i}{d\epsilon}\right] = \frac{dL}{d\epsilon}, \tag{3.3}$$

and thus the term in the brackets is conserved if the Lagrangian is invariant under the transformation. The trick is figuring out symmetry transformations that leave the Lagrangian unchanged.

Example 3.1: Linear Momentum

Consider a particle moving in a gravitational field near the surface of the earth:

$$L = \frac{1}{2}m\dot{x}^2 + \frac{1}{2}m\dot{y}^2 - mgy.$$

What is the conserved Noether quantity for horizontal translations?

Solution: The Lagrangian is invariant under the translation

$$x \to x + \epsilon,$$

since there is no explicit or implicit x dependence in the Lagrangian. Thus, the conserved Noether quantity is

$$p_x = \frac{\partial L}{\partial \dot{x}}\frac{\partial x}{\partial \epsilon} = m\dot{x}.$$

Noether's theorem immediately (and unsurprisingly) yields conservation of linear momentum.

Example 3.2: Angular Momentum

Consider the motions of a planet in the potential of the sun, or any other central potential:

$$L = \frac{1}{2}m\left[\dot{r}^2 + r^2\dot{\theta}^2 + r^2\sin^2\theta\dot{\phi}^2\right] - U(r),$$

where $U(r)$ is some unknown central potential. Under a central potential, the transform is

$$\phi \to \phi + \epsilon.$$

Find the conserved Noether quantity.

Solution: The transformation leaves the Lagrangian unchanged, since ϕ enters the Lagrangian only in the time derivative. Thus,

$$p_\phi = \frac{\partial L}{\partial \dot{\phi}} \frac{\partial \phi}{\partial \epsilon}$$

$$= mr^2 \sin^2 \theta \, \dot{\phi},$$

which is simply the angular momentum in the z-direction.

While the form of the Lagrangian may remain invariant under translations in time, the *value* of the Lagrangian likely will not. Thus,

$$\frac{dL}{dt} \neq 0.$$

In that case, equation (3.3) produces the conservation relation

$$\frac{d}{dt}\left[\sum_i \frac{\partial L}{\partial \dot{q}_i} \dot{q}_i - L \right] = 0. \tag{3.4}$$

You may recognize the term in the square brackets as the **Hamiltonian**, the classical energy of the system.

Energy and time invariance are very intimately related. You may even have anticipated a relationship of this form from the time-dependent Schrödinger equation,

$$i\partial^0 = \hat{H},$$

just as spatial invariance and momentum are connected via the operator

$$i\partial^i = \hat{p}^i.$$

It's not chance that the time derivative operator and the energy operator are one and the same in quantum mechanics.

Noether's theorem for particles produces a number of important conservation laws:

- Symmetry in time translation → conservation of energy
- Symmetry in spatial translation → conservation of linear momentum
- Symmetry in rotation → conservation of angular momentum

But the true power of Noether's theorem comes in exploring the symmetries of field Lagrangians. In those cases, there aren't necessarily any simple Newtonian analogs.

3.2 Noether's First Theorem

Conservation laws for fields are somewhat more complicated than for individual particles. Under Newtonian mechanics, in particular, we can simply specify some quantity— energy, momentum, and so on—and note that it doesn't change over time. But that *privileges* time over the spatial coordinates, and our insistence on Lorentz invariance in all things suggests that conservation laws need to have a somewhat more general form.

Rather than using simple numbers, we'll develop a theory of **conserved currents**, generically written as J^μ. The most famous of these currents, the electromagnetic charge density and three-dimensional current density, can be written

$$J^\mu = \begin{pmatrix} \rho \\ J^x \\ J^y \\ J^z \end{pmatrix}.$$

A conserved current (whether electromagnetic or otherwise) is such because

$$\partial_\mu J^\mu = \dot{\rho} + \nabla \cdot \vec{J} = 0. \tag{3.5}$$

This form is Lorentz invariant (note the matching indices!), but more important, it immediately yields the familiar statement of conservation of mass or energy in fluid dynamics. While the form itself is Lorentz invariant, for many practical calculations, it makes sense to separate the time from the other coordinates.

Integration over an arbitrary volume yields

$$\int dV \left(\dot{\rho} + \nabla \cdot \vec{J} \right) = 0.$$

Defining a charge Q (electrical or otherwise) as the integral over ρ and invoking the divergence theorem for the second term yields a very compact relationship:

$$\dot{Q} = -\oint_S \vec{J} \cdot d\vec{S}. \tag{3.6}$$

The charge within a region decreases or increases only when a current carries it out or in across a surface. Supposing that surface is the entire universe, or any sufficiently isolated region of space, Q will be a conserved quantity.

We note that the definition of a conserved quantity becomes a bit more complicated in curved spacetime. Under those circumstances, care needs to be taken with the volume element. We consider just such a situation in the context of an expanding universe in §3.6. But for now, we're confronted with a more fundamental question: How are these currents generated in the first place?

3.2.1 Symmetries and Conservation Laws

We showed for a single classical particle that if the Lagrangian (or the action, generally) is invariant under some transformation, then there is a conserved value. Now, we're going to do something more general. Noether's first theorem really states

> If a system has a continuously symmetric action, then there is a corresponding conserved current.

Having motivated with a single-particle Lagrangian, we consider coordinate transformations on a Lagrangian density on a group of fields:

$$\delta\mathcal{L} = \frac{\partial\mathcal{L}}{\partial\phi_a}\delta\phi_a + \frac{\partial\mathcal{L}}{\partial(\partial_\mu\phi_a)}\partial_\mu(\delta\phi_a)$$

$$= \partial_\mu\left(\frac{\partial\mathcal{L}}{\partial(\partial_\mu\phi_a)}\right)\delta\phi_a + \frac{\partial\mathcal{L}}{\partial(\partial_\mu\phi_a)}\partial_\mu(\delta\phi_a)$$

$$= \partial_\mu\left(\frac{\partial\mathcal{L}}{\partial(\partial_\mu\phi_a)}\delta\phi_a\right), \tag{3.7}$$

where the fields and coordinates vary with respect to the transformation parameter ϵ. We will see how, in short order, but in general, we may suppose that the symmetry transform is described by a parameter ϵ such that

$$\delta\phi_a = \frac{\partial\phi_a}{\partial\epsilon}\epsilon.$$

The variation of the field with the transformation parameter can be a bit tricky, but it will become more intuitive with a few worked examples, as we'll see in the next section.

If the Lagrangian is unchanged on the transformation, then our work is done. Comparison of equation (3.7) with equation (3.5) suggests that we have already found a conserved current.

Noether's theorem is useful, in large part, because the *form* of the Lagrangian is often patently symmetric upon a particular transformation. The Klein-Gordon Lagrangian (and all other Lagrangians found in nature) has no explicit position or time dependence, and thus displacements in spacetime leave the form of the Lagrangian unchanged.

The same cannot be said for the *value* of the Lagrangian. The Lagrangian is also a function, and thus transformations—especially spatial transformations—mean that the Lagrangian needs to be evaluated in the transformed coordinates, and this means that it will have a different value.

How can the Lagrangian vary? Unlike a field, which can be complex or can represent vectors or other complicated objects, the Lagrangian must be a real-valued scalar. Thus, we tend to care only about coordinate transformations; that is,

$$\frac{\partial\mathcal{L}}{\partial\epsilon} = \partial_\mu\mathcal{L}\frac{\partial x^\mu}{\partial\epsilon}. \tag{3.8}$$

Supposing that the divergence of the coordinate on the transform is zero (as is the case for coordinate rotations and displacements),

$$\partial_\mu\frac{\partial x^\mu}{\partial\epsilon} = 0,$$

which yields a particularly nice result:

$$\frac{\partial\mathcal{L}}{\partial\epsilon} = \partial_\mu\left(\frac{\partial x^\mu}{\partial\epsilon}\mathcal{L}\right). \tag{3.9}$$

If the perturbation to the Lagrangian is expressible in terms of a *total* derivative, then equation (3.7) produces a conserved current density

$$\boxed{J^\mu = \frac{\partial \mathcal{L}}{\partial(\partial_\mu \phi_a)} \frac{\partial \phi_a}{\partial \epsilon} - \frac{\partial x^\mu}{\partial \epsilon} \mathcal{L},} \tag{3.10}$$

where there is an explicit sum over all the fields. If the form of the current isn't immediately intuitively obvious, don't worry. A few examples will help clarify things.

3.3 The Stress-Energy Tensor

We saw in our analysis of nonrelativistic particles that invariant Lagrangians upon displacement in space give rise to a conserved momentum, and symmetry in time give rise to a conserved energy. But special relativity suggests that the two displacements should be related to one another. Thus, consider a more general displacement in spacetime:

$$x^\nu \to x^\nu + \epsilon^\nu.$$

In this particular case, ϵ^ν represents four possible different transformation parameters. As a result, we expect to find four different (but related) conserved currents. Shifting our coordinates to the left (for instance) changes the value of various scalar fields at any given point in spacetime by

$$\phi_a(x) \to \phi_a(x) + \epsilon^\nu \partial_\nu \phi_a,$$

which may be expressed more compactly as

$$\frac{\partial \phi_a}{\partial \epsilon^\nu} = \partial_\nu \phi_a,$$

yielding a conserved Noether current:

$$T^\mu_{\ \nu} = \frac{\partial \mathcal{L}}{\partial(\partial_\mu \phi_a)} \partial_\nu \phi_a - \delta^\mu_{\ \nu} \mathcal{L}.$$

This is, in fact, a set of four conserved currents, known as the **stress-energy tensor**, by comparison with fluid mechanics (albeit extended into 3+1 dimensions).

Raising the lowered index yields the tensor form:

$$T^{\mu\nu} = \frac{\partial \mathcal{L}}{\partial(\partial_\mu \phi_a)} \partial^\nu \phi_a - g^{\mu\nu} \mathcal{L}. \tag{3.11}$$

The stress-energy tensor is, as we noted, a collection of four closely related currents, which may be interpreted as follows:

- T^{00} – The energy density of a field
- T^{i0} – The momentum density of a field in the ith direction
- T^{0i} – The energy flow through through a plane normal to the ith direction
- T^{ij} – The stress tensor as normally defined in fluid mechanics

The stress-energy tensor is most famous as the source term of general relativity. It provides the basis for the Einstein field equation [54],

$$G^{\mu\nu} = 8\pi G T^{\mu\nu}, \tag{3.12}$$

where $G^{\mu\nu}$ is known as the **Einstein tensor** and describes, roughly, the overall curvature of the universe. It's not just the energy density that produces gravitational fields: the pressure and bulk flow contribute as well.

3.3.1 The Stress-Energy of Real-Valued Scalar Fields

Consider a real-valued scalar field with an arbitrary potential $V(\phi)$:

$$\mathcal{L} = \frac{1}{2}\partial_\mu\phi\partial^\mu\phi - V(\phi). \tag{3.13}$$

The density component of the stress-energy tensor almost immediately yields

$$\begin{aligned}
\rho = T^{00} &= \dot{\phi}^2 - \mathcal{L} \\
&= \frac{1}{2}\dot{\phi}^2 + \frac{1}{2}(\nabla\phi)^2 + V(\phi).
\end{aligned} \tag{3.14}$$

The form of the energy density is almost identical (except for the gradient terms) with what you might naively expect from simply extrapolating from the energy of a particle in a one-dimensional potential. Unsurprisingly, the energy density is *not* Lorentz invariant, since it involves only a single component of a tensor.

Other terms in the stress-energy tensor have important physical interpretations. The diagonal spacelike terms, for instance, are equal to the pressure:

$$\begin{aligned}
T^{11} &= (\partial_1\phi)^2 + \mathcal{L} \\
&= \frac{1}{2}\dot{\phi}^2 + \frac{1}{2}\left[(\partial_1\phi)^2 - (\partial_2\phi)^2 - (\partial_3\phi)^2\right] - V(\phi).
\end{aligned}$$

It may take a little experimentation to verify that manipulating signs and positions of indices works, and after we've done that, we're left with a rather awkward-looking expression. However, this simplifies considerably if

$$(\partial_1\phi)^2 = (\partial_2\phi)^2 = (\partial_3\phi)^2 = \frac{1}{3}(\nabla\phi)^2,$$

as will be the case for an isotropic fluid. In that case, $T^{11} = T^{22} = T^{33}$ is the pressure:

$$P = \frac{1}{2}\dot{\phi}^2 - \frac{1}{6}(\nabla\phi)^2 - V(\phi). \tag{3.15}$$

The relationship between pressure and density will become particularly interesting when we consider the behavior of the inflaton field in the early universe.

3.3.2 Complex Scalar Fields

Next, consider a massive complex scalar field (equation 2.15). Following the example for a real field almost exactly, we can quickly compute the various components of the stress-energy tensor, and in particular, the energy density:

$$\rho = \dot{\phi}\dot{\phi}^* + \nabla\phi \cdot \nabla\phi^* + m^2\phi\phi^*. \tag{3.16}$$

The complex field can be thought of as a superposition of fields for an antiparticle pair, which we'll label *b particles* and *c particles* (as we motivated in equation 2.14). Further, we can express a general solution to the field as a plane-wave Fourier expansion of the Klein-Gordon solution (equation 2.12), albeit with a somewhat odd-looking normalization:

$$\phi(x) = \int \frac{d^3p}{(2\pi)^3} \frac{1}{\sqrt{2E_p}} \left[b_{\vec{p}} e^{-ip\cdot x} + c_{\vec{p}}^* e^{ip\cdot x} \right]. \tag{3.17}$$

The terms $b_{\vec{p}}$ and $c_{\vec{p}}$ are arbitrary complex coefficients with the surprising dimensionality of $[b_{\vec{p}}] = [c_{\vec{p}}] = [E]^{-3/2}$.

Most of our work in this book will be made enormously simpler by working in Fourier space, since plane waves are energy and momentum eigenfunctions. Linear operations, including time and space derivatives, are made especially simple, and because the fundamental basis involves the momentum and energy of individual particles, it will be especially simple to interpret Fourier modes physically.

Further, going from classical to quantum will prove especially simple in Fourier space. It should not surprise you to learn that when we ultimately quantize the scalar field, $b_{\vec{p}}$ and $c_{\vec{p}}$ will be operators, creating and annihilating the b and c particles of the appropriate momentum and energy. For now, they'll remain simple classical amplitudes.

We can operate on ϕ by taking time or spatial derivatives of equation (3.17). For instance,

$$\dot{\phi} = -i \int \frac{d^3p}{(2\pi)^3} \frac{E_p}{\sqrt{2E_p}} \left[b_{\vec{p}} e^{-ip\cdot x} - c_{\vec{p}}^* e^{ip\cdot x} \right].$$

But since the density is quadratic in the field, we need to integrate over a second dummy momentum. Thus, the first term in the energy-density relation (equation 3.16) is

$$\dot{\phi}\dot{\phi}^* = \iint \frac{d^3p\, d^3q}{(2\pi)^6} \frac{E_p E_q}{2\sqrt{E_p E_q}} \left[b_{\vec{p}} e^{-ip\cdot x} - c_{\vec{p}}^* e^{ip\cdot x} \right] \left[b_{\vec{q}}^* e^{iq\cdot x} - c_{\vec{q}} e^{-iq\cdot x} \right].$$

As we are treating superpositions of plane waves, on average, the field will represent a constant density and momentum of particles. That said, it will make sense to integrate over all space. Under those circumstances, our exponentials simplify dramatically. We will use the integral relationship

$$\int d^3x\, e^{ik\cdot x} = (2\pi)^3\delta(\vec{k}) \tag{3.18}$$

a lot. In that case, cross terms will produce very nice delta functions. For instance,

$$\int d^3x e^{i(p\mp q)\cdot x} = (2\pi)^3\delta(\vec{p}\mp\vec{q}).$$

Integrating over all space as well as d^3q yields

$$\int d^3x \dot{\phi}\dot{\phi}^* = \int \frac{d^3p}{(2\pi)^3} \frac{E_p^2}{2E_p} \left[b_{\vec{p}} b_{\vec{p}}^* + c_{\vec{p}} c_{\vec{p}}^* - b_{\vec{p}} c_{-\vec{p}} - b_{\vec{p}}^* c_{-\vec{p}}^* \right].$$

The last two terms might seem like a cause for concern, but be patient. Multiplying out the other two terms in the energy density (3.16) in similar (and unilluminating) fashion, we get

$$\int d^3x\, \rho = \int \frac{d^3p}{(2\pi)^3}\frac{1}{2E_{\vec{p}}}\left[\left(E_{\vec{p}}^2 + |\vec{p}|^2 + m^2\right)\left(b_{\vec{p}}b_{\vec{p}}^* + c_{\vec{p}}c_{\vec{p}}^*\right) + \left(-E_{\vec{p}}^2 + |\vec{p}|^2 + m^2\right)\left(b_{\vec{p}}c_{-\vec{p}} - b_{\vec{p}}^*c_{-\vec{p}}^*\right)\right]$$

$$= \int \frac{d^3p}{(2\pi)^3}\frac{1}{2E_{\vec{p}}}\left[\left(2E_{\vec{p}}^2\right)\left(b_{\vec{p}}b_{\vec{p}}^* + c_{\vec{p}}c_{\vec{p}}^*\right) + (0)\left(b_{\vec{p}}c_{-\vec{p}} - b_{\vec{p}}^*c_{-\vec{p}}^*\right)\right]$$

$$= \int \frac{d^3p}{(2\pi)^3}\, E_{\vec{p}}\left(b_{\vec{p}}b_{\vec{p}}^* + c_{\vec{p}}c_{\vec{p}}^*\right), \tag{3.19}$$

a very pleasing-looking result. A straightforward interpretation of this is that

$$N_c = \int \frac{d^3p}{(2\pi)^3}\, c_{\vec{p}}c_{\vec{p}}^*$$

represents the number of c's in the system. The result is similar for b particles.

Example 3.3: Calculate the linear momentum density $\int d^3x\, T^{i0}$ for a complex-valued scalar field as expanded in terms of Fourier modes.

Solution: The momentum is

$$\dot{\phi}\partial^i\phi^* + \dot{\phi}^*\partial^i\phi$$

with $g^{0i} = 0$. Computing the spatial derivative of equation (3.17), we obtain

$$\partial^i\phi(x) = -i\int \frac{d^3p}{(2\pi)^3}\frac{p^i}{\sqrt{2E_p}}\left[b_{\vec{p}}e^{-ip\cdot x} - c_{\vec{p}}^*e^{ip\cdot x}\right],$$

so

$$\dot{\phi}^*\partial^i\phi = \iint \frac{d^3p}{(2\pi)^3}\frac{d^3q}{(2\pi)^3}\frac{p^iE_q}{2\sqrt{E_pE_q}}\left[b_{\vec{p}}e^{-ip\cdot x} - c_{\vec{p}}^*e^{ip\cdot x}\right]\left[b_{\vec{p}}^*e^{iq\cdot x} - c_{\vec{p}}e^{-iq\cdot x}\right],$$

which, adding the complex conjugate and integrating over all space, quickly yields

$$\int d^3x\, T^{i0} = \int \frac{d^3p}{(2\pi)^3}\, p^i[b_{\vec{p}}b_{\vec{p}}^* + c_{\vec{p}}c_{\vec{p}}^*],$$

demonstrating that both b and c particles carry positive inertial mass.

3.4 Angular Momentum

As was the case in particle dynamics, rotation is also the generator for angular momentum in fields. To make matters more concrete, we will focus on rotations around the z-axis, which will, naturally enough, generate the z-component of angular momentum. Under those circumstances

$$\frac{\partial\phi_a}{\partial\epsilon} = (x^2\partial_1\phi_a - x^1\partial_2\phi_a).$$

The angular momentum is a conserved current l_z^μ, but we're really interested only in $\mu = 0$, since it corresponds to the angular momentum density, as opposed to the angular momentum flow. Thus, following the prescription for generating currents (equation 3.10), yields

$$\frac{\partial x^0}{\partial \epsilon} = 0,$$

and thus,

$$l_z^0 = \dot{\phi}\left[x^2 \partial_1 \phi_a - x^1 \partial_2 \phi_a\right], \tag{3.20}$$

which more or less immediately produces the relation

$$\vec{l} = \vec{r} \times \vec{p},$$

where $\vec{l}$ is the angular momentum density in each of the three spatial directions. This is, naturally, the expected result.

There's a slightly more intuitive way to describe the angular momentum density, one which doesn't require *quite* so many cross products, namely,

$$l_z = \frac{\partial \mathcal{L}}{\partial \dot{\phi}_a}\frac{\partial \phi_a}{\partial \theta}, \tag{3.21}$$

where θ is the angle of rotation around the z-axis. We'll find this form a bit more useful when we start considering the angular momenta of nonscalar fields later on.

3.5 Electric Charge

We've focused thus far exclusively on coordinate transforms, but many other symmetry transformations are possible. In the Standard Model, most of the interesting symmetries arise out of **internal symmetries**, which are perhaps simpler to explain by way of example than by a strict definition.

In quantum mechanics, since all measurable quantities (e.g., the probability density) are products of the wavefunction and its complex conjugate, (e.g., $\psi\psi^*$), changing the phase of the wave produces no change in observables; that is,

$$\psi\psi^* \to \left(e^{-i\theta}\psi\right)\left(e^{i\theta}\psi^*\right) = \psi\psi^*.$$

The same relation holds for complex-valued scalar fields. Since all terms in the Lagrangian are products of the field and its conjugate,

$$\mathcal{L} = \partial_\mu \phi \partial^\mu \phi^* - m^2 \phi\phi^*,$$

then transformations of the form

$$\phi \to \phi e^{-iq\theta}$$
$$\phi^* \to \phi^* e^{iq\theta}$$

will yield no change in either the form or the value of the Lagrangian. Here, q is an arbitrary coupling constant that we will associate with the electric charge of the particle.

Noether's theorem tells us exactly how to turn this symmetry into a conserved quantity. Under small changes in the parameter θ the fields change as

$$\frac{\partial \phi}{\partial \theta} = -iq\phi$$

$$\frac{\partial \phi^*}{\partial \theta} = iq\phi^*,$$

yielding an electromagnetic current:

$$J^\mu = iq \left[\phi \partial^\mu \phi^* - \phi^* \partial^\mu \phi \right]. \tag{3.22}$$

Calculating the current for the plane-wave expansion of the complex field (equation 3.17) gives

$$\int d^3x \, J^\mu = q \int \frac{d^3 p}{(2\pi)^3} \frac{p^\mu}{E_p} \left(b_{\vec{p}} b_{\vec{p}}^* - c_{\vec{p}} c_{\vec{p}}^* \right). \tag{3.23}$$

Several exciting results pop out immediately. For one, it is clear that the b and c particles have opposite charges, despite having the same mass, as we showed earlier. It's completely arbitrary which one we choose to call the particle and which one the antiparticle. It's merely historical accident (a convention owed to Benjamin Franklin) that we label the electron with a negative charge.

For the timelike component J^0, the total charge is

$$Q = q(N_b - N_c), \tag{3.24}$$

while for the spacelike components, the current is

$$I^i = q \langle v^i \rangle (N_b - N_c).$$

Since we haven't yet introduced any electromagnetism into our theory of fields, the very concept of charge is thus far meaningless. But in short order, we'll see how the same symmetry which gives rise to conserved charge will also give rise to Maxwell's equations themselves.

3.6 Digression: Inflation

As a final example of the centrality of conserved currents in understanding dynamical systems, we will explore how even a simple scalar field can exhibit complex behavior under the right circumstances. In particular, we will consider the dynamics of the very early universe, both as an interesting system in its own right, as well as a precursor to our work on the Higgs field in Chapter 9.

3.6.1 Fluids in an Expanding Universe

As Georges Lemaître predicted from general relativity in 1927 [101], and Edwin Hubble [90] observed two years later, the universe is expanding. Technically, Hubble simply noted that nearby galaxies are generally receding from us, with more distant galaxies receding more quickly, but the implication is the same.

Rewind the expansion, and the natural interpretation is that at some early time, the overall density and temperature of space must have been incredibly high. This is the foundation of the hot Big Bang model of cosmology. At the earliest moments (current estimates put it at around 10^{-36} s after the Big Bang), the dynamics of the universe were likely driven by a scalar field known as the **inflaton field** [85, 103]. We'll consider how the inflaton field and the expansion of the universe played off one another. But first, we'll need a little background physics.

Hubble noted that within the distance range that he could accurately measure, galaxy recession seemed to be linearly proportional to distance. Through the lens of general relativity, this is interpreted as the scale of the universe increasing monotonically by a dimensionless scale parameter $a(t)$. Expansion diffuses the density of a system, but some care must be used, since the expansion of a fluid under pressure does work on the universe. As a consequence, the density of a fluid scales as

$$\rho \propto a^{-3(1+w)}, \tag{3.25}$$

where w is the **equation of state** of the fluid, which in natural units may be written

$$w \equiv \frac{P}{\rho}. \tag{3.26}$$

In practice, w is a rational constant for many fluids ($\simeq 0$ for nonrelativistic matter, $1/3$ for radiation). For ordinary, nonrelativistic matter, the density is inversely proportional to the proper volume element, $\rho \propto a^{-3}$, as expected. Conversely, for relativistic fluids, including radiation, the expansion of the universe produces the somewhat more surprising relation $\rho \propto a^{-4}$. Not only does the density of photons decrease in an expanding universe, but the energy of each photon decreases as its wavelength stretches with the universe.

Most peculiar is the case of $w = -1$, wherein, counterintuitively, the density remains constant despite the expansion. A negative pressure corresponds to a tension, which we can generate with a wide range of scalar fields. Consider a Lagrangian with an as-yet-unspecified potential:

$$\mathcal{L} = \frac{1}{2} \partial^{\mu} \phi \partial_{\mu} \phi - V(\phi). \tag{3.27}$$

We may compute the density (equation 3.14) and pressure (equation 3.15) relations, and further simplify under the assumption of a homogeneous field:

$$w = \frac{P}{\rho} = \frac{\frac{1}{2}\dot{\phi}^2 - V(\phi)}{\frac{1}{2}\dot{\phi}^2 + V(\phi)}. \tag{3.28}$$

For systems where the potential term is larger than the kinetic term, $w \to -1$, and, following equation (3.25), the energy density of the field will remain a constant in time.

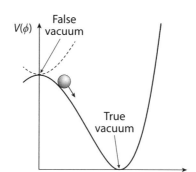

Figure 3.3. False vacuum.

3.6.2 The Friedmann Equation

Einstein's field equation (3.12) relates the stress-energy tensor to the time evolution of a gravitational system. During the 1920s, Alexander Friedmann and others found that the evolution of the expansion factor could be written as a fairly simple function of an integration constant and the energy density at any given time. This solution is known as the **Friedmann equation** [69] and can be used to compute the rate of expansion at any epoch in the universe:

$$\left(\frac{\dot{a}}{a}\right)^2 \equiv H^2 = \frac{8\pi G\rho}{3}, \tag{3.29}$$

where H is known as the **Hubble parameter**, after Edwin Hubble's observational discovery of the expanding universe. At present, the Hubble parameter has an approximate value of $(14\,\text{Gy})^{-1}$. Of particular relevance to our discussion of scalar fields is the inflaton field with $\rho = $ constant. In that case, equation (3.29) can be solved to produce

$$a \propto e^{Ht},$$

or, in other words, the universe exponentially expands without end. This is known as a **de Sitter universe**, and its dynamics seem to describe the first instant after the Big Bang. Classical inflation isn't *really* eternal; it came to an end roughly $10^{-34}\,\text{s}$ after it began (*nearly* an eternity in the context of the Big Bang). To understand how, we'll need to take a slightly closer look at the form of the inflaton potential.

3.6.3 The Inflaton Field

From the Planck time onward, the universe started to cool, but it wasn't until approximately $10^{-36}\,\text{s}$ after the Big Bang that the inflaton field began to become important. At that time, temperatures had cooled to energies of about $10^{16}\,\text{GeV}$.

Our picture of the inflaton field depends largely on the details of the inflaton potential $V(\phi)$, but we suppose a model with a false vacuum (where the field started) and a true, lower vacuum (where the field ended up), as seen in Figure 3.3.

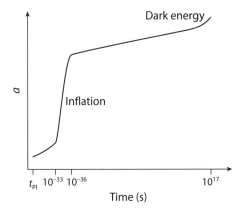

Figure 3.4. The growth of the universe during the inflationary epoch.

At an early time, the field starts near $\phi = 0$ (this is, in essence, what we mean by the vacuum). We have seen that the proper volume in an expanding universe requires a factor of a^3, and thus an appropriately normalized Lagrangian density must be written as

$$\mathcal{L} = a^3 \left(\frac{1}{2} \dot{\phi}^2 - V(\phi) \right).$$

Applying the Euler-Lagrange equation quickly yields

$$\ddot{\phi} + 3H\dot{\phi} = -\frac{\partial V}{\partial \phi}. \tag{3.30}$$

As the universe expands, the expansion itself acts as a friction term until ultimately the inflaton field reaches the true vacuum of the universe, and the inflationary epoch stops. The whole process takes, in most models, around 10^{-34} s, but during that time, the universe inflates on the order of 10^{100}.

Inflation blows up the universe to such extreme scales that any curvature that it may have had is essentially erased in the process, and inhomogeneities in the very early universe are smoothed out. While we haven't explored the quantum effects, quantum fluctuations in the inflaton field seed the universe with low-amplitude density perturbations—ultimately giving rise to cosmological structure. This discussion, however, is beyond the scope of this book.

Further, it's entirely possible (though difficult to swallow from a "naturalness" perspective) that a remnant of the inflaton field remains. The universe still has an energy component that seems to both accelerate expansion and remain constant in density over time. This **cosmological constant** or **dark energy** is small, about 10^{-120} the density of the inflaton field, but it is possible that the true vacuum of the inflaton potential wasn't identically zero.[1] It's only been in the last few billion years that the density of ordinary and **dark matter** has become sufficiently diffuse for dark energy to dominate [56,130],

---

[1] If the true vacuum isn't identically zero, then it differs only fractionally from zero, by 1 part in 10^{120}. This represents an incredible (in the most literal sense) **fine-tuning problem**, and barring some fundamental principle to enforce it, seems unlikely to have occurred naturally.

but dominate it does. In that sense, eternal exponential expansion describes the future evolution of the universe as well as the past.

Problems

3.1 A nonrelativistic particle moves in Cartesian coordinates with a potential energy

$$U(z) = mgz.$$

The Lagrangian will remain invariant under the transformation

$$x \rightarrow x + \epsilon y$$
$$y \rightarrow y - \epsilon x.$$

What is the corresponding conserved Noether quantity?

3.2 A particle is confined to move along the surface of an infinitely long cylinder of radius R. The equipotential curves on the surface makes a helical pattern:

$$U = U_0 \sin(\phi - \alpha z).$$

(a) What is the coordinate transformation on ϕ and z that leaves the Lagrangian invariant?
(b) What is the corresponding conserved quantity?

3.3 A particle of mass m and charge q in an electromagnetic field has a Lagrangian

$$L = \frac{1}{2}m\dot{\vec{r}}^2 - q(\Phi - \dot{\vec{r}} \cdot \vec{A}),$$

where Φ is the scalar potential, and $\vec{A}$ is the vector potential.
(a) Suppose (just for the moment) that the potential fields are not explicit functions of x. Use Noether's theorem to compute the conserved quantity of the electromagnetic Lagrangian (the canonical momentum).
(b) More generally, assume that the potential fields vary in space and time. What are the Euler-Lagrange equations for this Lagrangian corresponding to particle position x^i?
(c) Solve (and simplify) your previous solution explicitly for $m\ddot{x}$. Remember that the Euler-Lagrange equation uses a *total* time derivative (apply the chain rule!).
    Express the fields as combinations of components of $\vec{E}$ and $\vec{B}$ fields.

3.4 Consider a universe consisting of two real-valued scalar fields ϕ_1 and ϕ_2 with the total Lagrangian

$$\mathcal{L} = \mathcal{L}_1 + \mathcal{L}_2 + \mathcal{L}_{int},$$

where $\mathcal{L}_1$ and $\mathcal{L}_2$ are the free-field Lagrangians of the two fields:

$$\mathcal{L}_1 = \frac{1}{2}\partial_\mu\phi_1\partial^\mu\phi_1 - \frac{1}{2}m_1^2\phi^2; \quad \mathcal{L}_2 = \frac{1}{2}\partial_\mu\phi_2\partial^\mu\phi_2 - \frac{1}{2}m_2^2\phi_2^2.$$

There is also an interaction Lagrangian

$$\mathcal{L}_{int} = -\lambda\phi_1^2\phi_2^2,$$

with $\lambda > 0$.
(a) What is the dimensionality of λ?
(b) Compute the T^{00} component of the stress-energy tensor for the ϕ_1 and ϕ_2 free fields.
(c) Compute the T^{00} component of the stress-energy tensor for the interaction Lagrangian.
(d) Based on your previous answer, state whether the scalar fields attract or repel one another.

3.5 Compute all four components of $T^{1\mu}$ for a massive complex scalar field.

3.6 We've seen that a real-valued scalar field may be expanded as a plane-wave solution:

$$\phi(x) = \int \frac{d^3p}{(2\pi)^3} \frac{1}{\sqrt{2E_p}} [c_{\vec{p}} e^{-ip\cdot x} + c_{\vec{p}}^* e^{ip\cdot x}].$$

Compute the total anisotropic stress

$$\int d^3x\, T^{ij},$$

where $i \neq j$, for a real-valued field by integrating over the stress-energy tensor.

3.7 Consider the plane-wave solution for a complex scalar field (equation 3.17). The elements of the stress-energy tensor may all be expressed as

$$T^{\mu\nu} = \iint \frac{d^3p}{(2\pi)^3} \frac{d^3q}{(2\pi)^3} \frac{1}{\sqrt{4E_p E_q}} \left[B^{\mu\nu} b_{\vec{p}} b_{\vec{q}}^* e^{i(q-p)\cdot x} + B^{\mu\nu*} c_{\vec{p}}^* c_{\vec{q}} e^{-i(q-p)\cdot x} + \text{cross terms} \right].$$

(a) For the stress-energy components $T^{\mu 0}$ compute the coefficient $B^{\mu 0}$ explicitly.
(b) Now, consider the continuity equation

$$\partial_\mu T^{\mu 0} = 0.$$

Compute the coefficient corresponding to each term of this divergence.
(c) Combine the coefficients in the previous parts to demonstrate that energy is conserved for a complex scalar field.

3.8 We might suppose (incorrectly, as it happens) that a vector field has a Lorentz-invariant Lagrangian

$$\mathcal{L} = \partial_\mu A^\nu \partial_\nu A^\mu - m^2 A_\mu A^\mu.$$

(a) Compute the Euler-Lagrange equations for this Lagrangian.
(b) Assume a plane-wave solution for the vector field

$$A^\mu = \int \frac{d^3p}{(2\pi)} \epsilon^\mu \frac{1}{\sqrt{2E_p}} \left[a_p e^{-ip\cdot x} + a^* e^{ip\cdot x} \right],$$

where we haven't specified the polarization state(s) ϵ^μ explicitly.
   Develop an explicit relationship between the polarization, the momentum of the field, and the mass. What condition does this impose for a massless vector particle?
(c) What is the energy density of the vector field?

3.9 In subsequent chapters, we will often describe multiplets of scalar fields,

$$\Phi = \begin{pmatrix} \phi_1 \\ \phi_2 \end{pmatrix},$$

where ϕ_1 and ϕ_2 is each, in this case, a real-valued scalar field. For example,

$$\mathcal{L} = \frac{1}{2} \partial_\mu \Phi^T \partial^\mu \Phi - \frac{1}{2} m^2 \Phi^T \Phi$$

is a compact way of describing two free scalar fields with identical masses. This Lagrangian is symmetric under the transformation

$$\Phi \rightarrow (\mathbf{I} - i\theta\mathbf{X})\Phi,$$

where $\mathbf{X}$ is some unknown 2×2 matrix (to be determined), and θ is assumed to be small.

(a) What is the transformation of Φ^T (to first order in θ)? Show that $\Phi^T\Phi$ remains invariant under this transformation.

(b) What is the conserved current in this system?

(c) As we will see, for the particular case described in this problem (known as SO(2)) the elements of $\mathbf{X}$ are

$$\mathbf{X} = \begin{pmatrix} 0 & i \\ -i & 0 \end{pmatrix}.$$

Compute the conserved current in terms of ϕ_1 and ϕ_2 explicitly.

3.10 Consider a few field theory requirements of inflation.

(a) Following the Friedmann equation (3.29), at what equation of state w (equation 3.28) is the threshold between an accelerating universe ($\ddot{a} > 0$) and a decelerating one?

(b) Given the constraint on the previous problem, set a reasonable relationship on the "velocity" of the ϕ field ($\dot{\phi}$) such that inflation is ended.

(c) Assume that the change in ϕ is sufficiently slow that $\ddot{\phi} \simeq 0$. This is known as **slow roll inflation.** Using equation (3.30) with your previous result, give a constraint on where on the $V(\phi)$ curve inflation will end. You may assume the Hubble constant is, indeed, constant prior to this point.

Further Readings

• Goldstein, Herbert, Charles Poole, and John Safko. *Classical Mechanics*, 3rd ed. San Francisco: Addison Wesley, 2002. Goldstein is rightly considered the definitive text for advanced classical mechanics. The discussion of Noether's theorem, in Chapter 13, is particularly useful for supplementing the material presented here.

• Jackson, John David. *Classical Electrodynamics*. 3rd ed. New York: Wiley, 1999. Chapter 12 has a nice discussion of relativistic fields in the context of electromagnetism and, in particular, of the stress-energy tensor. This reading will anticipate our work in vector fields.

• Neuenschwander, Dwight E. *Emmy Noether's Wonderful Theorem*. Baltimore, MD: Johns Hopkins University Press, 2011.

• Peacock, John A. *Cosmological Physics*. Cambridge: Cambridge University Press, 1999. Chapter 11 has a very nice discussion of not only the inflationary paradigm from fields discussed here but also its relation to the growth of structure.

4 | Symmetry

Figure 4.1. Évariste Galois (1811–1832). Galois was a pioneer in the study of group theory.

We've been talking about symmetries of the Lagrangian in a very ad hoc sort of way. Our reliance on inspection and hairy algebraic calculations seems to contradict the idea that the Standard Model is *elegant*, *natural*, or *beautiful*, which are the sort of appellations that it's normally given.

In this chapter, we'll embark on a more formal discussion of symmetry, but we start by addressing why symmetry arguments are especially useful. Consider a uniform circular ring of electric charge for which we ask the question, what is the electric field at the center of the ring? It is, of course, possible to compute the field as an integral over every segment of the ring, but symmetry arguments in this case are far faster. There's no possible direction in which an electric field could be pointed that wouldn't break the circular symmetry of the system, and thus the field is zero.

Symmetry arguments don't so much allow us to solve new problems as they allow us to constrain the space of possible solutions. We have already seen, for instance, that the requirement for Lorentz invariance significantly reduces the number of potential Lagrangians describing physical interactions. Our goal will be to introduce new symmetries with the hope (informed by experiment and history) that they will limit the flexibility of theories to only a few parameters.

What we're really talking about here are **groups** of transformations that leave the Lagrangian unchanged. These symmetries, via Noether's theorem, ultimately give rise to conservation laws.

Groups will turn out to be the key to describing the contents of the particle zoo, and their interactions, and, more to the point, how the Standard Model can be said to be developed from a set of simple assumptions. But first, we need to say a few words about what groups are.

4.1 What Groups Are

The term *groupe* was first coined by Évariste Galois [73], a precocious French mathematician[1] whose work, along with that of Joseph-Louis Lagrange [94] is generally considered foundational to the development of group theory [68].

In terms of definition, groups are a fairly modest concept. A group is nothing more than a collection of operations, labeled $\{g_i\}$, that can be applied in sequence and which adhere to a few basic rules.

1. *Closure.* If g_i and g_j are part of the group, then

$$g_i \circ g_j = g_k \tag{4.1}$$

   must be as well. "Multiplication" (denoted by the $\circ$ symbol) in this context, must be contextually defined and need not correspond to what we'd normally think of as scalar multiplication. It simply means applying the transformations on some object, sequentially from right to left.

2. *Identity.* There must exist an element $\mathbf{I}$ such that

$$g_i \circ \mathbf{I} = \mathbf{I} \circ g_i = g_i \tag{4.2}$$

   for all members of the group. This is analogous to multiplying by one or adding zero in ordinary arithmetic.

3. *Inverse.* For every element g there exists an inverse $g_j = (g_i)^{-1}$ such that

$$g_j \circ g_i = g_i \circ g_j = \mathbf{I}$$

4. *Associativity.* Grouping doesn't matter. Thus,

$$g_i \circ (g_j \circ g_k) = (g_i \circ g_j) \circ g_k \, .$$

In general, elements of groups need not **commute**. As a reminder, the commutation operator is defined as

$$[\hat{A}, \hat{B}] = \hat{A}\hat{B} - \hat{B}\hat{A} \, . \tag{4.3}$$

---

[1] Galois exemplifies the truism that mathematicians do their best work at an early age. He was fatally wounded in a duel at age 20.

Table 4.1. The Multiplication Table of a Finite Group.

$\circ$	I	g_1	g_2	$\ldots$
I	I	g_1	g_2	
g_1	g_1	$g_1 \circ g_1$	$g_1 \circ g_2$	
g_2	g_2	$g_2 \circ g_1$	$g_2 \circ g_2$	
$\vdots$				

Note: The closure condition requires that all products are themselves members of the group.

Table 4.2. The Multiplication Table of the Z_2 Group.

$\circ$	1	-1
1	1	-1
-1	-1	1

For operations that commute (yielding a zero for the commutator) the order of operations doesn't matter. For instance, translations in space commute, since retracing your steps in reverse order gets you back to where you started. Such groups are called **Abelian** (after the Norwegian mathematician Neils Henrik Abel).

In many ways, *non-Abelian* groups are going to be of greatest interest to us. For instance, consider rotations of a sphere. A rotation through the equatorial axis followed by a rotation along the prime meridian will produce a different configuration than the operations applied in reverse order.

Taking our cues from quantum mechanics, we know that there are a number of important operators (momentum and position, for instance) that do not commute with one another. These commutation relations, in turn, provide the basis for ideas like the uncertainty principle. Most of the important symmetry groups in the Standard Model will behave similarly. We note by way of a teaser that the noncommutation of the symmetry groups tends to produce some interesting coupling effects in the electroweak and strong forces.

For the moment, however, we'll content ourselves with simply noting whether a group is or isn't Abelian and leave it at that.

4.2 Finite Groups

Finite groups are an easy foray into group theory, since in principle all their elements can be written in a multiplication table (Table 4.1).

One of the simplest finite groups is $\{1, -1\}$ under the operation of multiplication, often called Z_2. In that case, we can explicitly write out a multiplication table (Table 4.2).

Figure 4.2. The spiral galaxy M81 as imaged by the Hubble Space Telescope. Credit: NASA, ESA, and The Hubble Heritage Team (STScI/AURA)

Not only is this group finite, but it's **cyclical**, since every element appears exactly once in each row and column of the multiplication table. This group is also, incidentally, Abelian, since the multiplication table is symmetric. Don't worry if the parade of groups gets a bit confusing. Only a few groups are terribly important in the scheme of things, and we enumerate the important ones in Appendix D.

The Z_2 group looks for all the world like reflection or charge conjugation, since in each case, there are only two possible operations: flip or stay the same, and flipping twice is equivalent to staying the same. We will find that a number of groups are essentially equivalent to one another. The name for this property is **isomorphism**, which is another way of saying that there is a 1:1 mapping between the elements of one group and another.

Take a look at the spiral galaxy M81 in Figure 4.2. Apart from doing nothing (**I** in the language of groups), there is precisely one unique operation that will keep the appearance of the galaxy (essentially) the same: rotation by 180°. Applying this rotation twice returns the galaxy to its original configuration. This operation produces the exact same multiplication table as the Z_2 group and is thus an isomorphism.

Z_2 is the simplest nontrivial group and one that shows up surprisingly frequently in both physics and mathematics, but a higher-order group might shed a little more light on how finite groups (and by extension, infinite groups) work.

Example 4.1: The Equilateral Triangle (D_3)

Consider the symmetry transformations allowed on an equilateral triangle (also known as D_3 for the third dihedral group). For recordkeeping, we label the vertices of the triangle ABC as in Figure 4.3. What is the group of allowed transformations that leaves the appearance of the triangle unchanged?

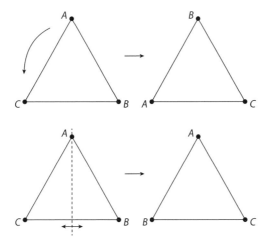

Figure 4.3. Some allowable symmetry transformations on an equilateral triangle, R_{-1} and F_1.

Table 4.3. The Multiplication Table of the Equilateral Triangle Group.

∘	I	R_1	R_{-1}	F_1	F_2	F_3
I	I	R_1	R_{-1}	F_1	F_2	F_3
R_1	R_1	R_{-1}	I	F_3	F_1	F_2
R_{-1}	R_{-1}	I	R_1	F_2	F_3	F_1
F_1	F_1	F_2	F_3	I	R_1	R_{-1}
F_2	F_2	F_3	F_1	R_{-1}	I	R_1
F_3	F_3	F_1	F_2	R_1	R_{-1}	I

Note: The members along the rows are to the left of the members above the columns.

Solution: There are six permutations of the vertices that leave the appearance of the triangle unchanged: ABC, BCA, CAB, ACB, CBA, BAC, and thus there must be six members of D_3:

- **I** : Do nothing
- R_1 : Rotate 1/3 of a clockwise turn
- R_{-1} : Rotate 2/3 of a turn or −1/3 of a turn
- F_1 : Flip through the vertical axis
- F_2 : Flip through the axis passing through the lower right corner
- F_3 : Flip through the axis passing through the lower left corner

This geometric object can be abstracted to its multiplication table (Table 4.3).

The asymmetry of the table immediately indicates that the group is non-Abelian.

Inspection of Table 4.3 indicates that there are a number of **subgroups** within D_3. For instance, {I, R_1, R_{-1}} form an Abelian subgroup of rotations, the cyclic group, Z_3. This can be seen by considering the terms in the table that contain only multiplications of $\mathbf{I}$, R_1, or R_{-1} and subsequently produce only those members. Z_3 is a subgroup of D_3.

The beauty of this formalism is that we can replace our triangle with another geometric shape or something abstract, and as long as the relative terms in the multiplication table remain the same, any conclusions about the group that hold for the triangle also hold for another object with the same symmetry group.

4.3 Lie Groups

Not all groups have a finite number of elements. For instance, the set of integers under the operation of addition has all the properties of a group. The inverse is simply the negative, and the identity is addition of zero.

In physics, we are typically much more concerned with continuous groups known as **Lie groups** (pronounced "Lee," named after the mathematician Sophus Lie [102], who did much of the groundbreaking work in their development). Unlike discrete groups, Lie groups are necessarily infinite and have one or more continuous parameters. Any unique value of the parameter in turn produces a unique group element.

4.3.1 Rotations and Notations

While mathematicians may find it useful to discuss members of a group as a pure abstraction, physicists tend to focus more on the representation of the group. We almost always operate on numbers or vectors or vector-like objects (bispinors, which we'll encounter in the next chapter), and so we'll generally find it most convenient to express members of Lie groups as square matrices, which we'll generically denote as $\mathbf{M}$.

Consider a set of points defined in two-dimensional Cartesian coordinates. We could imagine applying many different coordinate transformations: a uniform translation, for instance, or a Lorentz boost. But suppose we want to limit ourselves to only that set of transforms for which the distances between each point and the origin and each other remained fixed. Or, to put it another way, suppose we wish to find the set of transforms such that

$$r^2 = x_i x^i = \text{fixed}.$$

As we learned in our discussion of our discussion of metrics, the distance between the points and the origin will remain invariant upon a rotation. Having worked in plane geometry, the coordinate transformations will be written in the form

$$\mathbf{M}(\theta) = \begin{pmatrix} \cos\theta & \sin\theta \\ -\sin\theta & \cos\theta \end{pmatrix}. \tag{4.4}$$

This is the **irreducible** representation (because it's the smallest matrix that will describe it) of SO(2), which stands for "special orthogonal group in two dimensions," or, more simply, the "circle group."[2]

Because many of the matrices representing elements of important groups in the Standard Model have a number of things in common, some definitions are necessary to explain the shorthand. The most commonly used symbols are the following:

1. **Special (S)**: The determinant of the matrix is $+1$.
2. **Orthogonal (O)**: The inverse is equal to the transpose:

   $$\mathbf{M}\mathbf{M}^T = \mathbf{I}.$$

   As the determinants of the matrix and its transpose are equal, the determinant can be ± 1, and the matrix itself is real.
3. **Unitary (U)**: The inverse of a complex-valued $\mathbf{M}$ is equal to the Hermitian transpose:

   $$\mathbf{M}\mathbf{M}^\dagger = \mathbf{I}.$$

Inspection will show that rotations are both special and orthogonal, and as it can be represented by a 2×2 matrix, we label the group as SO(2).

We've seen rotational invariance before. We noted in the first chapter that a quantity that is completely independent of direction in 3-space will have an SO(3) symmetry— a designation that makes a little more sense now. The gravitational potential of the sun is completely isotropic, and thus the theory of Newtonian gravity should have an SO(3) symmetry. *However* (and this is a very important point) that doesn't mean that the solutions will be necessarily symmetric. As Kepler noted and Newton proved, the solutions to this symmetric potential are less-symmetric-looking ellipses. As David Griffiths [80] put it:

> The Greeks apparently believed that the symmetries of nature should be directly reflected in the motion of objects: Stars must move in circles because those are the most symmetrical trajectories. Of course, planets don't, and that was embarrassing (it was not the last time that naive intuitions about symmetry ran into trouble with experiments).

SO(2) and SO(3) are subgroups of a very important group that we assumed implicitly in our discussion of special relativity even though we didn't call it such—the **Poincaré group** [131], which consists of

- Linear translations in spacetime
- Rotations
- Boosts

The latter two transforms are a subgroup that constitutes the **Lorentz group**.

---

[2] A circle, naturally, because a circle is completely unchanged under a rotation.

Part of our working assumption in constructing realistic Lagrangians will be that they must be invariant under Poincaré group transformations. Fortunately, as near as we can tell, nature seems to have simplified our work by satisfying these symmetry conditions.

Returning to SO(2), we may consider subsequent applications of rotations. We'll leave it as an exercise for you to show that

$$\mathbf{M}(\theta_1)\mathbf{M}(\theta_2) = \mathbf{M}(\theta_1 + \theta_2) = \mathbf{M}(\theta_2)\mathbf{M}(\theta_1).$$

The members commute, and thus the SO(2) group is Abelian.

A *very* closely related group to SO(2) is U(1), but while the former is real valued, the latter is complex. As U(1) is parameterized by a single number, its members can be generated via

$$\mathbf{M}(\theta) = e^{-i\theta}.$$

It should be clear that both SO(2) and U(1) are isomorphic to one another, as they both represent a periodic rotation in one dimension.

4.3.2 Generators

We saw earlier that to calculate the Noether current, we need to compute the effect of arbitrarily small transforms. That is, both SO(2) and U(1) have a single continuous parameter θ. In the limit of $\theta \to 0$, both groups reduce to $\mathbf{I}$, the identity element.

We can expand any such transformation in terms of a Taylor series in the parameter; that is, defining a **generator X** for a group as

$$M = e^{-iX\theta}, \tag{4.5}$$

we can expand the transformation to first order as

$$\mathbf{M}(\theta) \simeq \mathbf{I} - i\mathbf{X}\theta. \tag{4.6}$$

For U(1),

$$e^{-i\theta} \simeq 1 - i\theta,$$

and thus the generator is simply $\mathbf{X} = 1$.

Example 4.2: What is the generator for SO(2)?

Solution: Taking the derivative of equation (4.4) with respect to θ,

$$\frac{\partial \mathbf{M}}{\partial \theta} = \begin{pmatrix} -\sin\theta & \cos\theta \\ -\cos\theta & -\sin\theta \end{pmatrix} = -i\mathbf{X},$$

and in the limit of $\theta \to 0$, $\sin \theta = 0$ and $\cos \theta = 1$. Thus, for SO(2),

$$\mathbf{X} = \begin{pmatrix} 0 & i \\ -i & 0 \end{pmatrix}, \tag{4.7}$$

which you may recognize as the negative of σ_2, the second Pauli spin matrix.

Once we've determined a complete set of generators for a symmetry group, we can return to equation (4.6) and suppose that it's merely the *first* term in a Taylor series. This makes sense, since by the closure requirement of groups, multiple small rotations must produce a larger rotation that is *also* a member of the symmetry group. This result is particularly useful because it means that when expressed as matrices, the elements' group can be generated via an exponential,

$$\mathbf{M} = e^{-i\vec{\theta} \cdot \vec{\mathbf{X}}}, \tag{4.8}$$

where we've used $\vec{\mathbf{X}}$, for instance, to represent an N-dimensional space vector composed of the Pauli generators. Thus, a dot product takes the form

$$\vec{\theta} \cdot \vec{\mathbf{X}} = \sum_i \theta^i \, \mathbf{X}_i \, .$$

You may note that generators play the same role that ordinary basis vectors do in ordinary geometry and trigonometry, where $\vec{\theta}$ represents a coordinate in a linear space. The combination of angles defines a set of coordinates, while generators define unit vectors in a **Hilbert space**, named after the mathematician David Hilbert [89, 147]. That we can think of group elements as being points in an N-dimensional space also means that our choice of basis vectors in that space isn't set in stone. We can combine them using the normal tricks of linear algebra, so long as there are N of them.

The linearity of the space, in turn, makes differentiation of the group elements a fairly trivial operation.

4.3.3 Generators and Noether's Theorem

Group theory formalism may just seem like circular reasoning (pun intended) for the U(1) and SO(2) groups, but it becomes important in other cases. Consider translational symmetry, introduced earlier:

$$x \to x + \epsilon.$$

We don't care so much about the coordinates but rather about a function defined by the now-changed coordinate

$$\phi(x + \epsilon) = \phi(x) + \epsilon \frac{\partial \phi}{\partial x},$$

which makes the generator for translations the operator

$$\mathbf{X} = -i\frac{\partial}{\partial x},$$

which looks *very* familiar. This starts to give us a deep insight into two seemingly unrelated facts:

1. The conserved Noether quantity for translational symmetry is momentum.
2. The momentum operator in quantum mechanics is the spatial derivative.

The reason that these are so important is that we will frequently want to take derivatives of our Lagrangian with respect to symmetry parameters. For instance, if we have a field ϕ_a operated on by a member of a group $\mathbf{M}$, then following Noether's theorem (equation 3.10), we'll often want to compute

$$\frac{\partial \phi_a}{\partial \theta^i} = -i\mathbf{X}_i \phi_a, \tag{4.9}$$

Knowing the generators means that we essentially get the currents instantly.

4.4 SU(2)

We close our discussion of group theory by introducing two very important Lie groups in particle physics, SU(2) and SU(3). The first is central not only in the weak interaction but also (and perhaps more familiarly) in the discussion of spin. The latter is the basis of our description of the strong force.

What we mean when we say something like "SU(2) *describes* the weak interaction"[3] is that the Lagrangian which describes the weak interaction will be invariant under SU(2) symmetry transformations. We can guess the existence of a group and then try to find the simplest Lagrangian that is invariant under those symmetry transformations, or we can do the whole thing the other way around (as we'll do now for pedagogical reasons).

4.4.1 The Members of SU(2)

We begin by imagining that the universe consists of a doublet of scalar particles, and for compactness we'll pair them together, and label the entire two-component object Φ:

$$\Phi = \begin{pmatrix} \phi_1 \\ \phi_2 \end{pmatrix}. \tag{4.10}$$

---

[3] Those familiar with the symmetries of the Standard Model may realize that we're fudging the truth somewhat at this stage. The underlying symmetry of the electroweak model is $SU(2)_L \otimes U(1)$, not SU(2). The weak interaction is necessarily left-handed, and the SU(2) symmetry by itself doesn't explain, for instance, massive mediators. We'll deal with these questions in Chapter 9.

These components don't correspond to time and space but are instead two components of a single field, each of which varies in time and space.

Let's further suppose that the Lagrangian contains only terms that look like

$$\Phi^\dagger \Phi = \phi_1^* \phi_1 + \phi_2^* \phi_2, \tag{4.11}$$

and corresponding derivatives that couple in the same way. We are primarily concerned with how two particles couple to one another, and we will generally focus on multiplets of particles that have identical interactions. For a transformation to be a symmetry, the preceding sum needs to remain unchanged. The Lagrangian satisfying these symmetry constraints would describe two particles that are identical but for their labels. The form of the terms in the Lagrangian looks almost exactly the same as a rotational invariance. From the perspective of an SU(2) theory, ϕ_1 and ϕ_2 particles are so utterly indistinguishable that swapping one for the other changes absolutely nothing in our toy universe.

It should be clear that whatever else we do to transform Φ, we need to do so from a 2×2 matrix, but the determinant of that matrix must be ± 1. Otherwise, the inner product of ϕ with itself wouldn't be conserved; that is,

$$\Phi \to \mathbf{M}\Phi$$

$$\Phi^\dagger \to \Phi^\dagger \mathbf{M}^\dagger$$

$$\Phi^\dagger \Phi \to \Phi^\dagger \mathbf{M}^\dagger \mathbf{M}\Phi,$$

so

$$\mathbf{M}^\dagger \mathbf{M} = \mathbf{I}, \tag{4.12}$$

which is the *definition* of the unitary group U (that is, the inverse of an element is the Hermitian conjugate). The reason that we have the *special* unitary group is that we already know the case where the determinant of $\mathbf{M}$ isn't 1,

$$e^{i\theta} \begin{pmatrix} 1 & 0 \\ 0 & 1 \end{pmatrix},$$

which is just a representation of U(1). Nothing prevents us from having a *combined* group of both SU(2) and U(1). Notationally, this is written as $SU(2) \otimes U(1)$ and is the union of the two groups. It is also, not coincidentally, the basis of electroweak unification.

For the moment, we'll confine ourselves to looking for the elements of SU(2). Expanding equation (4.12) yields

$$e^{i\vec{\theta}\cdot\vec{X}} e^{-i\vec{\theta}\cdot\vec{X}} = \mathbf{I},$$

which can be satisfied only if $\mathbf{X}_i^\dagger = \mathbf{X}_i$.

Counting the degrees of freedom (and eliminating the U(1) solution) produces three independent coordinates in Hilbert space and thus three unique generators. Fortunately, our studies of nonrelativistic quantum mechanics have not been in vain:

$$\sigma_1 = \begin{pmatrix} 0 & 1 \\ 1 & 0 \end{pmatrix}; \quad \sigma_2 = \begin{pmatrix} 0 & -i \\ i & 0 \end{pmatrix}; \quad \sigma_3 = \begin{pmatrix} 1 & 0 \\ 0 & -1 \end{pmatrix}, \tag{4.13}$$

which are a complete basis for describing any possible generator for SU(2) and also have the nice properties

$$\sigma_i \sigma_i = \mathbf{I} \quad \text{(no sum)} \tag{4.14}$$

and

$$[\sigma_i, \sigma_j] = 2i \sum_k \sigma_k \epsilon_{ijk}, \tag{4.15}$$

where the ϵ_{ijk} is known as the **Levi-Civita** symbol and yields 1 for indices in cyclic order (e.g., 231), -1 for indices in anticyclic order (e.g., 213), and 0 otherwise (e.g., 221).

4.4.2 Algebras

Owing to the closure requirement on groups, commutators of generators must take the form

$$[\mathbf{X}_i, \mathbf{X}_j] = \sum_k 2i f_{ijk} \mathbf{X}_k,$$

where the f symbols are known as the **structure constants** (sometimes defined without the factor i). The commutation relationships between the generators (and, by extension, between the group members themselves) define an **algebra**, which is simply a fancy way of describing how the vectors in the Hilbert space can be added and subtracted from one another.

Structure constants turn out to be exceedingly important. In some sense, the definition of a group (closure and associativity, in particular) requires that commutation relations between two generators produce a third generator. But these relations can be abstracted still further by introducing the **Baker-Campbell-Hausdorff** relations,

$$e^X e^Y = e^{X+Y+\frac{1}{2}[X,Y]} \tag{4.16}$$

and

$$e^X Y e^{-X} = Y + [X, Y], \tag{4.17}$$

where X and Y are assumed to be generators, operators, or matrices.

We've introduced U(1) and SU(2) because of their relative simplicity but also because they are of fundamental importance in the Standard Model. In 1941, Wolfgang Pauli recognized the significance of U(1) as the symmetry underlying electromagnetism [122], and 20 years later, Sheldon Glashow [78] proposed that the SU(2) symmetry, along with U(1), could unify the weak and electromagnetic forces. Understanding the underlying symmetries is going to be time well spent.

There is meaning behind the number of generators. Each generator will ultimately give rise to a new particle—a **mediating particle**—that communicates the interaction. U(1) has one generator, and thus electromagnetism has one particle, the photon. Likewise, SU(2) has three generators, and thus three mediators, W^+, W^-, and Z^0.

But before getting into the details of how a field moves between states, it's worth considering what the two "states" in an SU(2) theory correspond to. As it happens, this symmetry shows up in two different contexts.

4.4.3 Spin and Weak Isospin

In the most familiar context, the SU(2) symmetry group describes the spin of electrons. For spin-symmetric Lagrangians, converting all spin-up particles $|\uparrow\rangle$ to spin-down $|\downarrow\rangle$, and vice-versa, will have no effects on interaction energies, where we've used Paul Dirac's **bra-ket** notation [48] in anticipation of combining states with more than one particle.

Combinations of multiple particles lend an additional insight. We enumerate all possible combinations of two-particle systems:

$$|\uparrow\uparrow\rangle; \quad |\downarrow\downarrow\rangle; \quad \frac{1}{\sqrt{2}}\left(|\uparrow\downarrow\rangle + |\downarrow\uparrow\rangle\right) \quad \text{(triplet)} \tag{4.18}$$

and

$$\frac{1}{\sqrt{2}}\left(|\uparrow\downarrow\rangle - |\downarrow\uparrow\rangle\right) \quad \text{(singlet)}, \tag{4.19}$$

where the triplet of states is *symmetric* under change of spins (total spin of 1), and the singlet is *antisymmetric* (spin of 0). No amount of rotation in SU(2) will turn an element of the singlet state into a triplet, or vice-versa. Indeed, it's not a coincidence that there are three generators and three different ways of constructing symmetric states. The generators are the operations whereby the different elements of the triplets can be rotated into one another.

Part of the reason that Noether's theorem is so powerful is that it allows us to quickly relate the transformation properties of a unitary matrix to conservation laws. This can easily be seen by noting that for a symmetry transformation in general,

$$\mathbf{M}\hat{H}|\psi\rangle = \mathbf{M}E|\psi\rangle = E|\psi'\rangle = \hat{H}\mathbf{M}|\psi\rangle,$$

or, equivalently,

$$[\hat{H}, \mathbf{M}] = 0. \tag{4.20}$$

Symmetry operations commute with the Hamiltonian.

The advantage of the singlet-triplet classification scheme is that it allows us, almost without any additional work, to identify allowed processes which conserve energy or, in the case of spin, total angular momentum.

But there is a subtler manifestation of the SU(2) symmetry as well. Under weak interactions, particles can be sorted into doublets such as

$$\Psi = \begin{pmatrix} v_e \\ e \end{pmatrix}, \quad \begin{pmatrix} u \\ d \end{pmatrix}, \tag{4.21}$$

where v_e is the **electron neutrino**, e is the electron, u is the **up quark**, and d is the **down quark**. We haven't formally encountered most of these particles yet, but they have a centrality in physics (see Figure 1). Trios of quarks make up baryons, the lightest of which is the proton, followed closely by the neutron.[4]

The statement that Lagrangians of Ψ are invariant under SU(2) transformations means that as far as the weak interaction is concerned, the two particles in the doublet are interchangeable.[5] That is, turning *all* electrons to neutrinos and vice-versa will not alter the weak interaction calculation one iota, or even that

$$|ee\rangle; \quad |v_e v_e\rangle; \quad \frac{1}{\sqrt{2}}\left(|v_e e\rangle + |e v_e\rangle\right)$$

should all be energetically identical under the weak interaction.

Because the form of the Lagrangian requires terms like those found in equation (4.11), the mass of each particle in the doublet *should* be the same. As the electron ($m_e = 0.511\,\text{MeV}$) is far more massive than the nearly massless neutrino, this is clearly not the end of the story.

We hasten to also point out that that *doesn't* mean that electrons and neutrinos are identical as far as the other forces are concerned. Electromagnetism distinguishes very strongly between the two particles, as one (the electron) has charge and the other (the neutrino) does not. For what it's worth, the strong force will ignore both neutrinos and electrons equally.

4.5 SU(3)

Much of the development of the Standard Model hinges upon selecting possible symmetries for the underlying Lagrangian and exploring the fields and conserved currents generated by those symmetries. In 1964, Murray Gell-Mann [75] and George Zweig [170, 171] independently proposed that the strong force is described by the SU(3) group.

---

[4] It is curious that the neutron is heavier than the proton. The electromagnetic potential energy of the proton ought to make its mass the larger of the two. The massive neutron arises since a down quark is 2–4 MeV higher than an up quark, and neutrons comprise two downs and an up, while protons comprise two ups and a down. In problem 6.3, you'll have the opportunity to estimate the proton mass correction.

[5] We need to make the caveat (as often as possible) that the symmetry of the weak force isn't quite SU(2) for two reasons. First, the weak interaction is only a remnant of a broken symmetry of $SU(2)_L \otimes U(1)$, and second, the L corresponds to the fact that the weak interaction is left-handed. As a result, even absent electromagnetism, neutrinos and electrons are clearly distinguishable, if for no other reason than by their vastly different masses.

While we haven't discussed much about the strong force yet, we can make a few observations from simple inspection.

For instance, the elements of SU(3) are represented by a 3×3 matrix, which means that the symmetry operations are applied to a triplet. By analogy with our work with SU(2), there should be three possible states or particles that are completely symmetric with respect to the strong force. Unlike electrons and neutrinos, quarks respond to the strong force. Gell-Mann coined the word *quark* after a passage in *Finnegan's Wake* [76], and it caught on, in preference to Zweig's proposed "aces."

There are six different **flavors** of quarks—corresponding to a unique name and electrical charge—but those don't concern us here. What most concerns us is the color, the strong force equivalent to electrical charge. Every flavor of quark can come in "red," "green," or "blue":

$$\Psi = \begin{pmatrix} r \\ g \\ b \end{pmatrix}.$$

It should go without saying that the word "color" is simply a handy mnemonic for remembering the triplet. Indeed, as quarks can be probed only on the subfemtometer scale, color in the optical sense is meaningless.

We will ultimately derive the properties of the strong force interaction, but to give a sense of what color means, we consider a group of quarks of the same flavor. As with electromagnetism, there are some approximate rules of thumb:

- Quarks with different colors attract.
- Quarks with the same color repel.

These rules should look very similar to those of electromagnetism. In electromagnetism, flipping the signs of all charges doesn't change the forces at all. Likewise, changing all red quarks to green quarks and vice-versa will leave the energy of interaction in the system unchanged.

And what of the generators of the SU(3)?

In general, the SU(N) group will have

$$n_X = N^2 - 1 \tag{4.22}$$

generators, derivable entirely from the conditions that $\mathbf{X}^\dagger = \mathbf{X}$, and thus the trace of the generators must be zero (and consequently the determinant of the transform must be 1). In SU(2), this resulted in three generators, which we noted correspond to the three weak interaction mediating particles, $W^\pm$ and Z^0. Thus, in SU(3), we anticipate $3^2 - 1 = 8$ generators and thus eight mediators for the strong force. These, collectively, are known as **gluons**.

The generators are known as the **Gell-Mann matrices**, and while it's premature to delve too deeply into their role in the strong force, we can get a sense of them:

$$\lambda_1 = \begin{pmatrix} 0 & 1 & 0 \\ 1 & 0 & 0 \\ 0 & 0 & 0 \end{pmatrix}.$$

The seven others would prove equally unilluminating at this point. We'll get to them in due course. But first, we'll need to deal with fields more complex than scalar.

Problems

4.1 Consider a rectangle

 (a) List (and label) all the possible unique transformations that can be performed on the rectangle that will leave it looking the same as it did initially.
 (b) Construct a multiplication table for your set of transformations.
 (c) Does the set have the properties of a group?

4.2 Quaternions are a set of objects that are an extension of imaginary numbers, except that there are three of them, i, j, and k, with the relations

$$i^2 = j^2 = k^2 = ijk = -1.$$

 (a) Construct the smallest group possible that contains all the quarternions.
 (b) Compute the commutation relation $[j, i]$.
 (c) Construct a multiplication table for the quarternions.

4.3 Show that

$$\mathbf{M}(\theta_1)\mathbf{M}(\theta_2) = \mathbf{M}(\theta_1 + \theta_2) = \mathbf{M}(\theta_2)\mathbf{M}(\theta_1),$$

where $\mathbf{M}(\theta)$ are the elements of SO(2).

4.4 In quantum mechanics, SU(2) is used to describe the spin of electrons and other spin-1/2 fermions, including quarks. Consider a meson comprising a $u\bar{d}$ pair. Assume that the quarks have zero orbital angular momentum. Previously, we noted that there are a triplet of symmetric states and a singlet of antisymmetric states.
 (a) What is total spin for mesons in the symmetric and antisymmetric state? As a reminder, addition of spins yields the relationship

$$S^2 = s_1^2 + s_2^2 + 2\mathbf{s}_1 \cdot \mathbf{s}_2,$$

  with the quantum spin relation $S^2 = s(s + 1)$.
 (b) Starting with $|\uparrow\uparrow\rangle$ show that an SU(2) rotation of both spins results in transforming a symmetric spin state into a different superposition of symmetric states.

4.5 Consider a three quark system comprising three distinguishable flavors (e.g., uds). What are the possible total spin quantum numbers of the resulting baryon? Are there any states which would not be allowed if the three quarks were indistinguishable (e.g., uuu)?

4.6 Expand the series $e^{-i\theta\sigma_2}$ explicitly and reduce to common trigonometric, algebraic, or hypergeometric functions.

4.7 Using the Baker-Campbell-Hausdorff relations, compute
(a) $e^{i\theta^1\sigma_1}e^{i\theta^2\sigma_2}$.
(b) $e^{-i\sigma_1}\sigma_2 e^{i\sigma_1}$.

4.8 The SO(3) symmetry (rotation in three dimensions) has the generators

$$X_1 = \begin{pmatrix} 0 & 0 & 0 \\ 0 & 0 & -i \\ 0 & i & 0 \end{pmatrix}; \quad X_2 = \begin{pmatrix} 0 & 0 & i \\ 0 & 0 & 0 \\ -i & 0 & 0 \end{pmatrix}; \quad X_3 = \begin{pmatrix} 0 & -i & 0 \\ i & 0 & 0 \\ 0 & 0 & 0 \end{pmatrix}.$$

(a) Compute the commutator of $[X_1, X_2]$.
(b) Now consider a vector in this 3-D space:

$$V^i = \begin{pmatrix} 1 \\ 2 \\ 0 \end{pmatrix}.$$

Applying a small rotation, $\theta^3 \ll 1$, compute the first-order estimate of the transformed vector.
(c) In a short sentence (or an equation if you like), what about the vector (or other 3-D vectors) is going to be conserved in an SO(3) transformation?

4.9 Consider a 1+1 dimensional spacetime. We would like to compute transformations such that

$$\mathbf{M} \begin{pmatrix} V^0 \\ V^1 \end{pmatrix}$$

produce a fixed value of

$$(V^0)^2 - (V^1)^2.$$

Astute observers may recognize this as a restatement of Lorentz invariance. This dot product will be conserved if

$$\mathbf{M}^T \mathbf{g} \mathbf{M} = \mathbf{g},$$

where $\mathbf{g}$ is the metric, written as a matrix, and

$$\mathbf{M} = \begin{pmatrix} a & b \\ b & d \end{pmatrix},$$

where $\mathbf{M}$ is a symmetric, real-valued matrix.
(a) Multiply out the terms in equation 4.23, and determine b and d in terms of a based on the constraint, and rewrite $\mathbf{M}$ in terms of a explicitly.
(b) Compute the determinant of $\mathbf{M}$.
(c) Note that $a = 1$ corresponds to no boost (e.g., rest frame). More to the point, in general, $a = \gamma$. Expand $\mathbf{M}$ to first order in v (which will be the parameter of our group).
(d) Recalling the definition of the generator as

$$\mathbf{M} = I - ivX,$$

compute the generator of the group, X.

4.10 Consider a universe consisting of a complex field defined by two components:

$$\Phi = \begin{pmatrix} \phi_1 \\ \phi_2 \end{pmatrix}.$$

The Lagrangian takes the form

$$\mathcal{L} = \partial^\mu \Phi^\dagger \partial_\mu \Phi - m^2 \Phi^\dagger \Phi.$$

In some sense, there are *four* fields at work here: ϕ_1, ϕ_1^*, ϕ_2, ϕ_2^*. But for the purposes of this problem, you should generally think of Φ and $\Phi^\dagger$ as representing the *two* different fields. Since each is a 2-D vector, there are still four degrees of freedom.

(a) Consider a rotation in SU(2) in the θ^1 direction (σ_x). Expand **M** as an infinite series, and express as a 2×2 matrix of only trigonometric functions of θ^1.

(b) Verify numerically that your matrix (i) is unitary and (ii) has a determinant of 1.

(c) Compute a general expression for the current associated with rotations in θ^1.

4.11 The Gell-Mann matrices are the generators for the SU(3) group, the Lie group at the heart of the strong force. Two of the Gell-Mann matrices are

$$\lambda_1 = \begin{pmatrix} 0 & 1 & 0 \\ 1 & 0 & 0 \\ 0 & 0 & 0 \end{pmatrix}; \quad \lambda_4 = \begin{pmatrix} 0 & 0 & 1 \\ 0 & 0 & 0 \\ 1 & 0 & 0 \end{pmatrix}.$$

What is $[\lambda_1, \lambda_4]$?

Further Readings

- Cornwell, J. D. *Group Theory in Physics: An Introduction.* New York: Academic Press, 1997.
- Feynman, Richard P., and Robert B. Leighton. *The Feynman Lectures on Physics.* Reading, MA: Addison-Wesley, 1963. Chapter 52, Symmetry in Physical Laws, is especially good for giving a broad-brush overview of why symmetry is so important.
- Georgi, Howard. *Lie Algebras in Particle Physics: From Isospin to Unified Theories,* 2nd ed. Boulder, CO: Westview Press, 1999. The group theory neophyte will find chapters 1, 2, 3, and 7 especially instructive.
- Griffiths, David. *Introduction to Elementary Particles,* 2nd ed. Hoboken, NJ: Wiley-VCH, 2008. Chapter 4 has an excellent discussion of symmetries and elementary Lie groups.
- Weyl, Hermann. *Symmetry.* Princeton, NJ: Princeton University Press, 1952. Weyl doesn't explicitly deal in the language of group theory, but his discussion of symmetries in nature and mathematics is excellent.
- Zee, A. *Group Theory in a Nutshell for Physicists.* Princeton, NJ: Princeton University Press. 2016. Zee provides an important resource for mathematically minded students who want to learn more about group theory and its application in other areas of physics.

5 | The Dirac Equation

Figure 5.1. P.A.M. Dirac (1902–1984). Dirac's equation serves as the foundational description for relativistic fermions and thus for Lagrangians in the Standard Model.

Scalar fields are about as simple as fields can get, and our reliance on them has relieved us, so far, from being bogged down in the complexities of spin. However, the universe isn't obliging enough to allow us to ignore spin indefinitely. Most of the interesting *stuff* in our universe—everything we can directly interact with, in fact—comprises electrons and quarks, particles with intrinsic spin, and is collected under the umbrella of **fermions**.

The statistical properties of fermions had been understood since 1926, when Enrico Fermi [59] and Paul Dirac [44] independently found that particles like electrons failed to obey the classical specific-heat relationships predicted for an ideal gas, especially at low temperatures. Fermi and Dirac[1] hypothesized that the energy states of electrons are quantized, but antisocial; multiple electrons can't be found in the same quantum state. An explanation as to *why* fermions behave the way they do had to wait until the development

---

[1] Interestingly, and touchingly, both ascribed priority to the other.

of the **Dirac equation** in 1928 [46], which finds its origins in the unification of quantum mechanics and special relativity.

5.1 Relativity and Quantum Mechanics

5.1.1 The Form of the Field

There are many ways in which the quantum wave function and a classical field can be used almost interchangeably. There's not much difference in many applications between saying, "there is a 10% probability of finding a particle here," (quantum) and saying, "10% of the particles can be found here" (classical). In either case, we can construct a solution as a series of plane waves

$$\psi(x) \propto e^{-ip \cdot x}, \tag{5.1}$$

with the understanding that a measurement of intensity or probability will require the *square* of the wave.

Each mode represents a monoenergetic beam of coherent particles, but we so far lack a general dispersion relationship or a method for evolving ψ if we know only its initial conditions. In quantum mechanics (what's generally known as the **first quantization**) of the system is achieved by taking all the dynamical variables and turning them into operators,

$$\hat{p}^\mu \psi = i \partial^\mu \psi = p^\mu \psi, \tag{5.2}$$

and inserting the plane-wave solutions from equation (5.1). We have seen hints of this relationship in our discussion of Noether's theorem. Translational invariance (generated by spatial derivatives) gives rise to conserved momentum, while time invariance produces conserved energy. Consequently, the Hamiltonian $\hat{H}$ provides a mechanism for describing the time evolution of a wavefunction. The operator notation can be retroactively inserted into the Klein-Gordon equation (2.11):

$$\left(\hat{p}^\mu \hat{p}_\mu - m^2 \right) \phi = 0.$$

This is an elegant-looking equation, but it hides a bit of physical ugliness: the time-evolution component is second order in time.

A second-order time operator means that initial conditions for *both* ϕ and $\dot{\phi}$ must be specified to evolve the field. This situation may not, at first blush, seem that dire. It is not so different from the physics in the Newtonian framework, wherein we need to specify both initial positions and velocities to solve for future evolution.

But quantum fields are different. In particular, quantum mechanics is predicated on the idea that the probability density of particles integrated over space is a positive definite constant. Alas, second-order dynamic equations prevent such a simple interpretation of the field [80, 83].

If, however, we have a first-order equation of motion, then the initial state of a system tells us everything we need to know to evolve it indefinitely. That's precisely the state of affairs in *non*-relativistic quantum mechanics, wherein a freely propagating particle can be described by the Schrödinger wave equation:

$$\left(i\frac{\partial}{\partial t} + \frac{\nabla^2}{2m}\right)\psi = 0.$$

Unfortunately, as the Schrödinger equation privileges time over space, it's no simple matter to extend it to relativistic systems.

5.1.2 Electrons and Holes

When we quantize a classical field, energies turn into time derivatives. Thus, if we want an evolution equation that's first order in time, we'd like to turn the Klein-Gordon dispersion relation

$$(p^0)^2 - |\vec{p}|^2 - m^2 = 0 \tag{5.3}$$

into an equation that's linear in energy. Using a bit of middle-school algebra, we might be tempted to suppose that we can factor a quadratic equation into two linear equations. This approach seems especially promising if we consider the limiting (and patently frame-dependent) case of a particle at rest,

$$E^2 - m^2 = 0,$$

which quickly yields

$$E = \pm m,$$

either of which could be used to produce a differential wave equation linear in the time derivative. One of these solutions—the negative-energy one—seems a bit perplexing. Experimentally, there is no such thing as negative mass, and you might feel tempted to throw out the negative-energy solution as unphysical.

Common sense notwithstanding, Dirac was willing to follow this reasoning through to the end. As Richard Feynman put it [64]:

> When I was a young man, Dirac was my hero. He made a breakthrough, a new method of doing physics. He had the courage to simply guess at the form of an equation, the equation we now call the Dirac equation, and to try to interpret it afterwards.

Dirac [47] described his initial approach to negative energy particles ("holes") thus:

> It was proposed to get over this difficulty, making use of Pauli's Exclusion Principle which does not allow more than one electron in any state, by saying that in the physical world almost all the negative-energy states are already occupied, so that our ordinary electrons of positive energy cannot fall into them.

Dirac was referring to Pauli's recent work [120] wherein he explained the apparent ease with which atoms with odd numbers of electrons interact with each other compared with those carrying an even number. The noble gases, for instance, have even atomic numbers

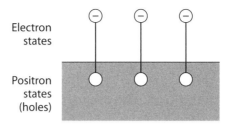

Electron states

Positron states (holes)

Figure 5.2. Dirac's negative energy "sea." As an electron is pulled out from the sea it leaves behind a negative-energy hole, what we now call a positron. Electrons and positrons are thus always created or annihilated in pairs.

(helium, $Z = 2$; neon, $Z = 10$; argon, $Z = 18$). Pauli introduced a quantum-numbering scheme, but it wasn't yet motivated as to *why* it worked. He simply supposed that multiple electrons couldn't occupy the same state. We'll justify the Pauli exclusion principle later in this chapter.

Dirac, initially, didn't believe that the negative-energy solutions corresponded to anything new, supposing that they might simply be protons.[2] But in 1931, he proposed the existence of a particle [47]:

unknown to experimental physics, having the same mass and opposite charge to an electron. We may call such a particle an antielectron.

The experimental discovery of the positron was almost immediate. Indeed, it technically *antedated* Dirac's prediction of the particle. The Soviet physicist Dmitri Skobeltsyn [151] presented a strange cloud chamber result in which particles seemed to be deflected in the direction opposite that predicted for an electron. This result was considered anomalous, and it wasn't confirmed until 1932 when Carl Anderson was given official recognition for discovery of the positron [14] in a cosmic ray shower, for which he won the Nobel Prize in Physics in 1936, which he shared with Victor Franz, for his discovery of cosmic radiation. It was his editor, incidentally, who coined the word "positron."

5.1.3 The γ-Matrices

To solve for the behavior of electrons and positrons generally (as opposed to those at rest), we attempt to factor equation (5.3) by assuming the heuristic form

$$p^\mu p_\mu - m^2 = (\beta^\kappa p_\kappa + m)(\gamma^\lambda p_\lambda - m) = 0, \qquad (5.4)$$

where, as always, the dummy indices are arbitrary.

---

[2] With the unfortunate complication that protons and electrons differ in mass by a factor of nearly 2000.

You're likely to encounter some difficulties if you assume β^{κ} and γ^{λ} are simply numbers. Expanding the factors in (5.4) explicitly yields

$$p^{\mu} p_{\mu} - m^2 = \beta^0 \gamma^0 p_0 p_0 + \beta^1 \gamma^1 p_1 p_1 + \ldots$$
$$+ (\beta^0 \gamma^1 + \beta^1 \gamma^0) p_0 p_1 + \ldots$$
$$+ (-\beta^0 + \gamma^0) m p_0 + \ldots$$
$$- m^2.$$

Since only the diagonal elements of the multiplication are coupled to one another, only the first and last lines of the expanded solution won't vanish. The third line immediately yields

$$\beta^{\kappa} = \gamma^{\kappa},$$

while the first and second can readily be combined to yield

$$\{\gamma^{\mu}, \gamma^{\nu}\} = 2g^{\mu\nu}\mathbf{I}. \tag{5.5}$$

The curly brackets denote the **anticommutation** operator

$$\{A, B\} \equiv AB + BA. \tag{5.6}$$

Each of the four elements of γ^{μ} represents a 4×4 matrix (Dirac found that nothing simpler would satisfy the anticommutation relation), but these γ-matrices are *not* unique. For instance, γ^0 in the **Dirac representation** is diagonal,

$$\gamma^{0(Dirac)} = \begin{pmatrix} \mathbf{I} & 0 \\ 0 & -\mathbf{I} \end{pmatrix},$$

while in **chiral representation** (also sometimes called **Weyl basis**) it can be written as

$$\gamma^0 = \begin{pmatrix} 0 & \mathbf{I} \\ \mathbf{I} & 0 \end{pmatrix}, \tag{5.7}$$

with the other three matrices expressible as

$$\gamma^i = \begin{pmatrix} 0 & \sigma_i \\ -\sigma_i & 0 \end{pmatrix}. \tag{5.8}$$

Here and throughout, we've used a block notation to describe the matrices for compactness. $\mathbf{I}$, for instance, represents the 2×2 identity matrix, and σ_i is the ith Pauli spin matrix (equation 4.13). We will use the chiral representation throughout in this book.

Example 5.1: Verify the anticommutation relation for γ^0 and γ^3.

Solution: Using the block notation, we get

$$\gamma^0 \gamma^3 = \begin{pmatrix} 0 & I \\ I & 0 \end{pmatrix} \begin{pmatrix} 0 & \sigma_3 \\ -\sigma_3 & 0 \end{pmatrix} = \begin{pmatrix} -\sigma_3 & 0 \\ 0 & \sigma_3 \end{pmatrix}$$

and

$$\gamma^3 \gamma^0 = \begin{pmatrix} 0 & \sigma_3 \\ -\sigma_3 & 0 \end{pmatrix} \begin{pmatrix} 0 & I \\ I & 0 \end{pmatrix} = \begin{pmatrix} \sigma_3 & 0 \\ 0 & -\sigma_3 \end{pmatrix},$$

which clearly add to zero, as required, since $g_{03} = 0$.

5.1.4 A Few Tricks with the γ-Matrices

After a while it will get tedious to multiply γ-matrices by hand. Fortunately, we don't always need to. For instance, the γ-matrices behave a *lot* like components of vectors, not least because they were derived in an effort to produce Lorentz invariance. As such, we may produce a one-form version of the matrices,

$$\gamma_\mu = g_{\mu\nu} \gamma^\nu, \tag{5.9}$$

which simply reverses the sign on the spacelike γ-matrices.

More significantly, we can contract the matrices with an ordinary 4-vector to produce a Lorentz-invariant term. For instance, for an arbitrary 4-vector a:

$$a_\mu \gamma^\mu = \slashed{a} \tag{5.10}$$

which is known as **Feynman's slash notation**. The slashed objects are *themselves* 4×4 matrices.

Example 5.2: Compute the components of $\slashed{p}$ for a particle moving in the z-direction with momentum p and energy E.

Solution: Writing the terms out explicitly, we get

$$\slashed{p} = E\gamma^0 - p\gamma^3$$

$$= E \begin{pmatrix} 0 & I \\ I & 0 \end{pmatrix} - p \begin{pmatrix} 0 & \sigma_3 \\ -\sigma_3 & 0 \end{pmatrix}$$

$$= \begin{pmatrix} 0 & 0 & E-p & 0 \\ 0 & 0 & 0 & E+p \\ E+p & 0 & 0 & 0 \\ 0 & E-p & 0 & 0 \end{pmatrix},$$

which may not look exciting, but upon taking the determinant, we find

$$det(p) = (E^2 - p^2)^2 = m^4,$$

a comfortingly Lorentz-invariant term.

From the definition of the slash relation, and from the anticommutation relation for the γ-matrices themselves, we quickly find

$$\{\not{a}, \not{b}\} = 2a \cdot b\mathbf{I}. \tag{5.11}$$

We will develop these sorts of relationships again on an as-needed basis (and in problems 5.2 and 5.3), but they will become *very* important when we encounter Casimir's trick two chapters hence. We also give some of the more commonly used relationships in Appendix A.

5.1.5 The Dirac Equation

The γ-matrices were necessitated by a desire to factor a quadratic dispersion relation to a linear relation,

$$\gamma^\lambda p_\lambda - m = 0,$$

from which we can quickly use the first quantization to yield

$$\boxed{i\gamma^\mu \partial_\mu \psi - m\psi = 0.} \tag{5.12}$$

Equation (5.12) is known the **Dirac equation**,[3] and a few things should be apparent from inspection. For one, ψ cannot be a simple number; it must, instead, be a four-component object:

$$\psi = \begin{pmatrix} \psi_1 \\ \psi_2 \\ \psi_3 \\ \psi_4 \end{pmatrix}.$$

This is *not* a 4-vector. This is a **Dirac Spinor**,[4] and it's going to have some interesting properties. While scalar fields were found to correspond to an ensemble of spin-0 particles, bispinor fields will be shown to correspond to spin-1/2 fermions: electrons, neutrinos, quarks, and other similar particles.

But take heed. Despite our use of quantum mechanical language, we have not yet generated a quantum field. While we are still in the realm of the classical, the number of particles is specified, but in quantum field theory (the so-called **second quantization**), even the number of particles cannot be said to be fixed. We will discuss quantization of

---

[3] Using our "slashed notation" from earlier makes it even more elegant: $(i\not\partial - m)\psi = 0$.
[4] Technically, it's a **bispinor**, since a spinor has two components, and the object in the Dirac equation has four.

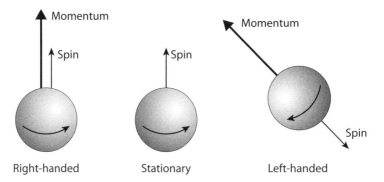

Figure 5.3. A right-handed particle.

the bispinor field at the end of this chapter, but for now, we allow ourselves the luxury of referring to the state of *a* particle.

5.2 Solutions to the Dirac Equation

The block form of the γ matrices suggests that the solution to the Dirac equation may be broken up into an upper and a lower pair of components in a bispinor field. That is, we may think of ψ as

$$\psi = \begin{pmatrix} \psi_L \\ \psi_R \end{pmatrix}, \tag{5.13}$$

where ψ_L and ψ_R have two elements each:

$$\psi_L = \begin{pmatrix} \psi_1 \\ \psi_2 \end{pmatrix}; \quad \psi_R = \begin{pmatrix} \psi_3 \\ \psi_4 \end{pmatrix}.$$

Very roughly speaking, the four components of the bispinor account for the two possible spin states of a fermion, along with the two possibilities of a particle or an antiparticle. We have also added a bit of suggestive notation in terms of subscripts, since the L and R refer to the handedness of a particle. Though we haven't yet derived the properties of spin, we will find that a right-handed particle's intrinsic angular momentum is in the same direction as the spin (as given by the right-hand rule, Figure 5.3), and that of a left-handed particle is the the reverse.

5.2.1 Stationary Particles

Consider a bispinor field that's entirely independent of position, in which case the Dirac equation reduces to

$$i\gamma^0 \dot{\psi} - m\psi = 0,$$

which the block form of the γ-matrices allows us to express as two coupled equations,

$$i\dot{\psi}_L - m\psi_R = 0$$

$$i\dot{\psi}_R - m\psi_L = 0,$$

which quickly yield four linearly independent solutions,

$$u_\pm(x) = \mathcal{N} \begin{pmatrix} 1 \\ 0 \\ 1 \\ 0 \end{pmatrix} e^{-imt}; \quad \mathcal{N} \begin{pmatrix} 0 \\ 1 \\ 0 \\ 1 \end{pmatrix} e^{-imt}$$

$$v_\mp(x) = \mathcal{N} \begin{pmatrix} 1 \\ 0 \\ -1 \\ 0 \end{pmatrix} e^{imt}; \quad \mathcal{N} \begin{pmatrix} 0 \\ 1 \\ 0 \\ -1 \end{pmatrix} e^{imt},$$

where $\mathcal{N}$ represents an as-yet-undetermined normalization.

Application of the Hamiltonian (equation 5.2) reveals the energy of the particles,

$$\hat{H}u_s = mu_s$$

$$\hat{H}v_s = -mv_s,$$

where s refers to the spin of the particle. We'll justify our choice of spin-up versus spin-down in §5.5.2.

For the v states, the energy of a stationary particle (the mass term) is *negative*. We saw earlier that these are the positrons, and they present their own particular problems. Our approach moving forward is to focus primarily on the u particles, and once we've got a firm intuition, to address the issues raised by the v particles.

5.2.2 Particles in Motion

We are now prepared to consider the free-field solutions to the Dirac equation. Assuming a generalized plane wave of the form in equation (5.1) yields

$$\left(\pm\gamma^\mu p_\mu - m\right)\psi(p) = 0,$$

where $\psi(p)$ is assumed to be spatially and time invariant. Multiplying out with the block form of the γ-matrices yields

$$\pm \begin{pmatrix} (p_0 - p^i\sigma_i)\psi_R \\ (p_0 + p^i\sigma_i)\psi_L \end{pmatrix} = m \begin{pmatrix} \psi_L \\ \psi_R \end{pmatrix}. \tag{5.14}$$

There is an implicit sign reversal in the spatial component of momentum when the indices are raised.

The block version of the Dirac equation may be solved by assuming the form of either the left- or right-handed component of the field and solving the coupled linear equation (5.14). We will save you some frustration by selecting only momentum states along the z-axis.

$$\psi_L = \mathcal{N} \begin{pmatrix} 1 \\ 0 \end{pmatrix},$$

which, along with the assumed negative in the exponent, generates

$$u_+(p, x) = \mathcal{N} \begin{pmatrix} 1 \\ 0 \\ \frac{E+p^z}{m} \\ 0 \end{pmatrix} e^{-ip \cdot x} \tag{5.15}$$

in general. We will often find it useful to isolate the propagating term in the solution, and thus we'll write

$$u_s(p, x) = u_s(p)e^{-ip \cdot x}. \tag{5.16}$$

We could solve for the three other basis states, but there's little to be gained in doing so until we answer a few questions. We haven't figured out a way to normalize or transform our solutions. We haven't yet determined a Lagrangian, which means that we can't compute any conserved currents, including the stress-energy tensor.

5.3 The Adjoint Spinor

Much as with the wavefunction in ordinary nonrelativistic quantum mechanics, we expect a quadratic combination of bispinor terms to yield a conserved scalar quantity. We define a new object, the **adjoint spinor**, as

$$\overline{\psi}(x) \equiv \psi^\dagger \gamma^0. \tag{5.17}$$

Just as ψ is a column vector, $\overline{\psi}$ can be thought of as a row vector.[5] A normalization will leave the product $\overline{\psi}\psi$ as a Lorentz-invariant scalar. We can, for instance, define the contraction of the basis spinors in terms of particle mass:

$$\overline{u}^{(+)}(p)u_+(p) = \mathcal{N}^2 \left[2\left(\frac{E+p^z}{m} \right) \right]$$
$$= 2m.$$

---

[5] Astute readers may wonder why we introduce the adjoint spinor at all, rather than simply use $\psi^\dagger$, which also produces a scalar upon contraction. The reason is that only the combination of $\overline{\psi}\psi$ produces a result that is Lorentz invariant.

The factor-of-2 convention, which will be useful in simplifying other expressions, produces the normalized basis spinors

$$
u_\pm(p) = \frac{m}{\sqrt{E+p^z}} \begin{pmatrix} 1 \\ 0 \\ \frac{E+p^z}{m} \\ 0 \end{pmatrix} \quad ; \quad \frac{m}{\sqrt{E-p^z}} \begin{pmatrix} 0 \\ 1 \\ 0 \\ \frac{E-p^z}{m} \end{pmatrix} \tag{5.18}
$$

$$
v_\pm(p) = \frac{m}{\sqrt{E-p^z}} \begin{pmatrix} 0 \\ 1 \\ 0 \\ \frac{-E+p^z}{m} \end{pmatrix} \quad ; \quad \frac{m}{\sqrt{E+p^z}} \begin{pmatrix} 1 \\ 0 \\ -\frac{E+p^z}{m} \\ 0 \end{pmatrix} , \tag{5.19}
$$

where E and M are defined to be positive numbers.

Example 5.3: Consider a spin-down Dirac particle moving at relativistic speeds $v \simeq 1$ in the $+z$-direction. What is the spinor representation?

Solution: In the relativistic limit, $m = 0$ and $p^z = E$. Noting that

$$
\frac{m}{\sqrt{E-p}} = \frac{m\sqrt{E+p}}{\sqrt{E^2-p^2}} = \sqrt{2E},
$$

we get a bispinor for a left-handed fermion:

$$
u_- = \begin{pmatrix} 0 \\ \sqrt{2E} \\ 0 \\ 0 \end{pmatrix}.
$$

Example 5.4: Consider a relativistic spin-up antiparticle in the $+z$-direction. What is the spinor representation?

Solution: The constraints are identical to those in the particle example but for v_+ rather than u_-. Substituting into equation 5.19 yields

$$
v_+ = \begin{pmatrix} 0 \\ \sqrt{2E} \\ 0 \\ 0 \end{pmatrix}.
$$

This is curious. A left-handed particle and a right-handed antiparticle, have an identical description so long as they are relativistic. This result will become even more important when we consider neutrinos.

We have explicitly chosen to set our coordinates along the direction of motion (the z-axis), as it's an arbitrary choice that makes our notation a bit neater. In principle, we could solve equation (5.14) with a velocity in an arbitrary direction. Regardless of the direction of motion, these basis spinors satisfy the generalized orthogonality and normalization conditions

$$\bar{u}_r(p)u_s(p) = 2m\delta_{rs}; \quad \bar{v}_r(p)v_s(p) = -2m\delta_{rs}; \quad \bar{u}_r(p)v_s(p) = \bar{v}_r(p)u_s(p) = 0. \tag{5.20}$$

Things are a bit more interesting for the antiparticle solutions, in which case the contraction yields a negative number. This difference in sign, as we've noted, is the difference between particles and antiparticles. Any wavefunction can be constructed as a linear series of $u_\pm(p)$ and $v_\mp(p)$, over any 4-momentum satisfying the relativistic dispersion relation (equation 5.3).

We may also compute the outer products, which produce a 4×4 matrix:

$$\sum_s u_s \bar{u}_s = \not{p} + m\mathbf{I}; \quad \sum_s v_s \bar{v}_s = \not{p} - m\mathbf{I}. \tag{5.21}$$

You can verify a particular case in problem 5.8.

Field theories are generally built upon Lagrangians, and the Dirac theory is no different. It is a simple matter to introduce a good candidate solution:

$$\boxed{\mathcal{L} = i\bar{\psi}\gamma^\mu \partial_\mu \psi - m\bar{\psi}\psi.} \tag{5.22}$$

By construction, the quantity $\bar{\psi}\psi$ is an invariant scalar, and based on the normalization, the dimensionality of the bispinor field must be $[\psi] = [E]^{3/2}$. This will come in handy when we do some dimensional analysis later. The Dirac equation (5.12) is trivial to derive from the Dirac Lagrangian, but it's important to have the Lagrangian; it is essential to determining the conserved Noether currents.

5.4 Coordinate Transformations

5.4.1 Rotations

Dirac spinors were developed as orthogonal solutions to the Dirac equation. Unlike scalars and vectors, spinors aren't simple tensorial objects, which means that we don't yet know how they transform under boosts and rotations.

Fortunately, our work in group theory has prepared us to delve headfirst into an unknown transformation. We will call the representation of rotations on spinors $\mathbf{M}_R$ and simply work through its properties,

$$\mathbf{M}_R = e^{-i\phi \mathbf{X}_R},$$

where the generator, $\mathbf{X}_R$ is to be determined, and where

$$\psi' = \mathbf{M}_R \psi; \quad \overline{\psi}' = \overline{\psi}\mathbf{M}_R^\dagger.$$

As a scalar, the mass term in the Dirac Lagrangian should be invariant under rotations,

$$m\overline{\psi}'\psi' = m\overline{\psi}\mathbf{M}_R^\dagger\mathbf{M}_R\psi,$$

demonstrating that $\mathbf{M}_R$ is unitary. Invariance in the Lagrangian also requires that vector terms transform according to Lorentz transforms:

$$\mathbf{M}_R^\dagger \gamma^\mu \mathbf{M}_R = \Lambda^{\overline{\mu}}_{\ \mu}\gamma^\mu. \tag{5.23}$$

For the particular case of a very small rotation around the z-axis (the form of the generator in equation 4.7), the coordinate rotation transformation may be written

$$\Lambda^{\overline{\mu}}_{\ \mu}\gamma^\mu = \begin{pmatrix} \gamma^0 \\ \gamma^1 - \delta\phi\gamma^2 \\ \gamma^2 + \delta\phi\gamma^1 \\ \gamma^3 \end{pmatrix} = e^{i\delta\phi\mathbf{X}_R}\gamma^\mu e^{-i\delta\phi\mathbf{X}_R}$$

to first order in $\delta\phi$. If we expand the exponentials using the Baker-Campbell-Hausdorff relations, the generator must satisfy the commutation relations

$$\left[\mathbf{X}_R, \gamma^1\right] = i\gamma^2$$

$$\left[\mathbf{X}_R, \gamma^2\right] = -i\gamma^1.$$

Thus, the generator for rotation of bispinors is

$$\mathbf{X}_R = \frac{1}{2}\begin{pmatrix} \sigma_3 & 0 \\ 0 & \sigma_3 \end{pmatrix}. \tag{5.24}$$

The factor of $1/2$ is a bit surprising and will have some interesting implications when we determine the intrinsic angular momentum of fermions. The generators of rotations around the other two principal axes follow fairly transparently.

Example 5.5: Consider a stationary spin-up electron. If you rotate the particle around the x-axis by an amount ϕ, what is the new state?

Solution: The initial state of the particle is

$$\psi = \sqrt{m}\begin{pmatrix} 1 \\ 0 \\ 1 \\ 0 \end{pmatrix}. \tag{5.25}$$

The transformation matrix can be written as

$$
\mathbf{M}_{R,x} = \exp\left[-i\frac{\phi}{2}\begin{pmatrix} \sigma_1 & 0 \\ 0 & \sigma_1 \end{pmatrix}\right] = \begin{pmatrix} \cos\frac{\phi}{2} & -i\,\sin\frac{\phi}{2} & 0 & 0 \\ -i\,\sin\frac{\phi}{2} & \cos\frac{\phi}{2} & 0 & 0 \\ 0 & 0 & \cos\frac{\phi}{2} & -i\,\sin\frac{\phi}{2} \\ 0 & 0 & -i\,\sin\frac{\phi}{2} & \cos\frac{\phi}{2} \end{pmatrix},
$$

where it is necessary to expand powers of the generator as a series to get the trigonometric form. This expansion yields

$$
\psi' = \mathbf{M}\psi = \sqrt{m}\begin{pmatrix} \cos\frac{\phi}{2} \\ -i\,\sin\frac{\phi}{2} \\ \cos\frac{\phi}{2} \\ -i\,\sin\frac{\phi}{2} \end{pmatrix}.
$$

Rotation by $\phi = \pi$ turns a spin-up electron, unsurprisingly, into spin-down.

5.4.2 Boosts

Our study of special relativity has primed us (so to speak) for the idea that rotations and boosts are intimately related. The arguments for constructing a generator for boosts are almost identical to those for a rotation, albeit with a different coordinate transform. To be explicit, we'll consider a boost in the $+z$-direction (following equation 1.26),

$$
\Lambda^{\overline{\mu}}_{\ \mu}\gamma^{\mu} = \begin{pmatrix} \gamma^0 + \delta v\gamma^3 \\ \gamma^1 \\ \gamma^2 \\ \gamma^3 + \delta v\gamma^0 \end{pmatrix},
$$

for small values of δv. We may follow the example of a rotation nearly exactly. A boost in the z-direction results in the commutation relations

$$
\left[\mathbf{X}_B, \gamma^0\right] = -i\gamma^3
$$
$$
\left[\mathbf{X}_B, \gamma^3\right] = -i\gamma^0,
$$

which are satisfied for

$$
\mathbf{X}_B = -\frac{i}{2}\begin{pmatrix} \sigma_3 & 0 \\ 0 & -\sigma_3 \end{pmatrix}, \tag{5.26}
$$

which is easily generalized to boosts along the x- and y-axes. The boost operator

$$
\mathbf{M}_B = e^{-i\theta\mathbf{X}_B}
$$

requires a bit of work to relate the parameter θ to a momentum or velocity in generality. Problem 5.11 focuses on this very question.

5.5 Conserved Currents

5.5.1 The Stress-Energy Tensor

We're now in for some real fun! We can compute the various conserved currents. Preeminent among them is the stress-energy tensor (equation 3.11), which may quickly be written as

$$T^{\mu\nu} = i\overline{\psi}\gamma^{\mu}\partial^{\nu}\psi - g^{\mu\nu}\mathcal{L}, \tag{5.27}$$

which gives

$$T^{00} = \overline{\psi}\left(-i\gamma^i\partial_i + m\right)\psi \quad \text{energy density;} \tag{5.28}$$

$$T^{i0} = i\overline{\psi}\gamma^i\partial^0\psi \qquad \text{momentum density.} \tag{5.29}$$

We'll get some additional clarity if we expand the Dirac field as a combination of plane waves,

$$\psi(x) = \sum_r \int \frac{d^3p}{(2\pi)^3} \frac{1}{\sqrt{2E_p}} \left[b_{\vec{p},r}u_r(p)e^{-ip\cdot x} + c_{\vec{p},r}^*v_r(p)e^{ip\cdot x}\right]. \tag{5.30}$$

a form that is almost identical (including normalization) to the plane-wave solution for a complex scalar field (equation 3.17) but with the inclusion of the bispinors. Given that $[\phi] = [E]^1$, and $[\psi] = [E]^{3/2}$, and $[u] = [E]^{1/2}$, the expansion makes sense at least on dimensional grounds.

The adjoint of the wave can be written

$$\overline{\psi} = \sum_s \int \frac{d^3q}{(2\pi)^3} \frac{1}{\sqrt{2E_q}} \left[b_{\vec{q},s}^*\overline{u}_s(q)e^{iq\cdot x} + c_{\vec{q},s}\overline{v}_s(q)e^{-iq\cdot x}\right],$$

where we have used a different dummy variable for both spin and the 4-momentum. Moving forward, we'll omit the explicit sum over spins, with the understanding that all spin states need to be included.

Integrating over all space we get a result *nearly* identical to what we found with the scalar field,

$$E = \int d^3x \, \overline{\psi}\left(-i\gamma^i\partial_i + m\right)\psi = \int \frac{d^3p}{(2\pi)^3} E_{\vec{p}} \left[b_{\vec{p},r}b_{\vec{p},r}^* - c_{\vec{p},r}c_{\vec{p},r}^*\right], \tag{5.31}$$

where—as a word to the wise if you're trying to replicate the result—some care must be taken to demonstrate that all cross terms in the Hamiltonian vanish.

All the same, this is an incredibly strange result. The b particles (we may as well call them *electrons*) have a positive energy at any speed, while the c particles (*positrons*) carry a negative energy, as we've already seen. No clever normalization will be able to square this discrepancy with the lack of negative-energy particles in the real physical universe.

The momentum, likewise, may be computed as

$$P^i = \int d^3x\, \overline{\psi}\left(i\gamma^i \partial_0\right)\psi = \int \frac{d^3p}{(2\pi)^3}\, p^i\left[b_{\vec{p},r}b^*_{\vec{p},r} - c_{\vec{p},r}c^*_{\vec{p},r}\right].$$

We'll resolve these negative-energy positrons before the end of the chapter, but for the moment, we have a few more conserved quantities to compute.

5.5.2 Angular Momentum

The general expression for angular momentum was developed during our discussion of Noether's theorem (equation 3.21):

$$l_z = \frac{\partial \mathcal{L}}{\partial_0 \phi_a}\frac{\partial \phi_a}{\partial \theta}.$$

For the Dirac Lagrangian, this yields

$$\vec{l} = \overline{\psi}\gamma^0 \vec{\mathbf{X}}_R \psi,$$

with the generator given by equation (5.24). Because we have selected our basis spinors as eigenstates of spin, the generator does double duty as the spin operator (hence the notation). Operating on the basis functions, the spin operator yields $+1/2$ for $u_+(0)$ and $v_-(0)$, while for $u_-(0)$ and $v_+(0)$ the angular momentum is $-1/2$.[6] The half-integer angular spin means that Dirac particles are fermions.

We can combine the spin operator with the momentum of a particle to produce a **helicity** operator:

$$\hat{h} = \frac{\vec{p}\cdot\hat{\vec{\mathbf{S}}}}{|\vec{p}|}. \tag{5.32}$$

The helicity operator produces $+1/2$ for a right-handed beam and $-1/2$ for a left-handed one.

Relativistic (and *only* relativistic) fermions are helicity eigenstates. If the spin and momentum are aligned, the particle is considered right-handed in the sense that a particle propagating in the direction given by your right-hand thumb will be spinning in the direction indicated by the curl of your fingers. The left handed case (u_- propagating in the $+z$-direction, for instance) is the same, but for your left hand (see Figure 5.3). This result justifies our labeling of ψ_L and ψ_R spinor components in the Weyl basis construction of the bispinor.

This distinction between left and right is an important one. In 1956, Tsung-Dao Lee and Chen-Ning Yang [99] did a literature review of the question of parity violation. While they found plenty of evidence that both electromagnetism and the strong force are ambidextrous, they didn't find any evidence one way or another regarding the weak

---

[6] It may appear surprising that v_+ has a spin of $-1/2$ rather than $+1/2$. This is yet another consequence of the fact that, classically, positrons have negative mass. The classical angular momentum of the positron may be negative, but its angular velocity is positive.

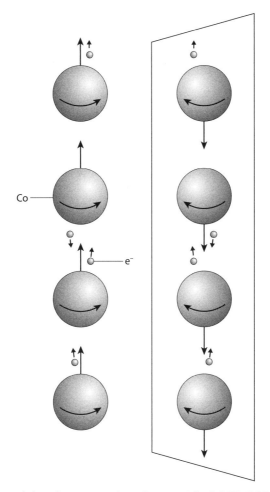

Figure 5.4. Cobalt-60 decay, as seen in a mirror. In 1957, C.-S. Wu [166] found that weak interactions are not symmetric with regard to parity transformations.

interaction. The following year, Chien-Shiung Wu and her collaborators tested the question by examining the weak nuclear decay of cobalt-60.

Wu found that the outgoing electrons were emitted preferentially opposite to the direction of the nuclear spin of the cobalt—a result which most definitely does *not* appear identical when seen in a mirror (see Figure 5.4). By performing the experiments at very low temperature, Wu was able to definitively show that neutrinos are always produced left-handed in weak reactions, while antineutrinos are always right-handed [166].

It is frequently helpful (and absolutely *required* once we get into the weak interaction) to decompose a beam into right-handed and left-handed components. Thus, we define the γ^5 matrix as

$$\gamma^5 \equiv i\gamma^0\gamma^1\gamma^2\gamma^3 = \begin{pmatrix} -\mathbf{I} & 0 \\ 0 & \mathbf{I} \end{pmatrix}. \tag{5.33}$$

Table 5.1. Bilinear Combinations of Dirac Spinors.

Bilinear		C	P
$\overline{\psi}_1 \psi_2$	Scalar	$-\overline{\psi}_2 \psi_1$	$\overline{\psi}_1 \overline{\psi}_2$
$\overline{\psi}_1 \gamma^\mu \psi_2$	Vector	$\overline{\psi}_2 \gamma^\mu \psi_1$	$-\overline{\psi}_1 \gamma^\mu \psi_2$
$\overline{\psi}_1 \gamma^\mu \gamma^5 \psi_2$	Axial Vector	$\overline{\psi}_2 \gamma^\mu \gamma^5 \psi_1$	$\overline{\psi}_1 \gamma^\mu \gamma^5 \psi_2$

Note: It should be understood that parity transformations on the field map $(t, \vec{x})$ to $(t, -\vec{x})$, even though this is not explicit in the table. Note that this table may differ from similar tables in other texts by a factor of -1 for charge conjugation. This is because we have not yet included the ad hoc anticommutation relation for the creation and annihilation operators that come from quantizing the field.

This matrix has the property of separating out spin; that is,

$$\frac{1}{2}(\mathbf{I} - \gamma^5)\psi = \begin{pmatrix} \psi_L \\ 0 \end{pmatrix} \tag{5.34}$$

$$\frac{1}{2}(\mathbf{I} + \gamma^5)\psi = \begin{pmatrix} 0 \\ \psi_R \end{pmatrix}. \tag{5.35}$$

We will find the γ^5 matrix especially useful when dealing with weak interactions, and it is essential in constructing axial vectors (see Table 5.1).[7]

5.5.3 Electric Charge

We've been talking about *electrons* and *positrons*, but we haven't discussed their most salient feature: electrical charge. The generator for the electromagnetic charge, as we've seen, is simply $\mathbf{X} = 1$, the generator for the U(1) group. For the Dirac Lagrangian, this yields a current density

$$J^\mu = q_e \overline{\psi} \gamma^\mu \psi, \tag{5.36}$$

where q_e is the charge of the electron.[8]

By now, the interpretation should be getting fairly familiar. Introducing the plane-wave solution (equation 5.30) yields a total current

$$I^\mu = \int J^\mu d^3 x = q_e \int \frac{d^3 p}{(2\pi)^3} E_p \left[b_{\vec{p},r} b^*_{\vec{p},r} + c_{\vec{p},r} c^*_{\vec{p},r} \right]. \tag{5.37}$$

Well, I'll be! According to our naive interpretation of Dirac fields, positrons contribute *negatively* to the energy density but *positively* to the charge! In some sense, this is less

---

[7] We encountered axial vectors before, the most familiar of which is the magnetic field, which is unchanged upon reflection.

[8] The fact that Franklin's convention makes q_e negative shouldn't overly concern us. What *will* concern is that in the classical interpretation, the electron and the positron will turn out to have the *same* charge.

of a problem than it might otherwise seem, since experimentally, we generally measure the mass/charge ratio [154] (which for the classical positron *will* be opposite that of the electron) rather than the mass and the charge individually.

However, we won't be able to get away with ignoring the problems with positrons indefinitely. Rather, we're going to need to be a bit more systematic about the difference between matter and antimatter.

5.6 Discrete Transforms

Noether's theorem allows us to determine conserved currents for our Lagrangians by use of continuous symmetries. But we have already encountered an important group of discrete transforms. We noted earlier (§2.5.2) that all theories in both the Standard Model and in general relativity are invariant upon transformations of successive applications of charge conjugation (C), reflection or parity (P), and time reflection (T), and all but the weak-interaction Lagrangian are invariant upon each of the three individually.

As simple as these transformations may first appear, they'll have a profound impact on the space of possible theories we can consider. But first, we need a sense of how fermions (and thus other particles) transform under C and P.

5.6.1 Charge Conjugation (C)

A charge conjugation transformation will turn positrons into electrons and vice-versa. We've already encountered the rather frightening proposition that positrons seem to have negative mass. All things being equal, we might naively suppose that electrons should quickly decay into positrons, an operation which evidently spoils the symmetry between them.

Ignoring this difficulty for just a few pages more, we consider what the charge conjugation operator looks like for Dirac fields:

$$\overline{\psi}\gamma^\mu\partial_\mu\psi \rightarrow \overline{\psi}\hat{C}^{-1}\gamma^\mu\partial_\mu\hat{C}\psi$$

This expression should be invariant upon the transformation. Both electromagnetism and the strong interaction exhibit C symmetry, but not the weak interaction. Since all neutrinos are left-handed and all antineutrinos are right-handed, charge conjugation violates symmetry in the weak interaction.

As to the form of the $\hat{C}$ operator, we may choose an answer and then show that it's right. Supposing

$$\psi^c = \hat{C}\psi = -i\gamma^2\psi^*, \tag{5.38}$$

we can quickly apply these to the basis spinors (equations 5.18, 5.19) to show that our form of the C operator yields

$$\hat{C}u_+ = v_+; \quad \hat{C}v_+ = u_+,$$

and similarly for the negative angular momentum states. Double application of charge conjugation returns a system to its original state, as expected.

Example 5.6: How does the scalar quantity $\overline{\psi}_1 \psi_2$ transform under charge conjugation?

Solution: Many important quantities can be described by quadratic combinations of Dirac fields (the so-called **bilinear covariants**). In addition to the scalar, there is the vector $\overline{\psi}\gamma^\mu \psi$, the axial vector $\overline{\psi}\gamma^\mu \gamma^5 \psi$, and several others. It is important to know how they change under discrete transformations.

Following equation (5.38), the transform of the adjoint may be computed as

$$\overline{\psi}_1^c = -i\psi_1^T \gamma^2 \gamma^0,$$

where we've introduced two minus signs by taking the conjugate of $-i$, as well as $(\gamma^2)^\dagger = -\gamma^2$. Thus,

$$\hat{C}(\overline{\psi}_1 \psi_2) = -\psi_1^T \gamma^2 \gamma^0 \gamma^2 \psi_2^*.$$

By construction γ^2 and γ^0 anticommute, and $\gamma^2 \gamma^2 = -I$, so

$$\hat{C}(\overline{\psi}_1 \psi_2) = -\psi_1^T \gamma^0 \psi_2^*.$$

Recalling the matrix relation $\mathbf{a}^T \mathbf{M} \mathbf{b} = \mathbf{b}^T \mathbf{M}^T \mathbf{a}$, and noting that γ^0 is symmetric (and thus its own transpose), we get

$$\hat{C}(\overline{\psi}_1 \psi_2) = -\overline{\psi}_2 \psi_1.$$

This and other bilinear covariant relations may be found in Table 5.1.

5.6.2 Reflection (P)

The parity transform turns all left-handed solutions to right-handed ones, in addition to reversing the $\vec{p} \cdot \vec{x}$ term in the exponent. For Dirac spinors, the operator may be represented by

$$\hat{P} = \begin{pmatrix} \sigma_1 & 0 \\ 0 & \sigma_1 \end{pmatrix}, \tag{5.39}$$

which reverses the spin and momentum of the basis Dirac spinors (and thus for Dirac fields generally). Again, this may be shown directly by applying the operator to the basis bispinors. The parity operator quickly yields

$$\hat{P}\hat{P} = \hat{I}.$$

Of course it is! Flip a mirror twice and you get the original!

Example 5.7: What is the result of a parity transformation of a spin-up electron propagating in the z-direction?

Solution: The parity transformation on a spin-up particle yields

$$\psi' = \frac{m}{\sqrt{E+p^z}} \begin{pmatrix} 0 & 1 & 0 & 0 \\ 1 & 0 & 0 & 0 \\ 0 & 0 & 0 & 1 \\ 0 & 0 & 1 & 0 \end{pmatrix} \begin{pmatrix} 1 \\ 0 \\ \frac{E+p^z}{m} \\ 0 \end{pmatrix} = \frac{m}{\sqrt{E+p^3}} \begin{pmatrix} 0 \\ 1 \\ 0 \\ \frac{E+p^z}{m} \end{pmatrix}.$$

We also need to explicitly include a reversal of momentum, $\vec{p} \to -\vec{p}$, which yields

$$\psi' = \frac{m}{\sqrt{E-p^z}} \begin{pmatrix} 0 \\ 1 \\ 0 \\ \frac{E-p^z}{m} \end{pmatrix},$$

which, as anticipated, is simply the spin-down electron.

Particles may be constructed as eigenstates of the parity operator:

$$\hat{P}\psi = \pm\psi.$$

Scalar fields, for instance, are necessarily unchanged on reflection and thus have a parity eigenvalue of $+1$. Vectors flip upon reflection, yielding a parity eigenvalue of -1. Axial vectors, as we've seen, are unchanged upon a parity transformation, as they are formed by the cross products of two vectors and thus have an eigenvalue of $+1$.

A spin-up electron, for instance, isn't a parity eigenstate. Seen in a mirror, spin-up appears spin-down. However, if we consider a *superposition* of stationary spin-up and spin-down electrons, in that case, the basis states are eigenstates for the parity operator; that is,

$$\hat{P}\left[\frac{1}{\sqrt{2}}(u_+ + u_-)\right] = \begin{pmatrix} 0 & 1 & 0 & 0 \\ 1 & 0 & 0 & 0 \\ 0 & 0 & 0 & 1 \\ 0 & 0 & 1 & 0 \end{pmatrix} \begin{pmatrix} \sqrt{m} \\ \sqrt{m} \\ \sqrt{m} \\ \sqrt{m} \end{pmatrix} = \begin{pmatrix} \sqrt{m} \\ \sqrt{m} \\ \sqrt{m} \\ \sqrt{m} \end{pmatrix}.$$

This is a $+1$ eigenstate of $\hat{P}$. A similar combination of $v_+ + v_-$ will reveal an eigenvalue of -1 (though the assignment of which is which is, naturally, completely arbitrary).

Many particles aren't eigenstates of parity or charge alone but, rather, combinations of C and P. **Mesons** are composite particles held together by the strong force (as are baryons) and composed of a quark and antiquark. For instance, a neutral **pion** comprises a superposition of up and down quarks and their antiparticles:

$$|\pi^0\rangle = \frac{1}{\sqrt{2}}\left(|u\bar{u}\rangle - |d\bar{d}\rangle\right). \tag{5.40}$$

Application of CP transformation quickly yields

$$\hat{C}\hat{P}|u\bar{u}\rangle = -|u\bar{u}\rangle,$$

with a similar relation for the down-antidown pair. Thus, pions have a CP of -1.

It may seem like a simple math trick to classify particles by their CP numbers, but the Dirac equation, along with the Lagrangians for all *three* of the fundamental forces are written in such a way that CP transformations should leave the Lagrangians intact. It is only the mass mixing that occurs from multiple generations (which we'll discuss in Chapter 10) that allows CP violation in the standard model, and thus it's worth considering the total CP number for complex and composite systems.

Fortunately, there's a simple rule for composite systems. You simply take the *product* of the constituents of the eigenvalue of the CP operator. Thus,

$$\pi^0 \rightarrow CP = -1$$
$$\pi^0 + \pi^0 \rightarrow CP = +1$$
$$\pi^0 + \pi^0 + \pi^0 \rightarrow CP = -1.$$

As we shall see in Chapter 10, conservation of CP *should* strictly limit the decay products that are possible. While the strong and electromagnetic force quite obligingly enforce CP symmetry, the weak interaction is not quite so cooperative.

5.7 Quantum Free-Field Theory

We've gone about as far as we can with the free-field classical Dirac Lagrangian. No amount of sweeping under the rug will solve the problem of negative-mass antimatter,

$$\overline{\psi}\psi \xrightarrow{C} -\overline{\psi}\psi, \tag{5.41}$$

as we've seen throughout. Our derivation in example 5.6 was necessarily classical. We used simple matrix relations to reverse the order of two fields. But order matters in quantum mechanics, and to get a sense of *how* order comes into it, we'll take a few steps back and revisit the simple mass on a spring.

5.7.1 The Quantum Harmonic Oscillator

Because it will serve as a template for field quantization, we'll do a simple outline of quantization of a familiar example: the quantum harmonic oscillator. The simple harmonic oscillator Lagrangian (equation 2.3) yields a solution of the form

$$q(t) = \sqrt{\frac{1}{2m\omega}}\left[a\,e^{-i\omega t} + a^*\,e^{i\omega t}\right], \tag{5.42}$$

which is guaranteed to be real-valued, with a dimensionless amplitude a. For a nonrelativistic particle, the momentum can be written

$$p(t) = m\dot{q} = -i\sqrt{\frac{m\omega}{2}}\left[a\,e^{-i\omega t} - a^*\,e^{i\omega t}\right]. \tag{5.43}$$

The transition from classical to quantum should be familiar. We simply "promote" the variables q and p into operators $\hat{q}$ and $\hat{p}$ on wavefunctions. Quantizing the position yields

$$\hat{q} = \sqrt{\frac{1}{2m\omega}}\left[\hat{a}\,e^{-i\omega t} + \hat{a}^\dagger\,e^{i\omega t}\right], \tag{5.44}$$

where $\hat{a}^\dagger$ and $\hat{a}$ are known as the "raising" and "lowering" operators, respectively; that is,

$$\hat{a}^\dagger|n\rangle = \sqrt{n+1}|n+1\rangle$$
$$\hat{a}|n\rangle = \sqrt{n}|n-1\rangle, \tag{5.45}$$

and where $n = 0$ corresponds to the ground-state energy of the system. While the basis states are not eigenstates of $\hat{a}$ or $\hat{a}^\dagger$ individually,

$$\hat{a}^\dagger\hat{a}|n\rangle = n|n\rangle.$$

These operations are particularly important by analogy, since the quantum field versions of the raising and lowering operators will be the **creation** and **annihilation operators**, which when operating on the state of a field will add or destroy a particle, respectively. You will have the opportunity to check the validity of these relations in problem 5.13.

In 1927, Werner Heisenberg [87] noted that momentum and position form a conjugate pair—that you can't measure either without increasing the uncertainty of the other. The order in which the operators are applied clearly matter. In other words,

$$[\hat{q}, \hat{p}] = i,$$

which is simply a mathematically sophisticated way of saying that if we're measuring both position and momentum, the order in which we measure them matters. The commutation relation for the momentum and position quickly produces a similar relation for the raising and lowering operators,

$$[\hat{a}, \hat{a}^\dagger] = 1, \tag{5.46}$$

justifying our initial choice of normalization in equation (5.42). We can use exactly the same argument when trying to figure out quantum perturbations of a scalar field.

5.7.2 *Quantized Scalar Fields*

We developed classical fields by direct analogy from our work with springs, so it makes sense to mimic our work with quantum harmonic oscillators when quantizing fields. In equation (3.14), we generated an energy density for a real-valued scalar field as

$$\rho = \frac{1}{2}\dot{\phi} + \frac{1}{2}(\nabla\phi)^2 + V(\phi),$$

where the field may be expanded as a series of plane waves:

$$\phi(x) = \int \frac{d^3 p}{(2\pi)^3} \frac{1}{\sqrt{2E_p}} \left[a_{\vec{p}} e^{-ip\cdot x} + a_{\vec{p}}^* e^{ip\cdot x} \right] \tag{5.47}$$

Recognizing that quantization is in our future, we must pay particular attention to the order of the coefficients $a_{\vec{p}}$ and $a_{\vec{p}}^*$. Expanding the energy in this form yields

$$E = \int d^3 x \, \rho(x) = \int \frac{d^3 p}{(2\pi)^3} E_p \left[\frac{1}{2} a_{\vec{p}} a_{\vec{p}}^* + \frac{1}{2} a_{\vec{p}}^* a_{\vec{p}} \right]. \tag{5.48}$$

The notation here is telling. Just as with the simple harmonic oscillator, there is a strong temptation to promote the coefficients $a_{\vec{p}}$ to operators, where

- $\hat{a}_{\vec{p}}^{\dagger}$ creates a ϕ particle with momentum $\vec{p}$,
- $\hat{a}_{\vec{p}}$ annihilates a ϕ particle with momentum $\vec{p}$,

and where, following the example of the raising and lowering operators of the quantum harmonic oscillator,

$$[\hat{a}_{\vec{p}}, \hat{a}_{\vec{q}}^{\dagger}] = (2\pi)^3 \delta(\vec{p} - \vec{q}). \tag{5.49}$$

In that context, the amplitude of the field ϕ may also be written as an operator:

$$\hat{\phi}(x) = \int \frac{d^3 p}{(2\pi)^3} \frac{1}{\sqrt{2E_p}} \left[\hat{a}_{\vec{p}} e^{-ip\cdot x} + \hat{a}_{\vec{p}}^{\dagger} e^{ip\cdot x} \right]. \tag{5.50}$$

The Hamiltonian is constructed by quantizing equation (5.48) and invoking the commutation relation (equation 5.49):

$$\hat{H}_{Scalar} = \int \frac{d^3 p}{(2\pi)^3} E_p \left[\hat{N}(p) + \frac{1}{2} (2\pi)^3 \delta^{(3)}(0) \right]. \tag{5.51}$$

That last term is somewhat nasty. The expectation value of the energy in a particular state can computed via

$$\langle E \rangle = \langle n | \hat{H} | n \rangle, \tag{5.52}$$

where $|n\rangle$ represents some generalized state of particles flying through space. Even in a vacuum, $n = 0$, equation (5.51) implies that there's a nonzero energy.

While the delta-function formally integrates to infinity, over a finite volume

$$(2\pi)^3 \delta^{(3)}(0) \to V, \tag{5.53}$$

a relationship which will serve us well later. Thus,

$$\langle \rho \rangle_{vacuum} = \frac{1}{2} \int \frac{d^3 p}{(2\pi)^3} E_p. \tag{5.54}$$

Assuming this expression can be integrated up to the Planck energies, it yields an astonishing density of

$$\langle \rho \rangle_{vacuum} = \frac{E_{Pl}^4}{4\pi^2}.$$

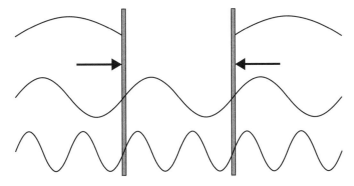

Figure 5.5. The Casimir effect.

This is approximately 10^{120} times the current best estimates of the total energy density of the universe [56, 130]. Since general relativistic gravity is supposed to see *all* the energy in the universe, this enormous constant is something of an embarrassment.

The energy perturbations from a quantized scalar field resemble, in a way, a pot of boiling water. Bubbles arise out of seemingly nowhere, and pop in short order. While there's no saying where a particular bubble will arise at any given time, *on average* we can say that there are going to be so many per unit area.

These perturbations are realized by the creation and annihilation of particles. Recognizing that our model of our scalar field is ultimately only an infinite array of simple harmonic oscillators allows a shortcut to figuring out the mathematics of a more formal theory of quantized scalar fields.

The analog of this vacuum energy has been experimentally detected in the photon field. In 1948 [35], Hendrik Casimir proposed that quantum fluctuations of the electromagnetic field should produce a force of attraction between two nearby conducting plates.

Since the electromagnetic field necessarily vanishes on the plates themselves, the closer the plates are placed to each other, the fewer allowable modes of the vacuum state and, hence, the lower the energy. In effect, the **Casimir effect** predicted the attraction of these conducting plates as evidence of a nonzero electromagnetic vacuum state. It also showed, not incidentally, that the electromagnetic energy density has a positive, rather than negative, zero point. This effect has been observed experimentally down to the submicron scale [95, 152].

5.7.3 Quantizing the Fermionic Fields

We quantized our scalar field without encountering anything *too* weird, but fermionic fields present problems almost immediately. As with scalar fields, the underlying issues will become clearest if we begin by quantizing the plane-wave expansion of a fermionic field (equation 5.30),

$$\hat{\psi}(x) = \sum_r \int \frac{d^3 p}{(2\pi)^3} \frac{1}{\sqrt{2E_p}} \left[\hat{b}_{\vec{p},r} u_r(p) e^{-ip \cdot x} + \hat{c}^{\dagger}_{\vec{p},r} v_r(p) e^{ip \cdot x} \right], \tag{5.55}$$

where, following both the simple harmonic oscillator and the quantized scalar field, we can interpret our newly found operators in the following manner:

- $b_{\vec{p},r}^{\dagger}$ creates an electron of 4-momentum p and spin state r.
- $b_{\vec{p},r}$ annihilates an electron.
- $c_{\vec{p},r}^{\dagger}$ creates a positron.
- $c_{\vec{p},r}$ annihilates a positron.

The Hamiltonian operator may be computed, following equation (5.31), as

$$\hat{H}_{Dirac} = \int \frac{d^3p}{(2\pi)^3} E_p \left[\hat{b}_{\vec{p},r}^{\dagger} \hat{b}_{\vec{p},r} - \hat{c}_{\vec{p},r} \hat{c}_{\vec{p},r}^{\dagger} \right]. \tag{5.56}$$

We've made no assumptions about the commutation relations between $\hat{c}$ and $\hat{b}$ operators and have thus simply written them out in order. There is a minus sign difference between the electrons (b) and the positrons (c), but they also differ in the order in which they operate.

We'll now assert (but not prove) that fermionic fields obey an *anticommutation* relation of the form

$$\{b_{\vec{p},r}, b_{\vec{q},s}^{\dagger}\} = \{c_{\vec{p},r}, c_{\vec{q},s}^{\dagger}\} = (2\pi)^3 \delta^{rs} \delta(\vec{p} - \vec{q}), \tag{5.57}$$

and all other anticommutators for the creation and annihilation operators vanish. Thus,

$$\hat{c}_{\vec{p},r} \hat{c}_{\vec{p},r}^{\dagger} = -\hat{c}_{\vec{p},r}^{\dagger} \hat{c}_{\vec{p},r} + (2\pi)^3 \delta^{(3)}(0).$$

And hence,

$$\hat{H}_{Dirac} = \int \frac{d^3p}{(2\pi)^3} E_p \left[\hat{N}_{b,r}(p) + \hat{N}_{c,r}(p) - (2\pi)^3 \delta^{(3)}(0) \right], \tag{5.58}$$

where the number operator is drawn by exact analogy from the scalar field for each species of particle. We note two important results:

1. When properly quantized, positrons really do have positive energy. We leave it as an exercise for the reader to show that the charge of the positron is opposite to that of the electron.

2. Not only is the contribution of fermions to the vacuum energy negative but the contribution per each of the four modes is the same as the positive energy contribution from a scalar field with one independent mode. This distinction between fermions and bosons will be important later on when we consider supersymmetry, but for now, and indeed, in most of our discussion, we simply ignore the vacuum energy entirely.

5.7.4 The Pauli Exclusion Principle

Suppose we start with empty space $|0\rangle$ and create an electron with momentum $\vec{p}$. Mathematically, the new state can be generated by

$$|\psi(p)\rangle = \hat{b}_{\vec{p},r}^{\dagger} |0\rangle.$$

We add a second particle with momentum $\vec{q}$, and the new state is

$$\left|\psi(p,q)\right\rangle = \hat{b}^{\dagger}_{\vec{q},s}\hat{b}^{\dagger}_{\vec{p},r}|0\rangle.$$

Consider what happens if we switch the order in which we label the particles; that is,

$$\left|\psi(q,p)\right\rangle = \hat{b}^{\dagger}_{\vec{p},r}\hat{b}^{\dagger}_{\vec{q},s}|0\rangle.$$

In many ways, $\left|\psi(p,q)\right\rangle$ and $\left|\psi(p,q)\right\rangle$ should be observationally identical. They should, for instance, have the same energy, the same charge, the same momentum, the same number of particles, and so forth. Practically speaking, calculations like

$$\langle\psi(p,q)|\hat{O}|\psi(p,q)\rangle = \langle\psi(q,p)|\hat{O}|\psi(q,p)\rangle$$

should be identical to one another for any operator, $\hat{O}$.

But that does not mean that the states themselves are identical. That is, we may imagine an exchange operator:

$$\hat{\mathcal{X}}\left|\psi(p,q)\right\rangle = \left|\psi(q,p)\right\rangle. \tag{5.59}$$

Application of the exchange operator twice in a row will leave a system unchanged, and thus the only possible eigenvalues are ± 1. But we can see that based on the anticommutation relation, fermions have an eigenvalue of -1. That is, since

$$\{\hat{b}^{\dagger}_{\vec{q},s},\hat{b}^{\dagger}_{\vec{p},r}\} = 0,$$

then the exchange operator on a pair of fermions yields

$$\left|\psi(q,p)\right\rangle_{\text{fermion}} = -\left|\psi(p,q)\right\rangle_{\text{fermion}}. \tag{5.60}$$

Bosons, particles with integer spin (including spin-0 scalar fields) obey a commutation relation,

$$\left|\psi(q,p)\right\rangle_{\text{boson}} = \left|\psi(p,q)\right\rangle_{\text{boson}}, \tag{5.61}$$

which produces the Bose-Einstein statistical distribution [31].

The relationship between particle exchange and spin is known as the **spin-statistics theorem** and has deep significance to the structure of matter. We've argued from a position of mere common sense that for antiparticles to have opposite charge and positive mass, Dirac fields must somehow anticommute upon replacement. But where does this conclusion come from?

This is not a simple question. In the 1960s, Richard Feynman posed the question in his *Feynman Lectures on Physics* [65] by noting:

> An explanation has been worked out by Pauli from complicated arguments of quantum field theory and relativity. ... We have not been able to find a way of reproducing his arguments on an elementary level. ... This probably means that we do not have a complete understanding of the fundamental principles involved.

But experimentally it is the case that fermions anticommute and bosons commute. The anticommutation relation is at the center of the Pauli exclusion principle. If the exchange of two identical particles in the same state can produce a negative sign, then the natural consequence is

$$\hat{\mathcal{X}}|\psi(p,\,p)\rangle = |\psi(p,\,p)\rangle = -|\psi(p,\,p)\rangle,$$

which can be true only if the state is zero. In other words, fermionic particles of identical type cannot be in the same state and position as one another. No such relation exists for bosonic particles.

The exclusion principle gives rise to the simple orbital relations of the periodic table and, as a consequence, virtually all chemistry; it's the mechanism that prevents the ultimate collapse of white dwarves and neutron stars, which explode in the form of supernova and produce the heavy elements that make complex life possible.

Problems

5.1 Evaluate:
(a) $\{\gamma^0, \gamma^0\}$
(b) $\gamma^2\gamma^0\gamma^2$ (which arose during the transformation of the Lagrangian mass term under charge conjugation).
(c) $[\gamma^1, \gamma^2]$

5.2 Simplify the anticommutation relation, $\{\not{a}, \not{b}\}$, where a and b are two different 4-vectors.

5.3 While the generalization of this problem can be found in Appendix A, compute the various traces of combinations of γ-matrices explicitly (that is, multiply them out).
(a) $Tr(\gamma^\mu \gamma^\nu)$ for the specific cases of:

  i. $Tr(\gamma^0\gamma^0)$
  ii. $Tr(\gamma^1\gamma^1)$
  iii. $Tr(\gamma^1\gamma^0)$

(b) Compute $Tr(\not{a}\not{b})$, for two arbitrary 4-vectors a and b.

5.4 Using only the quadratic trace relationship $Tr(\gamma^\mu\gamma^\nu) = 4g^{\mu\nu}$, which you derived in problem 5.3 compute:
(a) $Tr(\gamma^0\gamma^0\gamma^1\gamma^1)$
(b) $Tr(\gamma^0\gamma^1\gamma^0\gamma^1)$
(c) $Tr(\gamma^0\gamma^0\gamma^1\gamma^2)$

5.5 In the text, we have chosen a spin basis for our bispinors that aligns with the momentum axis. Instead, consider a particle propagating in the x-direction. Find the $u_+(p)$ eigenstate of the Dirac equation.

5.6 Show $\bar{u}_+(p)v_-(p) = 0$ explicitly. Do not simply assume the orthogonality condition.

5.7 In quantum field theory calculations, we will often find it useful to compute products like

$$\left[\bar{u}(1)\gamma^\mu u(2)\right],$$

where the 1 corresponds to the spin, mass, and 4-momentum of one particle state, and 2 corresponds to the similar quantities for a second particle.

For particle 1,

$$m = m_1; \quad \vec{p} = 0; \quad s = +1/2,$$

and for particle 2,

$$m = 0; \quad \vec{p} = p_z \hat{k}; \quad s = +1/2.$$

(a) Calculate the vector of values $\left[\bar{u}(1)\gamma^\mu u(2)\right]$ for the states listed.
(b) Calculate the vector of values for $s_2 = -1/2$.

5.8 Compute the *outer* product over the positron bispinors

$$\sum_s v_s(p)\bar{v}_s(p)$$

for a particle propagating in the $+z$ direction. Compare your answer with equation (5.21).

5.9 We found that the generator for spinor rotations in the ith direction may be written as

$$\mathbf{X}_R^i = \frac{1}{2}\begin{pmatrix} \sigma_i & 0 \\ 0 & \sigma_i \end{pmatrix}.$$

Calculate the commutation relation

$$[\mathbf{X}^1, \mathbf{X}^2].$$

5.10 As a follow-up to example 5.5, what happens to an electron bispinor upon rotation of 2π? How many rotations are required to return the electron to its initial condition?

5.11 Starting with a spin-up electron at rest, use the generalized form of the boost relation to compute the spinor for a spin-up particle moving in the $+x$-direction. Using the relation $\bar{\psi}\gamma^1\psi = 2p^1$, derive a general relationship between momentum and the boost parameter θ.

5.12 For the single-particle Dirac equation Hamiltonian

$$\hat{H} = -i\gamma^i \partial_i + m$$

(a) Compute the commutator of the Hamiltonian operator with the z-component of the angular momentum operator $[\hat{H}, \hat{L}_z]$, where

$$\hat{\vec{L}} \equiv \vec{r} \times \hat{\vec{p}}.$$

(b) Now consider the spin operator

$$\hat{\vec{S}} = \frac{1}{2} = \begin{pmatrix} \sigma & 0 \\ 0 & \sigma \end{pmatrix}.$$

Compute the z-component of $\hat{S}^z u_-(p)$.
(c) Compute $[\hat{H}, \hat{S}^z]$.
(d) Comparing your answers to parts a and c, derive a conserved quantity for free fermions. Conserved quantities commute with the Hamiltonian.

5.13 Starting with equation 5.45 compute:
(a) $[\hat{a}, \hat{a}^\dagger]$ for the quantum harmonic oscillator
(b) Using this result, and the definition of the quantized momentum and position, compute $[\hat{q}, \hat{p}]$.

5.14 Consider two identical fermions, combined to produce a boson, in states p and q, respectively. Construct a wavefunction for the boson which is symmetric under interchange of p and q.

Further Readings

- Duck, Ian, and E.C.G. Sudarshan. *Pauli and the Spin-Statistics Theorem.* New York: World Scientific, 1998. Duck and Sudarshan attempt to address the question posed by Feynman as to whether anyone has derived an elementary derivation of the spin-statistics theorem, and discuss many of the physical and astrophysical results of the theorem.
- Feynman, Richard P., Robert B. Leighton, and Matthew L. Sands. *The Feynman Lectures on Physics.* New York: Addison-Wesley, 1970. The quote about the spin-statistics theorem, is drawn from Lecture 4, on identical particles, and is particularly helpful.
- Griffiths, David J. *Introduction to Quantum Mechanics,* 2nd ed. New York: Pearson, 2004. Griffiths's descriptions of the time-independent Schrödinger equation (chapter 2) and of identical particles (chapter 5) serve as an extremely good supplement to the material in this chapter.
- Gross, Franz. *Relativistic Quantum Mechanics and Field Theory.* Hoboken, NJ: Wiley-VCH, 1993. Chapter 5 has an excellent discussion of the Dirac equation (albeit in Dirac basis rather than the chiral basis used herein).
- Neuenschwander, Dwight. The spin-statistics theorem and identical particle distribution functions. *Radiations,* Fall 2013, p. 27.
- Susskind, Leonard, and Art Friedman. *Quantum Mechanics: The Theoretical Minimum.* New York: Basic Books, 2014. Susskind and Friedman have a very nice discussion of unitary operators in quantum mechanical systems in chapter 4.
- Tong, David. "Lectures on Quantum Field Theory." 2006. http://www.damtp.cam.ac.uk/user/tong/qft. html. Cambridge University, Tong has a very nice discussion in section 4 on the Dirac equation.

6 | Electromagnetism

Figure 6.1. Hermann Klaus Hugo Weyl (1885–1955). Weyl developed one of the early gauge theories in an effort to unify general relativity and electromagnetism. Source: ETH-Bibliothek Zürich, Bildarchiv.

Electromagnetism has the advantage over the other fundamental forces of being evident on human scales.[1] James Clerk Maxwell unified the laws into his famous equations in 1862, and within a half century, Einstein [53] used Maxwell's equations as a launching point for special relativity. The weak and strong interactions are constructed on the model of electromagnetism, which means that we're going to have to do a bit of ahistorical analysis to put Maxwell's equations into a useful formulation for our future work.

6.1 A Toy Model of Electromagnetism

While electromagnetism is familiar, the principles for generating the electromagnetic Lagrangian are a bit more sophisticated. Instead, we will begin with a toy theory of scalar fields which consist of two species of spin-zero particles:[2]

- ϕ's, which roughly (very roughly) correspond to electrons;
- η's, which will stand in for photons.

---

[1] Gravity excepted, as it always is in the case of the Standard Model.
[2] Both electrons and photons have spin, and that's a complication that can wait for a few pages.

Figure 6.2. A mass of stationary ϕ field placed at the origin.

This theory has been unapologetically pulled out of thin air. In the next section, we will motivate how assumptions of internal symmetries inevitably give rise to particle interactions, but for now, we'll simply take the scalar Lagrangian as a given.

We've previously seen Lagrangians for free particles only in unending and unswerving streams with nothing to create or destroy particles or to alter their trajectories. Every Lagrangian we've seen to this point has been second order and has produced linear equations of motion. If we want to mimic the physics of the real world, we propose a third-order Lagrangian (with the requirement that it satisfy Lorentz invariance) of the form

$$\mathcal{L} = \frac{1}{2}\partial_\mu\phi\partial^\mu\phi - \frac{1}{2}m_\phi^2\phi^2 + \frac{1}{2}\partial_\mu\eta\partial^\mu\eta - \frac{1}{2}m_\eta^2\eta^2 - \lambda\phi^2\eta. \tag{6.1}$$

The Lagrangian itself may look a bit intimidating, but the first two pairs of terms on the right are comfortably familiar. They simply represent the free-field Lagrangians of two different species of scalar particles. The final term, parameterized by an interaction strength λ is an interaction between the two fields (and has dimensionality, in this theory, of $[E]^1$) which contains all the new physics.

As with a classical case, we can compute the Euler-Lagrange equations, but contrary to the Klein-Gordon equation, our toy model has generated source terms:

$$(\Box + m_\phi^2)\phi = -2\lambda\eta\phi;$$

$$(\Box + m_\eta^2)\eta = -\lambda\phi^2.$$

The dynamics of each of the free fields contains influences from the other. Most notably, a universe filled with only ϕ fields will spontaneously start to produce η fields!

Suppose we place a *lump* of ϕ at the origin (Figure 6.2). From dimensional analysis, we may approximate the lump as a delta function:

$$\phi^2 \sim \frac{1}{m_\phi}\delta(\vec{x}).$$

Provided the ϕ field doesn't evolve during timescales of interest, the oscillating term in the Euler-Lagrange equation for η will die out quickly, leaving the relation

$$(\nabla^2 - m_\eta^2)\eta \sim \frac{\lambda}{m_\phi}\delta(\vec{x}). \tag{6.2}$$

This is interesting—very interesting. It looks very much like the Poisson equation for electromagnetism and gravity, but with a pesky mass term. Writing $\eta(\vec{x})$ as a Fourier series,

$$\eta(\vec{x}) \sim \int \frac{d^3 p}{(2\pi)^3} e^{i\vec{p}\cdot\vec{x}} \eta_p,$$

and invoking the relation

$$\int d^3 x e^{i\vec{p}\cdot\vec{x}} = (2\pi)^3 \delta(\vec{p}) \tag{6.3}$$

allows us to quickly integrate over all space to get

$$(|\vec{p}|^2 + m^2)\eta_p \sim -\frac{\lambda}{m_\phi}.$$

Transforming back to real space, we get

$$\eta(\vec{x}) \sim -\frac{\lambda}{m_\phi} \int \frac{d^3 p}{(2\pi)^3} \frac{e^{i\vec{p}\cdot\vec{x}}}{|\vec{p}|^2 + m_\eta^2}.$$

This integral looks ugly, but it's not too bad. After all, the η field is constructed to be isotropic around the origin. We may utilize the exact analytic relation

$$\int \frac{d^3 p}{(2\pi)^3} \frac{e^{i\vec{p}\cdot\vec{x}}}{|\vec{p}|^2 + m_\eta^2} = \frac{e^{-m_\eta r}}{4\pi r}, \tag{6.4}$$

where r is the radial coordinate.

Placing a second lump of ϕ at position $\vec{r}$ with respect to the origin allows us to compute the interaction energy between the two as

$$E_{\text{Int}} = \int d^3 x \, \mathcal{H}(x)$$

$$\sim -\frac{\lambda^2}{m_\phi^2} \frac{1}{4\pi r} e^{-m_\eta r}, \tag{6.5}$$

which is very similar to the form discovered by Hideki Yukawa in the 1930s [168].[3]

Though we've simplified our work by using a toy scalar model, the general form of the potential—at least at large separations and thus low energies—will be the same for most interactions that we'll encounter. This result is quite remarkable for a number of reasons. First, starting with only two isolated ϕ particles, we found that a **mediating field** η arose spontaneously between the two and created an interaction energy.

Second, the potential energy is negative, which means the forces between the two ϕ's are attractive, regardless of the sign of the coupling constant λ (which enters quadratically). For the scalar theory, alike charges *attract*. Newtonian gravity (as opposed to general relativistic gravity) can be thought of as a classical theory mediated by a scalar field. Indeed, in that theory like charges attract one another.

Surprisingly, the sign of the interaction is a consequence of the spin, of the mediator particle. If the mediator has an even spin, then charges with like sign will attract (e.g., s = 0 for the scalar theory, s = 2 for a graviton, if a consistent quantum theory of gravity

---

[3] Yukawa explicitly solved for Dirac fields, as we will in due course.

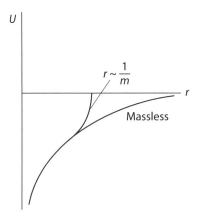

Figure 6.3. The Yukawa potential for a massive mediator.

is ever formulated[4]). On the other hand, in true electromagnetism, which uses a spin-1 photon as a mediator, like charges *repel*.

Third, the strength of the interacting potential field varies as $1/r$ on scales smaller than $r \ll 1/m_\eta$. The mediator effectively gets tired over large distances (e.g., Figure 6.3) and gives up. An important instance of this is the weak interaction, which has extremely massive mediators. The W^- boson, which mediates interactions between neutrinos and charged particles, has a range of approximately 10^{-18} m, smaller than the canonical size of an atomic nucleus (approximately a femtometer). The photon, however, is massless, which is another way of saying that electromagnetism is a long-range force. By the same token, as gravity seems to exhibit an inverse-square force on large scales, we expect the graviton to be massless as well.

6.2 Gauge Transformations

Our scalar version of electromagnetism was clearly ad hoc. Interactions in the Standard Model are based on a much more fundamental principle of symmetry. Though Maxwell's equations are likely familiar to most readers, there's a trove of symmetry hidden within the equations—in particular, **gauge symmetries**. Gauge symmetries, as the name suggests, center around the question of measurement. As they are going to be central to our discussion of the Standard Model, a working definition is in order:

A gauge transformation is a change in variables which leaves all measurable quantities unaltered.

A gauge symmetry is indicative of a system in which the Lagrangian has one or more redundant degrees of freedom. To pick a particularly simple example, consider potential

---

[4] The "charge" in gravity is the mass. Like charges in this case means that all particles have positive mass and are thus gravitationally attractive.

energy in classical mechanics:

$$\vec{F} = -\nabla(U + C) = -\nabla U.$$

The addition of a constant will have no effect on the net force.

6.2.1 Local Gauge Invariance

In electromagnetism, the gauge freedoms are more complicated than an additive constant but can still be generated by an extremely simple principle. Just as Noether currents are generated from continuous symmetries in a Lagrangian, we'll find that those same symmetries can give rise to the fundamental forces and mediators themselves.

For electromagnetism, a current was generated using the U(1) symmetry. As a reminder, the generator is $\mathbf{X} = 1$, which means that a Dirac field transforms under the **global gauge transformation**

$$\psi \to \psi e^{-iq_e\theta}, \tag{6.6}$$

where the coupling constant q_e is the dimensionless charge of an electron. The symmetry parameter θ is completely irrelevant, but it must be assumed to be constant over all space and time. That is what makes the transformation global, one that leaves the Dirac Lagrangian

$$\mathcal{L} = \overline{\psi}(i\gamma^\mu \partial_\mu - m)\psi$$

completely unchanged. We've already shown that the conserved Noether current for this transformation is

$$J^\mu = q_e\overline{\psi}\gamma^\mu\psi\,,$$

but we haven't demonstrated that what we're calling electric charge has any connection to what is produced by rubbing a balloon on your hair.

Local gauge invariance takes the idea of symmetry a step further. A system which is locally invariant contains a parameter $\theta(x)$ which can vary arbitrarily over space. It's not obvious why local gauge invariance should exist in physical laws, but it seems to be a governing property of our universe. As with an additive constant in a potential energy field, the gauge field $\theta(x)$ (and its derivatives) shouldn't appear in the final form of the Lagrangian or any other measured quantity at all.

A local gauge transformation in U(1),

$$\psi \to \psi e^{-iq_e\theta(x)} \qquad \overline{\psi} \to \overline{\psi}e^{iq_e\theta(x)}, \tag{6.7}$$

introduces some interesting complications. Transforming the Dirac equation yields

$$\mathcal{L} = \overline{\psi}(i\gamma^\mu \partial_\mu - m)\psi \to \overline{\psi}(i\gamma^\mu \partial_\mu - m)\psi + q_e\overline{\psi}\gamma^\mu\psi\partial_\mu\theta(x).$$

Oh no! The local U(1) gauge transformation introduced a new set of terms into the Lagrangian! As written, the Dirac Lagrangian is not invariant under U(1) local gauge transformations.

Hermann Weyl [158] initially proposed that gauge theory might be used to reconcile general relativity and electromagnetism, but when that failed, he suggested that a local gauge theory might be used to generate classical electromagnetism [122,160]:

> This principle of gauge invariance is quite analogous to that previously set up by the author, on speculative grounds, in order to arrive at a unified theory of gravitation and electricity. But I now believe that this gauge invariance does not tie together electricity and gravitation, but rather electricity and matter in the manner described above.

The gauge field $\theta(x)$ in this case, can be made to disappear entirely by introducing a *new* vector field $A_\mu(x)$. The A-field will obey the Gauge transformation

$$A_\mu \rightarrow A_\mu + \partial_\mu \theta \tag{6.8}$$

in addition to the transformation of the fermionic fields (equation 6.7). We have chosen the normalization in anticipation of the final version of our Lagrangian, which gives $[A^\mu] = [E]^1$ (the same as for a scalar field).

A Lagrangian with derivative terms like

$$\partial_\mu \psi \rightarrow D_\mu \psi = (\partial_\mu + i q_e A_\mu)\psi \tag{6.9}$$

will produce no explicit dependence on $\theta(x)$, provided that *both* ψ and A_μ are transformed simultaneously. D_μ is known as the **gauge covariant derivative** and is basically expected to replace the ordinary derivative in the Lagrangian. Thus, the semifinal form of the Lagrangian looks like

$$\mathcal{L} = \overline{\psi}\left[i\gamma^\mu\left(\partial_\mu + i q_e A_\mu\right) - m\right]\psi, \tag{6.10}$$

which is invariant under both global and *local* gauge transformations.

Example 6.1: Show that equation (6.10) is invariant under a U(1) gauge transformation.

Solution: We may ignore the mass term. Instead, we note

$$i\overline{\psi}e^{iq_e\theta(x)}\gamma^\mu\partial_\mu e^{-iq_e\theta(x)}\psi = i\overline{\psi}\gamma^\mu\partial_\mu\psi + q_e\overline{\psi}\gamma^\mu\partial_\mu\theta\psi,$$

and

$$-q_e\overline{\psi}\gamma^\mu\left[A_\mu + \partial_\mu\theta\right]\psi = -q_e\overline{\psi}\gamma^\mu A_\mu\psi - q_e\overline{\psi}\gamma^\mu\partial_\mu\theta\psi.$$

The final terms in the two expressions exactly cancel for arbitrary functions of $\theta(x)$, demonstrating gauge invariance. However, to get to this result, we needed to introduce the additional term

$$\mathcal{L}_{\text{Int}} = -A_\mu q_e\overline{\psi}\gamma^\mu\psi, \tag{6.11}$$

which is nothing more than the electromagnetic field (A_μ) interacting with the conserved U(1) current.

Multiplying out the gauge-invariant Lagrangian produces a free-field component for the Dirac field, while the gauge term represents an interaction which can be simplified as

$$\mathcal{L}_{\text{Int}} = -J \cdot A,$$

where J is the electromagnetic current (equation 5.36). Finally! Our decision to call the U(1) Noether current "electrical current" is starting to bear some fruit.

6.2.2 The Vector Potential

We're not quite done, since, as written, there is no free-field Lagrangian for A^μ. We may as well ruin the surprise now and say that A^μ is the electromagnetic potential,

$$A^\mu = \begin{pmatrix} \Phi \\ A^x \\ A^y \\ A^z \end{pmatrix}, \tag{6.12}$$

where Φ, in this context, represents the scalar potential, and $\vec{A}$ is the vector potential. We don't yet have a sense of what the free-field Lagrangian for this 4-potential look like, but we have some clues:[5] The free-field Lagrangian must be second order in A^μ and its derivatives, and it must be invariant under U(1) transformations (else it would produce additional terms that would need to be swallowed).

This requirement demands that terms like $M^2 A_\mu A^\mu$ can't appear in the Lagrangian, since under a gauge transformation they would produce terms like

$$M^2 A_\mu A^\mu \rightarrow M^2 A_\mu A^\mu + 2 M^2 A_\mu \partial^\mu \theta + M^2 \partial_\mu \theta \partial^\mu \theta,$$

which is manifestly *not* gauge invariant. The free-field Lagrangian for our A-field can't have a mass term, which means that the A-field describes a massless particle.

Likewise, we anticipate a dynamic term in the Lagrangian that contains first derivatives of the potential. Though it might be counterintuitive to do so, it is often useful to decompose tensors into a symmetric component,

$$G_{\mu\nu} = \partial_\mu A_\nu + \partial_\nu A_\mu,$$

and an antisymmetric one, known as the **Faraday tensor**,

$$F_{\mu\nu} = \partial_\mu A_\nu - \partial_\nu A_\mu, \tag{6.13}$$

such that

$$\partial_\mu A_\nu = \frac{1}{2} \left(G_{\mu\nu} + F_{\mu\nu} \right).$$

---

[5] And thanks to a century and a half of experience with Maxwell's equations, we'll know if we get the right answer.

The symmetric term can't appear in the free-field Lagrangian because it transforms under a local gauge transformation as

$$G_{\mu\nu} \to G_{\mu\nu} + 2\partial_\mu \partial_\nu \theta,$$

which is, again, not gauge invariant. Only the Faraday tensor is gauge invariant, which means that the Lagrangian has to be composed of quadratic combinations; that is, it must take the form

$$\mathcal{L}_{A, Free} = \text{Const} \times F_{\mu\nu} F^{\mu\nu}.$$

Since the normalization doesn't matter, to meet with normal convention, the constant is taken to be $-1/4$, which gives us everything we need to know to come up with a theory of electrodynamics:

$$\mathcal{L} = \overline{\psi}(i\gamma^\mu \partial_\mu - m)\psi - A_\mu J^\mu - \frac{1}{4} F_{\mu\nu} F^{\mu\nu}. \tag{6.14}$$

It may not be obvious from such a humble equation, but this Lagrangian implicitly contains all of Maxwell's equations, the energy relation for a free radiation field, the Lorentz force law, and much more.

6.3 Interpreting the Electromagnetic Lagrangian

6.3.1 What the Faraday Tensor Means

The Faraday tensor $F^{\mu\nu}$ is antisymmetric, which means that the diagonal components are all zero (by definition), and the top six components are just the negative of the bottom six. What are these six unique components? They are simply the electric and magnetic fields, combinations of first derivatives of time and space of the vector and scalar potentials:

$$\vec{E} = -\nabla \Phi - \dot{\vec{A}}; \tag{6.15}$$

$$\vec{B} = \nabla \times A. \tag{6.16}$$

We haven't shown that it has the properties that you've developed in earlier coursework, but we'll demonstrate the legitimacy of our labeling in short order.

Example 6.2: What is the F_{01} component of the Faraday tensor?

Solution:

$$F_{01} = -\dot{A}^x - \partial_x \Phi = E^x.$$

We've simply labeled this as the x-component of the electric field; we haven't (yet) shown that this is the same field that shows up in Gauss's and Lorentz's law. At this stage, we're simply *defining* the $\vec{E}$- and $\vec{B}$-fields in terms of various first derivatives of the potentials. The other F_{0i} components follow transparently.

The space-space terms of the Faraday tensor are a bit trickier.

Example 6.3: What is the F_{12} component of the Faraday tensor?

Solution:

$$F_{12} = -\partial_x A_y + \partial_y A_x$$
$$= \partial_x A^y - \partial_y A^x$$
$$= (\nabla \times \vec{A})^z$$
$$= B^z.$$

Writing out the Faraday tensor explicitly, we obtain

$$F^{\mu\nu} = \begin{pmatrix} 0 & -E^x & -E^y & -E^z \\ E^x & 0 & -B^z & B^y \\ E_y & B_z & 0 & -B^x \\ E^z & -B^y & B^x & 0 \end{pmatrix}, \tag{6.17}$$

where the first index denotes the column, and the second denotes the row.

Even the composition of the Faraday elements tells a story. Knowing nothing else about component fields except how they are generated, we may say

$$\partial_\lambda F_{\mu\nu} + \partial_\mu F_{\nu\lambda} + \partial_\nu F_{\lambda\mu} = 0. \tag{6.18}$$

Equation (6.18) is known as a **Bianchi identity**, which we leave as an exercise for the reader (problem 6.7) to prove. Though the relation represents 64 possible identities (four values each for μ, ν, λ), in reality, there are far fewer, since permutations of the three indices produce identical relations. Further, there are lots of trivial results, as is the case when any two indices repeat. The only interesting identities are those for which all three indices are different.

Example 6.4: Expand the Bianchi identity explicitly for the case in which $\mu = 0$, $\nu = 1$, $\lambda = 2$.

Solution:

$$\partial_y E^x + \dot{B}^z + \partial_x(-E^y) = 0,$$

or

$$(\nabla \times \vec{E})^z = -\dot{B}^z.$$

Combining all the nonvanishing Bianchi identity terms yields

$$\boxed{\nabla \times \vec{E} = -\dot{\vec{B}}.}\ \text{Faraday's law of inductance} \tag{6.19}$$

Likewise, $\mu = 1$, $\nu = 2$, $\lambda = 3$ yields

$$\partial_z B^z + \partial_x B^x + \partial_y B^y = 0,$$

or in the more familiar form,

$$\boxed{\nabla \cdot \vec{B} = 0.}\ \text{Gauss's law for magnetism} \tag{6.20}$$

The *form* of the Faraday tensor law immediately generated two of Maxwell's equations.

6.3.2 The Dynamics of the Free-Field Potential

We obtain the other two Maxwell's equations by computing the Euler-Lagrange equations for the components of the vector field in our electroweak Lagrangian (equation 6.14) and treating each component as a separate field. This may look daunting, but by expanding the Lagrangian and grouping explicitly, it is straightforward to show that

$$\frac{\partial \mathcal{L}}{\partial(\partial_\mu A_\nu)} = -F^{\mu\nu},$$

which inserted into the Euler-Lagrange framework yields

$$\boxed{\partial_\mu F^{\mu\nu} = J^\nu.} \tag{6.21}$$

The Euler-Lagrange equation is beautifully Lorentz invariant. We might even have guessed the form up to a sign or a constant multiple, but it contains an amazing amount of information. Multiplying out for $\nu = 0$,

$$\boxed{\nabla \cdot \vec{E} = \rho,}\ \text{Gauss's law} \tag{6.22}$$

and for the spacelike terms,

$$\boxed{\nabla \times \vec{B} = \dot{\vec{E}} + \vec{J}.}\ \text{Ampère's law} \tag{6.23}$$

We can also immediately verify conservation of electric charge. Since the Faraday tensor is antisymmetric, it is easy to show

$$\partial_\nu \partial_\mu F^{\mu\nu} = 0,$$

which, from equation (6.21), immediately yields

$$\partial_\nu J^\nu = 0,$$

as, of course, it must.

6.3.3 Gauges, Again

All this work, and we have rediscovered Maxwell's equations and their relationships to the 4-potential (equations 6.15 and 6.16). As with gravitational fields, we never measure the potentials directly, and thus the choices of Φ and $\vec{A}$ are not unique. Rather, we measure the $\vec{E}$ and $\vec{B}$ fields by their effect on charged particles.

Consider what will happen if we make the following transformation:

$$\Phi \rightarrow \Phi + \dot{\theta};$$

$$\vec{A} \rightarrow \vec{A} - \nabla\theta.$$

This is nothing more than a rewriting of the local U(1) gauge symmetry from which we developed electromagnetism in the first place.

Since the curl of a gradient is zero, the $\vec{B}$ field is unchanged. Likewise, the U(1) transformation creates two canceling terms in the $\vec{E}$ field, which subsequently remains unchanged. To be clear, *any* particular solution to the gauge field $\theta(x)$ is physically valid, but some solutions are more mathematically useful than others. We can impose additional constraints on the 4-potential. For example, we may set

$$\partial^\mu A_\mu + \Box\theta = 0,$$

from which we may solve the Poisson equation and find a solution such that

$$\partial_\mu A^\mu = 0.$$

This is known as **Lorenz gauge**.[6]

We can take this a step further, since solutions of the form

$$\partial_\mu \partial^\mu \theta = 0$$

are not unique. We may select one that allows us to break Lorentz invariance and split the Lorenz condition into two parts,

$$\nabla \cdot \vec{A} = 0 \qquad \dot{\Phi} = 0,$$

for free fields. As it's time-invariant, we're free to set the scalar potential however we like, and it'll remain constant. This choice is known as **Coulomb gauge**.

It's fine for the time derivative of the scalar potential to vanish, as it doesn't explicitly appear in the Faraday tensor, which means that we should, in general, be able to calculate Φ directly from $\vec{A}$. In the absence of sources, we have the Poisson relation

$$\nabla^2 \Phi = -\nabla \cdot \dot{\vec{A}}, \tag{6.24}$$

meaning that Φ can be derived (up to a constant) from the 3-potential. In the case where $\vec{A}$ is divergentless (or where the divergence is constant in time), Φ can be said to disappear entirely. This property will turn out to be extremely useful in generating the free-field solution for the photon field.

---

[6] Named after Ludvig Lorenz, not Henrik Lorentz, and frequently misattributed. It's a bit confusing because it arises in the context of Lorentz invariance.

6.3.4 The Stress-Energy Tensor

Even in the absence of nearby source terms, a freely propagating electromagnetic field carries energy. We may compute the stress-energy tensor (equation 3.11) for the electromagnetic field:

$$T^{\mu\nu} = -F^{\mu\lambda}\partial^{\nu}A_{\lambda} - g^{\mu\nu}\mathcal{L}.$$

The term $\partial^{\nu}A_{\lambda}$ is a bit troubling. The expectation is that only electric and magnetic fields will appear in any final calculation. But we can get a deeper insight by expanding this troublesome derivative in terms of

$$\partial^{\nu}A_{\lambda} = F^{\nu}{}_{\lambda} + \partial_{\lambda}A^{\nu},$$

which yields

$$T^{\mu\nu} = -F^{\mu\lambda}F^{\nu}{}_{\lambda} - \frac{1}{2}g^{\mu\nu}\left(|\vec{E}|^2 - |\vec{B}|^2\right) + \delta T^{\mu\nu},$$

where the $\delta T^{\mu\nu}$ is defined as

$$\delta T^{\mu\nu} = -F^{\mu\lambda}\partial_{\lambda}A^{\nu} = \partial_{\lambda}(F^{\lambda\mu}A^{\nu})$$

and where the last step may be proven by expanding the component of the Faraday matrix explicitly under the assumption of a free field. Thus,

$$\partial_{\mu}\delta T^{\mu\nu} = \partial_{\mu}\partial_{\lambda}\left(F^{\lambda\mu}A^{\nu}\right) = 0.$$

This term vanishes because a permutation of μ and λ yields a minus sign, canceling the whole thing. Since energy isn't transferred into or out of the δT term, the unperturbed stress-energy tensor is also conserved, regardless of gauge choice. For the energy component:

$$T^{00} = \frac{1}{2}\left(|\vec{E}|^2 + |\vec{B}|^2\right) - \vec{E}\cdot\nabla\Phi.$$

We can make this last term disappear entirely by an appropriate choice of gauge, yielding an energy density for a free electromagnetic field of

$$\rho_{EM} = \frac{1}{2}\left(|\vec{E}|^2 + |\vec{B}|^2\right). \tag{6.25}$$

Although this is a happy ending, there's something a little suspicious about this. While the gauge-dependent energy term doesn't show up in electromagnetic interactions, we're encountering the same problem we did with the Casimir effect. Gravity is supposed to see *all* energy. Is the gauge term real or not?[7]

---

[7] This question is meant to be rhetorical. Since the gauge term produces no additional effects within electromagnetism, we will ignore it from here on out. Nevertheless, it should be treated as an open question within the model.

6.3.5 The Lorentz Force Law

There's another useful application of our newly simplified stress-energy tensor:

$$T^{\mu\nu}_{EM} = -F^{\mu\lambda}F^{\nu}{}_{\lambda} - g^{\mu\nu}\mathcal{L}.$$

We can ask what happens when this tensor interacts with a current density distribution. In particular, we want to compute

$$\partial_{\mu}T^{\mu\nu}_{EM} = -F^{\nu}{}_{\lambda}\partial_{\mu}F^{\mu\lambda} - F^{\mu\lambda}\partial_{\mu}F^{\nu}{}_{\lambda} + \frac{1}{2}F_{\alpha\beta}\partial^{\nu}F^{\alpha\beta}$$

$$= -F^{\nu}{}_{\lambda}J^{\lambda} - F^{\mu\lambda}\partial_{\mu}F^{\nu}{}_{\lambda} + \frac{1}{2}F_{\alpha\beta}\partial^{\nu}F^{\alpha\beta}$$

$$\partial_{\mu}T^{\mu\nu}_{EM} + F^{\nu\lambda}J_{\lambda} = \frac{1}{2}F_{\alpha\beta}\left[\partial^{\nu}F^{\alpha\beta} - 2\partial^{\alpha}F^{\nu\beta}\right]$$

$$= \frac{1}{2}F_{\alpha\beta}\left[\partial^{\beta}F^{\alpha\nu} + \partial^{\alpha}F^{\beta\nu}\right]$$

$$= 0,$$

where the term on the right identically cancels because inside the brackets an exchange of α and β does nothing, while outside, an exchange reverses the sign of $F_{\alpha\beta}$, and we used the Bianchi identity between steps 3 and 4.

Thus, we find

$$\partial_{\mu}T^{\mu\nu}_{EM} = -F^{\nu\lambda}J_{\lambda}. \tag{6.26}$$

Momentum and energy are flowing into the EM field, but from where? The only possibility is from the Dirac field, yielding

$$\partial_{\mu}T^{\mu\nu}_{\psi} = F^{\nu\lambda}J_{\lambda}. \tag{6.27}$$

This is the Lorentz force law in disguise. Consider $\nu = 1$, since T^{01} represents the momentum density in the x-direction. Thus,

$$\dot{T}^{01}_{\psi} + \partial_i T^{i1}_{\psi} = \rho E^x + J^y B^z - J^z B^y.$$

The nonmomentum terms on the left simply represent the pressure (T^{11}) and the anisotropic stress. Ignoring those as being irrelevant to EM interaction, extending to the three spatial directions, and recalling that $\dot{\vec{p}} = \vec{F}$ yields

$$\vec{F} = q(\vec{E} + \vec{v} \times \vec{B}), \tag{6.28}$$

where we've gone from density to an individual charge to make the equation appear more familiar.

6.3.6 A Moment of Reflection

It's worth stepping back for a moment to consider what we've accomplished. Asserting nothing more than a first-order time evolution, Lorentz invariance, and the assumption of U(1) local gauge invariance, we were able to

- Develop a Lagrangian that predicts the existence of spin-1/2 particles and antiparticles.
- Predict the existence of a massless vector field (photons) that mediates the interaction between charged fermions.
- Derive all of Maxwell's equations from first principles.
- Compute the energy density of a freely propagating electromagnetic field.
- Compute the Lorentz force law.

Be careful patting yourself on the back, though. Symmetry arguments can go only so far. For instance, we needed to enter the mass of the electron as well as the coupling term (electric charge) into the Lagrangian by hand. No further amount of clever analysis yet known is going to explain these terms. This marks one of the limits of the Standard Model.

6.4 Solutions to the Classical Free Field

We haven't *really* established any new results about electromagnetism apart from demonstrating that the fundamental equations can be generated more or less from first principles (no small feat!). But now we'll explore the implications of our field theoretical approach with an eye toward quantizing the field. For a free electromagnetic field, the Euler-Lagrange equations yield

$$\partial_\mu F^{\mu\nu} = 0, \tag{6.29}$$

or equivalently,

$$\partial^\mu \partial_\mu A^\nu - \partial^\nu \partial_\mu A^\mu = 0.$$

Here's where gauge choice comes into play. In particular, using Lorenz gauge ($\partial_\mu A^\mu = 0$) causes the second term in our Euler-Lagrange equation to drop out entirely. This yields (again, specifically in Lorenz gauge) a dynamical equation for the photon field:

$$\partial^\mu \partial_\mu A^\nu = 0. \tag{6.30}$$

This looks like four duplicates of a massless Klein-Gordon equation. We can thus expand it as

$$A^\mu(x) = \sum_{r=0}^{3} \int \frac{d^3 p}{(2\pi)^3} \frac{1}{\sqrt{2E_p}} \varepsilon^{(r)\mu}(\vec{p}) \left[a_{\vec{p},r} e^{-ip\cdot x} + a^*_{\vec{p},r} e^{ip\cdot x} \right], \tag{6.31}$$

where $\varepsilon^{(r)\mu}(\vec{p})$ represents the unit vectors for four possible polarization states.

As the photon is described by a vector field, we know its rotation properties. Vectors require a full rotation to return to their original state, which, following the example from the Dirac particles means that photons are spin-1, and thus bosonic.

In quantum mechanics you likely derived a general relationship for the degeneracy of angular momentum. For particles with a spin s the degeneracy is expected to be

$$n_m = 2s + 1$$

possible spin states. For the spin $-1/2$ Dirac particles, this expression accurately yielded two possible states.

By this argument, we might naively expect photons to have three possible polarization states. However, our gauge constraint allows us to set the scalar potential to zero (equation 6.24). In that case, there are two independent polarization modes which satisfy the Lorenz condition,

$$\varepsilon_1 = \begin{pmatrix} 0 \\ 1 \\ 0 \\ 0 \end{pmatrix} \qquad \varepsilon_2 = \begin{pmatrix} 0 \\ 0 \\ 1 \\ 0 \end{pmatrix}, \tag{6.32}$$

as the particular solutions for a photon field propagating along the z-direction. For any momentum, there will be two polarization modes defined perpendicular to the direction of motion. More generally, we have

$$\varepsilon^{(r)} \cdot p = 0. \tag{6.33}$$

Even though we computed them, there may be some confusion as to why there are only two polarization states, rather than the expected three. The third is swallowed by the masslessness of the photon. The missing spin state is $m = 0$, which implies an observational frame in which the photon is at rest. Since there is no such frame, there are only two polarization states. You will demonstrate the implication of mass on the third state in problem 6.9.

6.5 The Low-Energy Limit

6.5.1 Dirac Spinors at Low Energy

Since we've explored the free fields for both Dirac particles and photons, it's finally worth a little discussion of how the two interact with one another. We'll focus on the electron (positive energy solutions) so as to not have to worry about negative signs. We begin by separating the Dirac equation under electromagnetism (6.14) into spacelike and timelike terms, as well as into block components of the Dirac spinor:

$$(i\partial_0 - q_e \Phi)\psi_R + \sigma^i(i\partial_i - q_e A_i)\psi_R - m\psi_L = 0; \tag{6.34}$$

$$(i\partial_0 - q_e \Phi)\psi_L - \sigma^i(i\partial_i - q_e A_i)\psi_L - m\psi_R = 0.$$

Where we have written $A_0 = \Phi$ as the electromagnetic scalar potential.

In the limit of particles moving slowly, $|\vec{p}| \ll m$, in the z-direction, we may approximate the two basis spinors (equations 5.18) as

$$u_{\pm}(p, x) \simeq \sqrt{m} \begin{pmatrix} 1 - \frac{v}{2} \\ 0 \\ 1 + \frac{v}{2} \\ 0 \end{pmatrix} e^{-imt} e^{i\vec{p}\cdot\vec{x}}; \qquad \sqrt{m} \begin{pmatrix} 0 \\ 1 + \frac{v}{2} \\ 0 \\ 1 - \frac{v}{2} \end{pmatrix} e^{-imt} e^{i\vec{p}\cdot\vec{x}} \qquad (6.35)$$

to first order in v.

Noting the form of equations (6.34), it is clear that sums and differences of the ψ_L and ψ_R will make a more convenient basis than the two modes themselves. That is, we may define

$$\phi \equiv \frac{1}{2\sqrt{m}}(\psi_L + \psi_R)e^{imt}; \qquad \chi \equiv \frac{1}{2\sqrt{m}}(\psi_L - \psi_R)e^{imt}.$$

This yields

$$\phi^{(+)}(p, x) = \begin{pmatrix} 1 \\ 0 \end{pmatrix} e^{i\vec{p}\cdot\vec{x}}; \qquad \phi^{(-)}(p, x) = \begin{pmatrix} 0 \\ 1 \end{pmatrix} e^{i\vec{p}\cdot\vec{x}},$$

with the components of χ defined similarly. The split Dirac equations (6.34) can quickly be rewritten in terms of ϕ and χ:

$$(i\partial_0 - q\Phi)\phi e^{-imt} - \sigma^i(i\partial_i - q_e A_i)\chi e^{-imt} - m\phi e^{-imt} = 0;$$

$$(i\partial_0 - q\Phi)\chi e^{-imt} - \sigma^i(i\partial_i - q_e A_i)\phi e^{-imt} + m\chi e^{-imt} = 0.$$

Expanding the time derivative of the exponential yields

$$(i\partial_0 - q\Phi)\phi - \sigma^i(i\partial_i - q A_i)\chi = 0;$$

$$(i\partial_0 - q\Phi)\chi - \sigma^i(i\partial_i - q A_i)\phi + 2m\chi = 0.$$

In the second equation, the time derivative $i\partial_0$ represents a kinetic energy term (much smaller than m), and $q\Phi$ is the electrostatic potential (again, assumed to be much smaller than m in the weak field limit). Simplifying, we get

$$\chi \simeq \frac{1}{2m}\sigma_i(i\partial_i - q A_i)\phi,$$

which we can then insert into the first equation of the preceding pair to yield

$$i\dot{\phi} - q\Phi\phi - \frac{1}{2m}\sigma^i(i\partial_i - q A_i)\sigma^j(i\partial_j - q A_j)\phi = 0.$$

This is a first-order differential equation in time and second order in space. It's starting to look a *lot* like the Schrödinger wave equation. It is worth noting that the σ^i's commute with derivatives and A_i's. For $i = j$, $\sigma^i \sigma^i = I$,

$$i\dot\phi = \left[\frac{1}{2m}(-i\nabla - q\vec{A})^2 + q\Phi + \frac{1}{2m} \sum_{i \neq j} \sigma^i \sigma^j (i\partial_i - q A_i)(i\partial_j - q A_j) \right] \phi.$$

The sum term will vanish by symmetry if there are two derivatives or two vector potentials. Only the terms

$$\sum_{i \neq j} \sigma^i \sigma^j (i\partial_i - q A_i)(i\partial_j - q A_j) = iq \sum_{i \neq j} \sigma^i \sigma^j (\partial_i A_j - \partial_j A_i)$$

remain. You may recognize the derivatives as components of the $\vec{B}$ field, since

$$\sigma^i \sigma^j = i \sum_k \sigma^k \epsilon_{ijk}.$$

Thus,

$$i\dot\phi = \left[\frac{1}{2m}(-i\nabla - q\vec{A})^2 + q_e \Phi - \left(\frac{q_e \vec{\sigma}}{2m} \right) \cdot \vec{B} \right] \phi. \tag{6.36}$$

Amazing! This is the Schrödinger equation of a nonrelativistic electron in an electromagnetic field. As, of course, it must be.

6.5.2 The Spin of an Electron

In many ways, we're not all that interested in low-energy interactions. Most of the interesting new results will come from evaluating them at high energy. But there are a few secrets from the first quantization that are still worth uncovering. Consider an electron at rest in a magnetic field B_0 in the positive z-direction. The magnetic term in the Schrödinger equation yields an interaction Hamiltonian:

$$\hat{H}_{\text{Int}}\phi = -\frac{q_e B_0}{2m} \begin{pmatrix} 1 & 0 \\ 0 & -1 \end{pmatrix} \begin{pmatrix} \phi_1 \\ \phi_2 \end{pmatrix}. \tag{6.37}$$

The emission of photons that would allow the electron to go from spin-up (energy of $e B_0/2m$) to spin-down (the negative of that). Classically, a loop of current produces a moment $\vec{\mu}$ with an interaction energy

$$U = -\vec{\mu} \cdot \vec{B}.$$

For a classically rotating mass of charges, the magnetic moment and the angular momentum may be related via

$$\mu = \frac{q}{2m} L.$$

But we already know the angular momentum of an electron ($S = 1/2$), which means that the Dirac equation revealed a nonclassical prediction,

$$\mu = g_e \frac{q_e}{2m_e} S,$$

where g_e (also known as the **dimensionless magnetic moment**) is

$$g_e = 2.$$

In practice, this is only an approximation. We'd need to invoke quantum field theory to get a better estimate of g_e. Incredibly, this result can be measured experimentally to about 1 part in a *trillion* and is consistent with the quantum field theory theoretical result [118]:

$$\frac{g-2}{2} = 1.15965218086 \times 10^{-3} \pm 7.6 \times 10^{-13}.$$

The existence of a magnetic moment was known well before Dirac derived the properties of fermions. In 1897, Pieter Zeeman [169] noted that electronic transitions in magnetic fields exhibited a split in the spectral line. Unlike orbital line-splitting (which produces an odd number of energy levels), the spin splitting the number of further doubles states, producing an even number of energy levels. The **Zeeman effect** was, in fact, one of the original motivations for Dirac's work.

6.6 Looking Forward

Electromagnetism serves as a template to developing other fundamental interactions. In 1954, Chen-Ning Yang and Robert Mills [167] proposed that this basic mechanism could be applied to other symmetries besides U(1), by assuming a local gauge transformation of the form

$$\mathbf{M} = \exp(-i\theta^i(x)\mathbf{X}_i),$$

where $\mathbf{X}_i$ are the various generators for the symmetry operation. Conserved currents can then be generated via Noether's theorem,

$$J_{(i)}^\mu = \overline{\psi}\gamma^\mu \mathbf{X}_i \psi \tag{6.38}$$

as the ith current for a particular symmetry. Following the model from electromagnetism, the Dirac Lagrangian can be adjusted to include a covariant gauge derivative:

$$\partial_\mu \to D_\mu = (\partial_\mu + i\mathbf{X}_i A_\mu^{(i)}). \tag{6.39}$$

That's not to say that the weak and strong interactions won't introduce a number of interesting new twists. For one, the U(1) group is Abelian (obviously, since there's only one generator). SU(2) and SU(3), are non-Abelian groups, which quickly results in self-interactions of mediators.

Masses also present some problems. The gauge symmetry approach naturally produces massless mediators, regardless of the symmetries. In electromagnetism, this poses no problem (and, in fact, is quite desirable, since the photon *is* massless), but the $W^\pm$ and the Z^0, the mediators for the weak interaction, are not.

We'll see how to deal with these nontrivial complications in due course.

Problems

6.1 At some time $t = 0$, a $\phi - \eta$ system (equation 6.1) is initialized such that

$$\phi(x) = Ax \qquad \eta(x) = \dot{\eta}(x) = 0,$$

where x in this case is the x^1 spatial coordinate.
(a) What is the evolution of $\eta(x, t)$ for early times? You may assume λ is small.
(b) Based on your previous result, compute the time evolution of ϕ at early times.

6.2 Suppose, contrary to our work in this chapter, that the photon had a very small mass, $\sim 10^{-4}$ eV. What would the effective range of the electromagnetic force be? Express your answer in meters. Approximately how light (in kilograms) would the photon need to be such that earth-scale magnetic fields would still be measurable?

6.3 The neutron has a mass approximately 1.3 MeV larger than that of the proton, largely because it is made of two down quarks and an up as opposed to two up quarks and a down. While the quark masses aren't well constrained, a reasonable estimate is that $m_d - m_u \simeq 3$ MeV, in excess of the measured difference between protons and neutrons.

Estimate an approximate contribution of the proton electric field energy assuming a characteristic size of 10^{-15} m. Is this value consistent with the quark and baryon mass difference?

6.4 Starting with the Lagrangian for a complex scalar field,

$$\mathcal{L} = \partial_\mu \phi \partial^\mu \phi^* - m^2 \phi \phi^*,$$

assume a local gauge invariance of the form

$$\phi \to \phi e^{-iq\theta(x)}.$$

(a) Compute the additional Lagrangian terms arising from the gauge transformation.
(b) Now include a vector field A_μ which obeys the transform in equation (6.8). Find a gauge-invariant Lagrangian in terms of the scalar and vector fields. Expand all terms.
(c) Calculate the Euler-Lagrange equation for ϕ^*.
(d) Compute the equation of motion for ϕ in the limit of weak electromagnetic fields ($A \cdot A \simeq 0$).

6.5 Prove the Bianchi identity (equation 6.18) is necessarily true given the definition of the antisymmetric Faraday tensor.

6.6 In classical electrodynamics, radiation is propagated along the Poynting vector,

$$\vec{S} = \vec{E} \times \vec{B},$$

an ordinary 3-vector. Express the components of S^i in terms of components of $F^{\mu\nu}$ in as simplified a form as possible.

6.7 Consider a free-field solution of the electromagnetic 4-potential,

$$A^\mu = \begin{pmatrix} 0 \\ A\cos(\omega(z-t)) \\ A\sin(\omega(z-t)) \\ 0 \end{pmatrix},$$

where ω and A are positive, real-valued constants.
(a) What gauge condition(s) are satisfied by the potential vector?
(b) Compute the nonzero components of the Faraday tensor for this field, construct $\vec{E}$ and $\vec{B}$ explicitly, and verify $\vec{E} \cdot \vec{B} = 0$.

(c) Based on conservation of current, $\partial_\mu F^{\mu\nu} = J^\nu$. What is the local current density?

(d) Compute $\vec{E} \times \vec{B}$.

6.8 The quantity $Q = \vec{B}^2 - \vec{E}^2$ is measured in one frame. What is the value of Q in a frame boosted a speed v in the $+x$-direction?

6.9 In developing the two polarization-states model for the photon we relied upon U(1) gauge invariance, which in turn depends upon a massless photon. We know that spin-1 particles are supposed to have three spin states, but we claimed that the third state was swallowed by the Coulomb gauge condition. Let's approach the question of three states by assuming that the photon *does* have mass and obeys the Lagrangian

$$\mathcal{L} = -\frac{1}{4} F_{\mu\nu} F^{\mu\nu} + \frac{1}{2} M^2 A^\mu A_\mu.$$

(a) Write the Euler-Lagrange equation for the massive photon field.

(b) Let the photon field take the form of a single plane wave:

$$A^\mu = \varepsilon^\mu e^{-ip\cdot x}.$$

Express the Euler-Lagrange equation as dot products of p and ε with themselves and each other. Show that the transverse wave condition (equation 6.33) drops out of the dispersion relation regardless of whether the field has mass.

For a particle propagating in the z-direction, this immediately demonstrates that equation (6.32) gives two possible polarization states for a massive or massless photon field.

(c) What is the third possible polarization state for a massive photon propagating in the z-direction?

(d) What are the electric and magnetic fields of the massive photon field in this third polarization state? What happens to those fields for $m = 0$?

6.10 Consider an electron in a spin state

$$\phi = \begin{pmatrix} a \\ b \end{pmatrix}$$

in a magnetic field B_0 oriented along the z-axis. We will calculate the **Larmor frequency** by which the electron precesses.

(a) Turn the interaction Hamiltonian (equation 6.37) into a first-order differential equation in time.

(b) Solve the differential equation(s) in part (a). What is the frequency of oscillation of the phase *difference* between the two components?

Further Readings

- Cottingham, W. N., and D. A. Greenwood. *An Introduction to the Standard Model of Particle Physics.* Cambridge: Cambridge University Press, 1998. Much of the discussion relating the Dirac equation to the Schrödinger wave equation follows Cottingham and Greenwood's discussion of low-energy phenomenology in chapter 7. The entire chapter is well worth a read.
- Griffiths, David J. *Introduction to Electrodynamics,* 2nd ed. New York: Addison-Wesley, 2012. There are few (if any) introductory electrodynamics textbooks that can match Griffith's discussion of Gauss's laws.
- Jackson, J. D., and L. B. Okun. Historical roots of gauge invariance. *Rev. Mod. Phys.* 73:663 (2001). Available on the arXiv preprint server; hep-ph/0012061. Jackson and Okun present an excellent review of the historical basis for the development of gauge theory and, in particular, the priority of the discovery of U(1) in electromagnetism.
- Jackson, J. D. *Classical Electrodynamics,* 3rd ed. Hoboken, NJ: Wiley, 1998. Jackson is *the* classic text for intermediate/advanced electromagnetism. In particular, students may benefit from chapter 12 and the discussion of the electromagnetic stress-energy tensor.

7 | Quantum Electrodynamics

Figure 7.1. Richard Feynman (1918–1988). Feynman elucidated a number of important connections in quantum mechanics and particle physics, and developed both the diagrams and path integral formalism that bear his name. Photo copyright Tamiko Thiel 1984.

The distinction between particle and field is an important one. Einstein's successful argument for the particle nature of light, for instance, earned him the 1921 Nobel Prize in Physics. While our work won't be quite so grandiose, we're going to need to leave the realm of classical fields and instead consider quantum particles.

We knew this was coming. We've seen, for instance, that in a purely classical model we get the charge and mass of the positron completely wrong. The time has come to treat particles as individual objects, but fortunately, we've already developed the machinery to do just that.

In this chapter, we'll introduce a visual and computational device known as **Feynman diagrams** as a way of quantizing classical theories. Though we'll use Feynman's formalism, his approach is mathematically equivalent to other approaches, in particular, those of Julian Schwinger [148] and Sin-Itiro Tomonaga [155], with whom Feynman shared

Table 7.1. Some Well-Known Decay Processes and Timescales.

Decay	ΔE	Interaction	τ
$p \to ?$	?	GUT?	$\gtrsim 10^{34}$ years
$n \to p + e^- + \bar{\nu}_e$	0.782 MeV	Weak	881.5 s
$\mu^- \to e^- + \bar{\nu}_e + \nu_\mu$	105.1 MeV	Weak	2.2×10^{-6} s
$\pi^- \to \mu^- + \bar{\nu}_\mu$	33.9 MeV	Weak	2.6×10^{-8} s
$2p \to 1s$; (Lyman α)	10.2 eV	EM	1.6×10^{-9} s
$\pi^0 \to \gamma + \gamma$	135 MeV	Strong	8.5×10^{-17} s
$\Delta^+ \to p + \pi^0$	160 MeV	Strong	6×10^{-24} s

Note: All data taken from the Particle Data Group Live, `http://pdg.lbl.gov/`. Some of these particles may be unfamiliar, as they haven't yet come up in our discussion. Pions ($\pi^\pm$, π^0), for instance, are composed of a quark-antiquark pair, and the delta particle Δ^+ is a spin-3/2 version of the proton.

the 1965 Nobel Prize [50]. We will revisit Schwinger's and Tomonaga's contributions in our discussion of renormalization in Chapter 11, but for the moment, we will approach quantum field theory from Feynman's perspective.

7.1 Particle Decay

In a quantum universe, it's impossible to say that a particular system *will* do something in a particular time, but we aim for the next best thing: quantifying the probability of what *might* happen [87]. Reactions can be expressed in terms of an expected rate,

$$\Gamma \equiv \frac{1}{\tau}, \tag{7.1}$$

where τ is the characteristic timescale of an interaction. Statistically, decay is a Poisson process, whereby the probability that a decay won't have occurred after a time t is

$$P(> t) = e^{-t/\tau} .$$

In Table 7.1 we provide a number of different reactions and their experimentally derived decay times [119] in an effort to figure out why some are fast and some are slow. Even without a detailed analysis, we see that, all things being equal, strong decays are much faster than electromagnetic ones, which in turn, are much faster than weak decays. Likewise, decays that liberate more energy tend to be much faster than those which are only barely energetically favorable. At the extreme, proton decay has *no* energetically allowed channel, since it is the lightest of the baryons; proton decay has never been observed. Detection of a proton decay would indicate important new physics beyond the Standard Model.

7.1.1 The Interaction Hamiltonian

Quantum field theories sound intimidating, but our approach will be quite similar to what you will have seen in ordinary nonrelativistic quantum mechanics:

1. Construct a Hamiltonian to evolve a wavefunction:

$$\hat{H}\psi = i\dot{\psi} \,.$$

2. Decompose the wavefunction into linear combinations of eigenstates of the Hamiltonian,

$$\psi = c_n |n\rangle,$$

with the one important difference: instead of describing, say, the energy state of an electron, in quantum field theories, the eigenstates of the Hamiltonian will represent multiple particles and their corresponding energies and momenta.

As a first step, we'll develop a quantum field theory for our toy scalar model (last seen in §6.1) and then assert, admittedly without rigorous proof, that simple extensions to the model will continue to hold as we move into complicated fermionic systems.

The scalar model consisted of an interaction term in the Lagrangian density of the form

$$\mathcal{L}_{\text{Int}} = -\lambda\phi^2\eta,$$

which results in a *classical* Hamiltonian of

$$H_{\text{Int}} = \lambda \int d^3x \eta(x)\phi(x)\phi(x), \tag{7.2}$$

where the reversal of sign mimics the sign reversal of the potential between the Lagrangian and the Hamiltonian. To get the total rather than the density, we need to integrate over all space.

To use all the machinery of quantum mechanics, we must promote the Hamiltonian to an operator. We must do the same for the η and ϕ fields. In §5.7 we found that quantization of the scalar ϕ field can be written as a combination of particle creation operators $\hat{a}_{\vec{p}}^{\dagger}$ and annihilation operators $\hat{a}_{\vec{p}}$:

$$\hat{\phi}(x) = \int \frac{d^3p}{(2\pi)^3} \frac{1}{\sqrt{2E_p}} \left[\hat{a}_{\vec{p}} e^{-ip\cdot x} + \hat{a}_{\vec{p}}^{\dagger} e^{ip\cdot x} \right].$$

We define the $\hat{\eta}$ operator identically but substitute $\hat{d}^{\dagger}$ and $\hat{d}$ for creation and annihilation of the η particles. Multiplying out the quantized Hamiltonian yields

$$\hat{H}_{\text{Int}} = \lambda \int d^3x \frac{d^3q_1}{(2\pi)^3} \frac{d^3q_2}{(2\pi)^3} \frac{d^3q_3}{(2\pi)^3} \frac{1}{\sqrt{8E_{q_1}E_{q_2}E_{q_3}}} \left(\hat{a}_{\vec{q}_3} e^{-iq_3\cdot x} + \hat{a}_{\vec{q}_3}^{\dagger} e^{iq_3\cdot x} \right)$$
$$\times \left(\hat{a}_{\vec{q}_2} e^{-iq_2\cdot x} + \hat{a}_{\vec{q}_2}^{\dagger} e^{iq_2\cdot x} \right) \left(\hat{d}_{\vec{q}_1} e^{-iq_1\cdot x} + \hat{d}_{\vec{q}_1}^{\dagger} e^{iq_1\cdot x} \right). \tag{7.3}$$

This 12-dimensional integral is, to put it lightly, kind of a bear, but at least it puts us on familiar footing. In ordinary quantum mechanics, we use the Hamiltonian to evolve a wavefunction $|\psi\rangle$, where the "state" in this case consists of a collection of particles with specified energies and momenta. We may integrate over time to produce a generalized

unitary evolution operator:

$$|\psi(t)\rangle = \hat{U}(t, t_0)|\psi(t_0)\rangle = \exp\left(-i\int_{t_0}^{t}\hat{H}_{\text{Int}}(t')dt'\right)|\psi(t_0)\rangle. \tag{7.4}$$

From our work with generators in group theory, we're well prepared to expand exponentials as infinite series:

$$\hat{U}(t, t_0) = \mathbf{I} - i\int_{t_0}^{t}\hat{H}_{\text{Int}}(t')dt' + \frac{1}{2}(-i)^2\int_{t_0}^{t}\int_{t_0}^{t}dt'dt''\hat{H}_{\text{Int}}(t')\hat{H}_{\text{Int}}(t'') + \dots. \tag{7.5}$$

There's an important assumption underlying this approach. We're assuming that the system is **perturbative**. That is, one term gives a bad approximation to the solution,[1] two terms give a better approximation, and so on.[2] The interaction Hamiltonian allows us to compute an amplitude for a particular transition known as the **S-matrix**, defined as:

$$S_{fi} \equiv \langle f|\hat{U}(\infty, -\infty)|i\rangle. \tag{7.6}$$

The S-matrix will serve as the central quantity in all our quantum field theory calculations, and much of the hard work for our scalar and real-world theories will involve getting a handle on this number.

7.1.2 The Decay Amplitude

As our first worked example, consider a decay of an η particle into a pair of ϕ's:

$$\eta_1 \to \phi_2 + \phi_3,$$

where the subscripts are written in anticipation of the labeling of momentum and energies of particles. Energy conservation, naturally, requires that $m_\eta > 2m_\phi$.

Having set the scene, we can ignore most of the terms in the Hamiltonian. If we start with an η and end with two ϕ's, then the only terms in the Hamiltonian that will contribute to the decay amplitude are those which annihilate an η and create two ϕ terms: $\hat{d}\hat{a}^\dagger\hat{a}^\dagger$. Keeping only the relevant terms in the Hamiltonian yields

$$\hat{H}_{\text{Int}} = \lambda\int d^3x \frac{d^3q_1}{(2\pi)^3}\frac{d^3q_2}{(2\pi)^3}\frac{d^3q_3}{(2\pi)^3}\frac{1}{\sqrt{8E_{q_1}E_{q_2}E_{q_3}}}\hat{a}_{\vec{q}_3}^\dagger\hat{a}_{\vec{q}_2}^\dagger\hat{d}_{\vec{q}_1}e^{-i(q_1-q_2-q_3)\cdot x}.$$

Inspection of the creation and annihilation operators tells a story. Once upon a time (the story goes), there was an η particle which got annihilated, and two ϕ particles were created in its place. Admittedly, it's not a particularly *interesting* story, but it's one we can visualize fairly simply, as we do in Figure 7.2. This is our first Feynman diagram.

As we've drawn it, the incoming particles appear at the bottom of the diagram, and the outgoing particles are at the top. This corresponds to time running "upward." Those

---

[1] As the first term is the identity, including only the first term would result in a completely static universe.

[2] There is an important caveat. The Hamiltonian can be written as $\hat{H}_0 + \hat{H}_{\text{Int}}$, where $\hat{H}_0$ is the "free-field" Hamiltonian that we're simply ignoring in this context. In general, $[\hat{H}_0, \hat{H}_{\text{Int}}] \neq 0$, so quadratic and higher terms in the evolution operator introduce permutation factors. While we will give the results of these complications, we won't dwell on them, and we won't derive them.

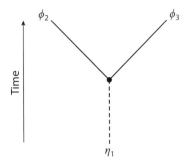

Figure 7.2. The lowest-order diagram for the decay of an η particle into two ϕ particles.

familiar with quantum field theory (QFT) may object to a vertical arrow of time. Many texts—most, in fact—draw the arrow of time horizontally, in an effort to mimic the direction of reading in many languages. We'll be using a vertical arrow for two reasons:

1. It is suggestive of the format of "spacetime diagrams" in relativity, wherein "up" is forward in time.
2. This was the format that Feynman himself used in his original formulation [61].

But please bear in mind that while the vertical axis represents time, the horizontal axis (while suggestive of space) is essentially meaningless, and the tilt of lines should not be used to infer the momentum of individual particles.

There's one more element to calculation of the amplitude, the initial and final states. In many nonrelativistic quantum systems (think of, for example, the hydrogen atom or a particle in a box), we may enumerate all possible states of the system and then simply figure out the transition amplitude between any two of them.

For quantum field theory calculations, states are described in terms of the number of particles of each type and momentum. This is most easily accomplished by starting with a vacuum and simply adding particles one at a time:

$$|i\rangle = \sqrt{2 E_{p_1}} \hat{a}_{\vec{p}_1}^\dagger |0\rangle. \tag{7.7}$$

A final state is simply the "ket,"

$$\langle f| = \langle 0| \hat{a}_{\vec{p}_3} \hat{a}_{\vec{p}_2} \sqrt{4 E_{p_2} E_{p_3}}, \tag{7.8}$$

where the normalization may look a bit odd but will be justified in due course.

However, we may get some insights from an appropriate normalization of the vacuum state itself,

$$\langle 0|0\rangle = 1,$$

along with the requirement that the annihilation operator $\hat{a}$ on the vacuum produces a multiplicative factor of zero. That, combined with equation (5.49) yields the relation

$$\langle 0| \hat{a}_{\vec{p}} \hat{a}_{\vec{q}}^\dagger |0\rangle = (2\pi)^3 \delta(\vec{p} - \vec{q}), \tag{7.9}$$

which can be used to evaluate the normalization of the initial state,

$$\langle i|i\rangle = 2E_{p_1}(2\pi)^3\delta^{(3)}(0);$$ (7.10)

the outgoing particles are evaluated similarly. Here, $\delta^{(3)}(0)$ is value of the three-dimensional Dirac delta function at zero. We will find a practical way of expressing this term in short order.

Our states are described as basis functions for the free-field Hamiltonian, which means that all the evolution comes in the form of an interaction Hamiltonian. This is known as the **interaction representation** of a system.[3]

Focusing, for the moment, on the simplest nontrivial term in the unitary operator (equation 7.5) suggests

$$\langle f|\hat{U}(t,t_0)|i\rangle = -i\int_{t_0}^{t}\langle f|\hat{H}_{\mathrm{Int}}(t')|i\rangle dt',$$

to first order. For a two-particle decay, estimating the S-matrix element is fairly straightforward:

$$S_{fi} = -i\lambda\sqrt{8E_{p_1}E_{p_2}E_{p_3}}\langle 0|\left[\int d^4x \frac{d^3q_1\,d^3q_2\,d^3q_3}{(2\pi)^9\sqrt{E_{q_1}E_{q_2}E_{q_3}}}e^{-i(q_1-q_2-q_3)\cdot x}\left(\hat{a}_{\vec{p}_3}\hat{a}_{\vec{q}_3}^\dagger\hat{a}_{\vec{p}_2}\hat{a}_{\vec{q}_2}^\dagger\hat{a}_{\vec{p}_1}\hat{a}_{\vec{q}_1}^\dagger\right)\right]|0\rangle$$

$$= -i\lambda(2\pi)^4\delta(p_1-p_2-p_3),$$ (7.11)

where we contracted the initial and final states via the orthogonality condition (equation 7.9) and the δ-function relation

$$\int d^4x e^{ip\cdot x} = (2\pi)^4\delta(p).$$ (7.12)

Astute readers will also note that we've rather arbitrarily coupled the creation of ϕ's with momentum $\vec{q}_2$ with the annihilation of $\vec{p}_2$ and $\vec{q}_3$ with $\vec{p}_3$, but another permutation is possible. That labeling represents an overcounting of possible outcomes, since the two particles are identical. We'll show how to correct for this in short order. For now, we ignore this issue.

We already have some very nice results. Because of the δ-function in equation (7.11), the decay amplitude is zero unless momentum and energy are exactly conserved. Further, we've found that the amplitude in this case is linearly proportional to the coupling constant, λ, suggesting the role that similar terms (e.g., the electric charge) will play in quantum field theory calculations.

---

[3] The interaction representation is not the only option. Nonrelativistic quantum mechanics is normally developed using the Schrödinger representation, wherein the waves are represented as combinations of eigenstates of the full Hamiltonian. In the Heisenberg representation, the states themselves evolve over time, even for unperturbed systems. Even our choice of approach is not unique. We will privilege time over the other coordinates. As another option, in the Feynman path integral formalism, all four spacetime coordinates are treated on equal footing throughout the calculation.

7.1.3 Fermi's Golden Rule for Decays

We've computed a decay amplitude for $\eta \to \phi + \phi$, but we haven't put that calculation in context. In nonrelativistic quantum mechanics, you may have encountered **Fermi's golden rule**, an approach to quantum transitions popularized by Enrico Fermi [60] but, like so much in quantum theory, originally developed by Dirac [45]. It begins with a basic statement of quantum mechanics wherein the appropriately normalized probability of transitioning from one state to another,

$$P = \frac{|S_{fi}|^2}{\langle i|i\rangle \langle f|f\rangle},$$

is the probability of a system making the transition from state $|i\rangle$ to $|f\rangle$ over some time $t - t_0$. The overall probability of a particular transition over a long but noninfinite time T can be used to compute a rate of decay,

$$\Gamma = V^2 \int \frac{d^3 p_2}{(2\pi)^3} \frac{d^3 p_3}{(2\pi)^3} \frac{|S_{fi}|^2}{T} \frac{1}{\langle i|i\rangle \langle f|f\rangle},$$

where we explicitly integrated over all possible momenta. The integral over all spacetime in the S-matrix is particularly interesting. We have seen an integral of this form before (equation 7.12). In this case, we're *guaranteed* to produce a conservation of momentum and energy hardwired into our amplitude.

However, a bit of a complication arises when we note that the decay rate is proportional to the S-matrix component *squared*, along with the state terms in the denominator. Both contain δ-functions.

We deal with this issue, as we've done before, with a bit of a trick. While formally, both integrals are over all space and time, we may instead consider one of the integrals to be over a suitably large box V and timescale T:

$$\lim_{p \to 0} \int d^4 x \, e^{ip \cdot x} = VT.$$

Comparison of this result with equation (7.12) yields a nice substitution:

$$\delta^{(4)}(p) \to (2\pi)^4 VT; \quad \delta^{(3)}(\vec{p}) \to (2\pi)^3 V \tag{7.13}$$

We're nearly there! The η-decay rate can thus be computed as

$$\Gamma = \lambda^2 \int (2\pi)^4 \frac{d^3 p_2}{(2\pi)^3} \frac{d^3 p_3}{(2\pi)^3} \frac{1}{8 E_{p_1} E_{p_2} E_{p_3}} \delta(p_1 - p_2 - p_3),$$

where all the volume terms drop out as if by magic!

This is marvelous! Getting here was a slog, but we need to do this calculation only once. Indeed, virtually everything in this relationship is a function of only the 4-momenta of the incoming and outgoing particles. The only theory-specific term is λ, the strength of the scalar interaction. In general, we define a **scattering amplitude** $\mathcal{A}$ such that the overall rate of decay can be be computed as

$$\boxed{\Gamma = S \int |\mathcal{A}|^2 (2\pi)^4 \delta(\Delta p) \frac{1}{2E_0} \prod_{\text{out states}} \left(\frac{d^3 p_i}{(2\pi)^3} \frac{1}{2E_i} \right),}$$ (7.14)

where E_0 is the initial energy (in the center-of-mass frame, simply the mass of the decaying particle), the delta function demands conservation of mass and energy, the ith state refers to the possible energies and momenta of outgoing particles, and S is the degeneracy factor, which for a decay into two identical particles produces a factor of $1/2$. Or, more generally,

$$S = \frac{1}{n!}$$

for every group of identical partners in the decay.

The decay rate calculation has two parts:

1. A purely kinematic term that normalizes the rate and guarantees conservation of momentum and energy, and
2. The scattering amplitude, which is specific to both the interaction and the expansion of the Hamiltonian. Comparison of equation (7.12) with our specific solution for η-decay suggests that in this theory $\mathcal{A} = \lambda$.

Example 7.1: What is the decay rate for

$$\eta \to \phi_2 + \phi_3$$

in the rest frame?

Solution: The δ-function in the decay rate equation (7.14) simplifies dramatically:

$$\Gamma = \lambda^2 (2\pi)^4 \iint \frac{d^3 p_2}{(2\pi)^3} \frac{d^3 p_3}{(2\pi)^3} \frac{1}{8 m_\eta E_2 E_3} \delta(m_\eta - E_2 - E_3) \delta(\vec{p}_2 + \vec{p}_3)$$

$$= \frac{\lambda^2}{(2\pi)^2} \int d^3 p_2 \frac{1}{8 m_\eta E_2 E_3} \delta(m_\eta - E_2 - E_3).$$

The particle decay is isotropic, so we may say

$$d^3 p_2 = 4\pi p^2 dp,$$

where we've dropped the index because the outgoing momenta of the daughter particles are equal and opposite. A six-dimensional integral has been reduced to one! The final δ-function is more challenging than it appears. The energies, E_2 and E_3 are explicitly a function of momentum.

In the rest frame of the η, each of the outgoing particles has energy $m_\eta/2$, yielding

$$p_F \equiv |\vec{p}_2| = \frac{\sqrt{m_\eta^2 - 4m_\phi^2}}{2}.$$

Figure 7.3. Some three-vertex diagrams for an η decay.

Noting

$$\int dx\, \delta(f(x)) = \frac{1}{|f'(x)|_{f(x)=0}} \tag{7.15}$$

allows us to write

$$\int \frac{p^2}{E_2 E_3}\, dp\, \delta(m_\eta - E_2(p) - E_3(p)) = \frac{p_F}{m_\eta}.$$

This gives a decay rate of

$$\Gamma = \lambda^2 \frac{p_F}{16\pi m_\eta^2}. \tag{7.16}$$

Example 7.1 lends several intriguing insights into decays in general:

- The outgoing particles have a unique energy in two-body decay. This will *not* be the case for decays of more particles, as you can test for yourself in problem 7.5.
- The larger the mass of the decaying particle (all things being equal), the slower the decay.
- The more energy liberated by the decay, the faster it occurs.

The scattering amplitude is *the* quantity that we're trying to extract from a Feynman diagram. As we noted earlier,

$$\mathcal{A} = \lambda,$$

though this result isn't exact. For small values of λ, a first-order diagram is fine, but the beauty of quantum field theory is that in principle we can take calculations to higher order by expanding the evolution operator appropriately. For instance, consider the operators that show up in the cubic term:

$$\hat{H}_{\text{Int}}^3 \propto \lambda^3 (\hat{a}\hat{a}\hat{a}^\dagger)(\hat{a}^\dagger \hat{a}\hat{a}^\dagger)(\hat{a}^\dagger \hat{a}^\dagger \hat{a}).$$

It looks horrid, to be sure. We've omitted subscripts, but the upshot is that internally, multiple ϕ and η particles are temporarily created and annihilated. The *net* effect, that of an η particle decaying into two ϕ's, remains the same. As with the first-order term, we also may draw three-vertex Feynman diagrams (Figure 7.3).

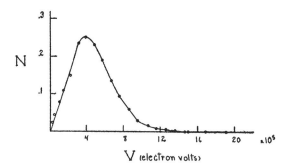

Figure 7.4. The energy spectrum of electrons emitted from the decay of radium E [150]. A characteristic feature of these spectra is a high-energy tail—a feature that does not appear in our toy scalar model and which is entirely attributable, as we'll see, to the fact that the particles involved are fermions.

What's remarkable about these diagrams is how much of the physics they capture. Each vertex carries with it a multiplicative factor λ, and each connects two ϕ lines with an η line. Even though we can draw diagrams with internal loops, evaluating them tends to produce some rather unpleasant infinities, but we may get a sense of how much new information they contribute (ignoring the infinities) from dimensional analysis. Since $\Gamma \propto |\mathcal{A}|^2$ and has units of $[E]^1$, for decay into massless η particles, the only dimensional term remaining is the mass of the ϕ. Thus, the contributions from interference between one-vertex and three-vertex diagrams will contribute corrections like

$$\Delta\Gamma_3 \sim \frac{\lambda^4}{m_\phi^5} \, .$$

The coefficient out front needs to be computed from the loop diagrams, to which we will return in Chapter 11.

7.1.4 A Foray into Beta Decay

Not all decays are monoenergetic. In 1914, James Chadwick observed the β-decay of radium B, which is similar in many ways to neutron decay. He, and a number of his contemporaries found, despite expectations, that the energy spectrum of electrons is continuous (The decay spectrum for radium E is shown in Figure 7.4).

In 1930, in a letter to a scientific meeting in Tübingen [121], Wolfgang Pauli offered up an explanation:

> I have hit upon a desperate remedy to save the … law of conservation of energy. Namely, the possibility that in the nuclei there could exist electrically neutral particles, which I will call neutrons,[4] that have spin 1/2 and obey the exclusion principle and that further differ from light quanta in that they do not travel with the velocity of light. The mass of the neutrons should

---

[4] As what we now call a neutron was discovered by James Chadwick in 1932, Enrico Fermi suggested the name "neutrino" to correspond to Pauli's hypothetical particle.

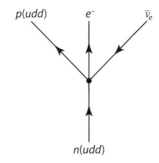

Figure 7.5. Fermi's model for neutron decay.

be of the same order of magnitude as the electron mass and in any event not larger than 0.01 proton mass. — The continuous beta spectrum would then make sense with the assumption that in beta decay, in addition to the electron, a neutron is emitted such that the sum of the energies of neutron and electron is constant.

The electron antineutrino was detected by Clyde Cowan, Frederick Reines, and their collaborators some 26 years later [40], a remarkable example of the existence of an entity theoretically predicted long before its discovery.

We can get a sense of the energetics of beta decays in general by considering neutron decay, a *staggeringly* slow weak interaction, with a mean lifetime of approximately 881.5 s:

$$n \rightarrow p + e^- + \bar{\nu}_e \, .$$

We can get a sense of why the process is so slow by considering how little energy there is to work with. The neutron ($m_n = 939.565$ MeV is only about 1.3 MeV more massive than the proton ($m_p = 938.272$ MeV). In a three-body decay, virtually all that energy goes into the electron and the antineutrino.

While we haven't explicitly talked about the weak interaction in any great depth, like all the forces in the fundamental model, it's mediated by a boson, in this case, the $W^\pm$. The $W^\pm$ bosons are massive enough that when Enrico Fermi was developing a model for the weak interaction in 1934 [163], he assumed the interaction occurred at a single point (Figure 7.5).

Example 7.2: While we aren't prepared to do a weak decay yet,[5] we may model a scalar version of neutron decay,

$$\phi_0 \rightarrow \phi_1 + \phi_2 + \phi_3 \, ,$$

where the masses of the four particles are equal to the particles in a neutron decay. What is the spectrum of "electron" energies? For this case, we will assume that the amplitude $\mathcal{A}$ is independent of energy.

---

[5] Additionally, both the proton and neutron are composite particles (and fermions at that).

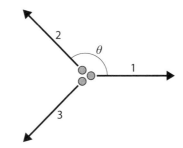

Figure 7.6. The geometry of a three-body decay.

Solution: Evaluating equation (7.14) under these assumptions yields

$$\Gamma = S \iiint |\mathcal{A}|^2 \frac{1}{16(2\pi)^5 M} \frac{d^3 p_1}{E_1} \frac{d^3 p_2}{E_2} \frac{d^3 p_3}{E_3} \delta(\Delta p) \ .$$

In a three-body decay, conservation of energy and momentum reduces the problem to two dimensions, uniquely determined by any angle and energy of the outgoing particles, as shown in Figure 7.6. We are particularly interested in processes resembling beta decays, that is,

$m_1 \ll M$ (the "electron");

$m_2 \simeq 0$ (the "antineutrino");

$M - m_3 \equiv \Delta E \ll M$ (the outgoing "proton").

Under those circumstances, the kinetic energy of the outgoing proton is very low, but the large mass is able to absorb the momentum of the outgoing particles. As shown in equation (B.6) in Appendix B, this yields a decay rate of

$$\Gamma = S \iint \frac{|\mathcal{A}|^2}{(4\pi)^3} \frac{\Delta E - E_e}{M^2} \sqrt{E_e^2 - m_e^2} d E_e d(\cos\theta), \tag{7.17}$$

where $M = m_n$, and $\Delta m = 1.3\,\text{MeV}$ for neutron decay. Assuming the scattering amplitude is independent (as is the case for our single-vertex scalar model), we can readily compute the spectrum of outgoing electrons,

$$f(E) \propto \frac{\Delta E - E_e}{M^2} \sqrt{E_e^2 - m_e^2},$$

which is plotted in Figure 7.7.

7.2 Scattering

We don't just have to wait for particles to decay; we can, instead, throw them together and see what pops out. In **scattering**, we're particularly interested in the **cross section** σ,

$$\Gamma = n\sigma v, \tag{7.18}$$

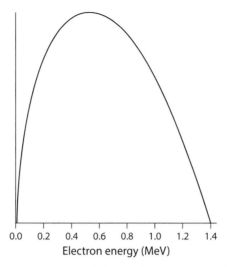

Figure 7.7. The energy spectrum of the outgoing electron in a simplified scalar model of neutron decay. The scalar model produces a spectrum which lacks the high-energy tail of observed beta decays (Figure 7.4), a detail that will be remedied when we revisit this problem using fermions.

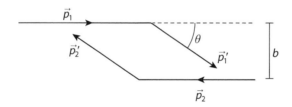

Figure 7.8. The $\phi-\phi$ scattering process in our toy scalar theory.

where n is the number density of target particles, and v is the relative speed of the incoming particles. Once again, by way of example, we'll use our simple scalar model and will focus on the process

$$\phi_1 + \phi_2 \to \phi_3 + \phi_4,$$

where the subscripts in this context represent unique identifiers of momentum and energy, rather than necessarily different species. A schematic of the scatter is shown in Figure 7.8.

7.2.1 Classical Cross Sections

Before considering quantum field scattering, it will be useful to consider the classical version of $\phi-\phi$ scattering. As we saw in the previous chapter (equation 6.5), the classical low-speed potential energy between two ϕ particles can be written as

$$U(r) \sim -\frac{\lambda^2}{m_\phi^2} \frac{1}{4\pi r} e^{-m_\eta r} \, .$$

We'll start by considering the limit of a massless η, which makes this problem identical to the classic gravitational scattering problem or, with a change of sign, to Rutherford scattering. There is a generalized form of Rutherford scattering which depends only on an inverse-square force law of the form

$$U = \frac{\kappa}{r}.$$

For electromagnetism and gravity, respectively the coefficient κ is

$$\kappa_{EM} = \frac{q_1 q_2}{4\pi}; \quad \kappa_G = -G m_1 m_2,$$

respectively. The different signs indicate that while electromagnetism is repulsive (for like charges), gravity is attractive.

Two particles approach each other as shown in Figure 7.8 and conserve energy and angular momentum at all times. After a little work, we get

$$\frac{d\sigma}{d\Omega} = \frac{\kappa^2 m^2}{16 p_0^4 \sin^4(\theta/2)}, \tag{7.19}$$

where the differential **solid angle** $d\Omega$,

$$d\Omega \equiv \sin\theta \, d\theta \, d\phi,$$

integrates to 4π over all directions in the unit sphere. In the familiar case of a massless mediator, the total cross section integrates over $d\Omega$ to infinity. In most experiments, there's more information to be gleaned from the differential cross sections than from the total.

7.2.2 Quantum Scattering: The Kinematic Component

Cross sections and decays are intimately related. Following the derivation of the decay exactly, we may calculate a collision rate of a particle scatter. We need compute only the relative speed of interacting particles. We leave it as an exercise for the reader to show that

$$v E_1 E_2 = \sqrt{(p_1 \cdot p_2)^2 - (m_1 m_2)^2}$$

is simply a Lorentz-invariant way of expressing the relative velocity. This yields, in the final analysis,

$$\sigma = S \int |\mathcal{A}|^2 (2\pi)^4 \delta(\Delta p) \frac{1}{4\sqrt{(p_1 \cdot p_2)^2 - (m_1 m_2)^2}} \prod \left(\frac{d^3 p_i}{(2\pi)^3 2 E_i} \right). \tag{7.20}$$

Example 7.3: What is the kinematic component of the scattering cross section calculation for the process $\phi_1 + \phi_2 \to \phi_3 + \phi_4$ in the center-of-mass frame?

Solution: For identical particles with energy E_0 and momentum p_0

$$\sqrt{(p_1 \cdot p_2)^2 - m^4} = 2 E_0 p_0$$

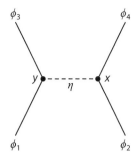

Figure 7.9. The t-channel diagram for scattering between two scalar particles.

in the center-of-mass frame. Thus,

$$\sigma = S \int |\mathcal{A}|^2 \frac{1}{32\, p_0\, E_0\, E_3\, E_4\, (2\pi)^2}\, d^3p_3\, d^3p_4\, \delta(p_1 + p_2 - p_3 - p_4)$$

$$= \frac{1}{2} \int |\mathcal{A}|^2 \frac{1}{32\, p_0\, E_0\, E_3^2\, (2\pi)^2}\, p_3^2 dp_3\, d\Omega\, \delta(2E_0 - 2E_3)$$

$$\frac{d\sigma}{d\Omega} = \frac{|\mathcal{A}|^2}{2(16\pi)^2\, E_0^2}. \tag{7.21}$$

We take this opportunity to give a very general result. For two-to-two particle scatters, in the center-of-mass frame:

$$\boxed{\frac{d\sigma}{d\Omega} = S\frac{|\mathcal{A}|^2}{(16\pi)^2\, E_0^2}\, \frac{|p_F|}{|p_I|}.} \tag{7.22}$$

7.2.3 The Propagator

The *real* challenge, as with decays, is computing the scattering amplitudes. We begin by drawing the lowest-order Feynman diagram (Figure 7.9).

This configuration—a simple two-vertex diagram in which the final and initial states of the same particles meet at a vertex—is common enough to be given a name: the **t-channel**. We will encounter a few other channels in due course, but the advantage of categorizing diagrams in this way is that conservation of 4-momentum guarantees that the η particle must carry a momentum of $p_1 - p_3$. This quantity (or, rather, its square), is known as a **Mandelstam variable**, and when we encounter more complicated diagrams we'll find useful to employ the shorthand

$$t = (p_1 - p_3)^2.$$

We are now faced with the challenge of solving for the scattering amplitude of a Feynman diagram with two vertices. The second-order term in the expansion of the unitary time operator takes the form

$$\hat{U}_{\text{Int}}^{(2)}(t, t_0) = \frac{1}{2}(-i)^2 \int_{t''}^{t} \int_{t_0}^{t} dt' dt''\, \hat{H}_{\text{Int}}(t')\, \hat{H}_{\text{Int}}(t'').$$

As with the scattering diagrams, Figure 7.9 also tells a story (which we may begin at either the left or right vertex):

1. A ϕ particle is annihilated at point y in spacetime, and an η and a ϕ are created.
2. Later, at another point, x in spacetime, the η particle is annihilated, along with a ϕ particle, and a ϕ particle is created.

All told, the S-matrix element becomes

$$S_{fi} = (-i\lambda)^2 \sqrt{16\, E_{p_1}\, E_{p_2}\, E_{p_3}\, E_{p_4}}\, \langle 0 | \hat{a}_{\vec{p}_3} \hat{a}_{\vec{p}_4}$$

$$\times \left[\int d^4x \frac{d^3k_4}{(2\pi)^3} \frac{d^3k_2}{(2\pi)^3} \frac{d^3q_2}{(2\pi)^3} \left(\hat{a}^\dagger_{\vec{k}_4} \hat{a}_{\vec{k}_2} \hat{a}_{\vec{q}_2} \right) \frac{1}{\sqrt{8\, E_{k_2}\, E_{q_2}\, E_{k_4}}} e^{-i(k_2 + q_2 - k_4)\cdot x} \right]$$

$$\times \left[\int d^4y \frac{d^3k_1}{(2\pi)^3} \frac{d^3k_3}{(2\pi)^3} \frac{d^3q_1}{(2\pi)^3} \left(\hat{a}^\dagger_{\vec{k}_3} \hat{a}^\dagger_{\vec{q}_1} \hat{a}_{\vec{k}_1} \right) \frac{1}{\sqrt{8\, E_{k_1}\, E_{q_1}\, E_{k_3}}} e^{-i(k_1 - q_1 - k_3)\cdot y} \right]$$

$$\times \hat{a}^\dagger_{\vec{p}_1} \hat{a}^\dagger_{\vec{p}_2} | 0 \rangle ,$$

where q_1 and q_2 correspond to the 4-momentum of the **virtual particle** (the η in the middle of the diagram[6]). This expression is downright terrifying! It is literally a 26-dimensional integral. However, we can be thankful for two things:

1. We are in the process of developing the Feynman rules, which means that we'll soon be able to calculate terms like this with no trouble, and
2. A great many of these integrals produce helpful δ-functions, yielding

$$k_i = p_i; \quad q_1 = q_2 .$$

Thus,

$$S_{fi} = (-i\lambda)^2 \int d^4x\, d^4y \frac{d^3q}{(2\pi)^3} \frac{1}{2E_q} e^{-i(p_2 + q - p_4)\cdot x} e^{-i(p_1 - q - p_3)\cdot y} . \tag{7.23}$$

This 11-dimensional integral might still prove a little daunting, especially as it includes calculations at different points in spacetime.

Quantum field theory, and the Standard Model generally, is based on the idea that forces are mediated by virtual particles. In electromagnetism, we never detect the photons that communicate the repulsion between two electrons. But to understand the dynamics of these mediators, we need to introduce a mathematical shorthand known as a **propagator.**

Take a good look at the spacetime dependence in equation (7.23). The only term that couples the spacetime events x and y is

$$\Delta(x - y) = \int \frac{d^3q}{(2\pi)^3} \frac{1}{2E_q} e^{-iq\cdot(x-y)} .$$

[6] A later simplification will establish that $q_1 = q_2$.

Propagators of this sort will show up in any calculation with more than one vertex. In the context of $\phi-\phi$ scattering, it represents the η particle propagating from y to x to carry the interaction signal. You may, in fact, find it reminiscent of the Green's function which you encountered in electromagnetism. This is no accident.

We need to be a bit careful. As we've described it, spacetime coordinate x occurs *after* y, and yet both terms are integrated over all spacetime. We shall assume that order *doesn't* matter (an effect we've accounted for by swallowing the factor of 2 in the Taylor series of the unitary time operator).

The propagator is particularly useful if we can rewrite it as an integral over the entire 4-momentum. We'll start with the solution and work backward:

$$\Delta(x-y) = i \int \frac{d^4 p}{(2\pi)^4} \frac{1}{p \cdot p - m^2} e^{-ip \cdot (x-y)}. \tag{7.24}$$

If you've mastered your Lorentz algebra, you'll recognize that the integral *should* blow up. If the mediator has the mass it's supposed to, then the denominator should be identically zero. However, because mediators are virtual particles, they aren't ever detected, and because quantum field theory is, well, *quantum*, it turns out that uncertainty rules the day. Mediators need not have the same mass as if they were in the wild (termed "on the mass shell").

We may evaluate the propagator by rewriting the denominator as

$$p \cdot p - m^2 = (p_0 - E_{\vec{p}})(p_0 + E_{\vec{p}}),$$

where $E_{\vec{p}}$ is the energy that a particle of mass m and momentum $\vec{p}$ is *supposed* to have. For convenience, for the rest of the analysis, we will *assume* that $x^0 > y^0$, and integrate over p_0:

$$\Delta(x-y) = \int \frac{d^3 p}{(2\pi)^3} e^{i\vec{p} \cdot (\vec{x}-\vec{y})} \int_{-\infty}^{\infty} \frac{dp_0}{(2\pi)} \frac{i}{(p_0 - E_{\vec{p}})(p_0 + E_{\vec{p}})} e^{-ip_0(x^0-y^0)}.$$

The integral over p_0 looks impossible. We have two roots, at $E_{\vec{p}}$ and $-E_{\vec{p}}$. But complex analysis gives us a trick. First, we introduce a slight change,

$$\int_{-\infty}^{\infty} \frac{dp_0}{(2\pi)} \frac{i}{(p_0 - (E_{\vec{p}} - i\varepsilon))(p_0 + E_{\vec{p}} - i\varepsilon)} e^{-ip_0(x^0-y^0)},$$

and do a semi-infinite semicircle integral from $p_0 = -\infty$ to ∞, and then around the complex part of the plane (Figure 7.10). In complex analysis—the Cauchy residue theorem to be exact —we learn that if you do a counterclockwise integral around a root,

$$\int \frac{dz}{z - z_0} f(z) = 2\pi i f(z_0).$$

In this case, there is a root at $p_0 = E_{\vec{p}} - i\varepsilon$. Since the contour integral is clockwise, this introduces a negative sign, and thus,

$$\int_{-\infty}^{\infty} \frac{dp_0}{(2\pi)} \frac{i}{(p_0 - E_{\vec{p}} + i\varepsilon)(p_0 + E_{\vec{p}} - i\varepsilon)} e^{-ip_0(x^0-y^0)} = \frac{2\pi}{2E_p}$$

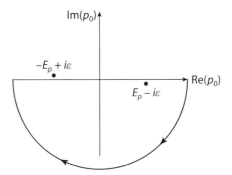

Figure 7.10. Contour-integral analysis of the propagator.

in the limit of $\varepsilon \to 0$. And so,

$$\Delta(x - y) = \int \frac{d^4 p}{(2\pi)^4} \frac{i}{p \cdot p - m^2 + i\varepsilon} e^{-ip\cdot(x-y)} = \int \frac{d^3 p}{(2\pi)^3} \frac{1}{2E_p} e^{-ip\cdot(x-y)} . \tag{7.25}$$

The Lorentz-invariant form is *very* important in computing the scattering amplitude. The virtual particle doesn't need to have the same mass as when it's at rest. The contribution of the amplitude is *maximized* on the mass shell, but the term may deviate. This also gives us an insight into why factors of $2E$ have so frequently shown up in our calculations.

7.2.4 The $\phi-\phi$ Scattering Amplitude

We are finally prepared to compute the amplitude from the Feynman diagram in Figure 7.9:

$$S_{fi} = -i\lambda^2 \int d^4 x d^4 y \frac{d^4 q}{(2\pi)^4} \frac{1}{q \cdot q - m_\eta^2} e^{-i(p_2 - p_4 + q)\cdot x} e^{-i(p_1 - p_3 - q)\cdot y}$$

$$= -i\lambda^2 (2\pi)^4 \frac{1}{(p_1 - p_3)\cdot(p_1 - p_3) - m_\eta^2} \delta(p_1 + p_2 - p_3 - p_3),$$

$$\mathcal{A}_t = \frac{\lambda^2}{(p_1 - p_3)\cdot(p_1 - p_3) - m_\eta^2} .$$

Plugging in the initial and final particle trajectories gives a Mandelstam variable,

$$(p_1 - p_3)\cdot(p_1 - p_3) = -2p_0^2(1 - \cos\theta) ,$$

which simplifies still further if we take a massless mediator $m_\eta = 0$:

$$\mathcal{A}_t = -\frac{\lambda^2}{2p_0^2(1 - \cos\theta)} . \tag{7.26}$$

We're not quite done. Because the outgoing particles are identical, it doesn't really matter which one we label as "3" and which one we label as "4." Classically, we'd say that "particle 1 and particle 3 are the same object, but at different times," but quantum

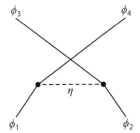

Figure 7.11. A "twisted" two-vertex diagram describing ϕ–ϕ scattering, representing the u-channel.

mechanically, they're not. The labeling of the outgoing particle is arbitrary, and we may just as easily say that it's particle 4 that meets particle 1 at a vertex, as shown in Figure 7.11.[7] The twisted configuration is known as the **u-channel**.

The approach is similar to that of the untwisted diagram but produces an amplitude,

$$\mathcal{A}_u = \frac{\lambda^2}{(p_1 - p_4) \cdot (p_1 - p_4) - m_\eta^2},$$

where

$$(p_1 - p_4) \cdot (p_1 - p_4) = -2 p_0^2 (1 + \cos \theta),$$

which is also known as the u Mandelstam variable. The rule with amplitudes (as with ordinary quantum mechanics) is that you add them *before* you compute the scattering cross section or decay rate. Thus, adding the amplitudes from Figures 7.9 and 7.11 yields

$$\mathcal{A} = \frac{\lambda^2}{2 p_0^2 (1 - \cos \theta)} + \frac{\lambda^2}{2 p_0^2 (1 + \cos \theta)} = \frac{\lambda^2}{p_0^2 \sin^2 \theta}.$$

We have one more diagram, although it's not one that shows up in electron-electron scattering. Figure 7.12 shows the so-called **s-channel** diagram, which quickly yields

$$\mathcal{A}_s = \lambda^2 \frac{1}{(p_1 + p_2) \cdot (p_1 + p_2)} = \frac{\lambda^2}{4 E_0^2}.$$

We now have a generalized solution to the toy scalar problem, but it's especially interesting in the nonrelativistic limit in which $p_0 \ll E_0$ and $E_0 \simeq m_\phi$, which essentially allows us to discard the last diagram entirely.

We may *finally* compute a QFT estimate of the scattering cross section:

$$\frac{d\sigma}{d\Omega} = \frac{\lambda^4}{2(16\pi)^2 p_0^4 m_\phi^2 \sin^4 \theta}. \tag{7.27}$$

---

[7] This issue of swapping particles is a subtle point that we first encountered in our discussion of the exchange operator in Chapter 5. The point is that all electrons (or up quarks or ϕ's) are made of the same "stuff," and thus from the perspective of measurement, there is *nothing* that uniquely identifies a particle at one point in time with a specific antecedent.

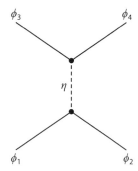

Figure 7.12. The s-channel contribution to $\phi-\phi$ scattering.

This is *very nearly* the same as the classical result (equation 7.19) but differs significantly at high energies (large scattering angle). It's worth noting that the overall scattering amplitude integrates to infinity. But that need not be the case. We find another interesting limit if $m_\eta \gg E_0$. In that case, all three contributions to the scattering amplitude are identical, and

$$\mathcal{A} = -\frac{3\lambda^2}{m_\eta^2} \, .$$

This produces an isotropic cross section, which may be integrated to yield

$$\sigma = \frac{9\lambda^4}{128\pi E_0^2 m_\eta^4} \, . \tag{7.28}$$

Having worked through a few examples, we're now prepared to come up with a generalized approach to computing amplitudes for our theory.

7.3 Feynman Rules for the Toy Scalar Theory

It's a bit cumbersome to start from scratch every time we want to calculate reaction rates or cross sections for some new process. But rest easy. The entire *point* of Feynman diagrams is that we can use them as a shortcut to figure out scattering amplitudes. The process of turning diagrams into amplitudes is known as **Feynman rules**. The approach should look comfortingly familiar, based on the work we've already down.

Throughout this book, whenever new quantum field theories are introduced, the Feynman rules will simply be appended to the following collected in Appendix C.

Preliminaries
Regardless of the quantum field theory, a few initial steps will set the scene for QFT calculations:

1. Compute and simplify (as much as possible) the kinematic components of the decay rate (equation 7.14) or cross section (equation 7.20).

2. Compute the degeneracy term $1/n!$ for identical types of outgoing particles.

3. Draw *all* possible Feynman diagrams to as high an order as precision as requirements demand. Each quantum field theory has a fundamental vertex which can be inferred directly from the classical interaction Lagrangian. A valid Feynman diagram can be twisted and rotated with abandon, so long as the appropriate number (and type) of particles go into and the appropriate number go out of each vertex.

4. Label each ingoing particle, p_1, p_2,... and each outgoing particle, p_{n+1}, p_{n+2},... For each draw an arrow (or assume a convention) which indicates the 4-momentum of the particles going into or out of a vertex.

The Amplitude

Having drawn the diagrams, we're now prepared to compute the scattering amplitude. These details are specific for the scalar theory, but we'll find some remarkably similar ones for quantum field theories for Dirac particles.

The "contribution" in the following list consists of multiplying all the terms together in a "sensible" order. For scalar theories, the order of multiplication doesn't matter, but for other theories, the wrong order will result in a nonscalar amplitude.

1 External Lines. The contribution for external scalar particles is trivial. We show only the ϕ particles for this calculation, but the contribution for η particles is identical (albeit drawn with a dashed line).

Particle	Contribution	Representation
Outgoing scalar	1	
Incoming scalar	1	

2 Vertex Factors. The scalar theory produces a multiplicative factor.

Vertex	Contribution $\times \left[(2\pi)^4 \delta(\Delta p)\right]$	Representation
Toy scalar theory	$-i\lambda$	

where the sum over 4-momenta represents the difference of the incoming and outgoing particles. For internal lines, especially, the "direction" of the 4-momentum (into or out of the vertex), while arbitrary, must be clearly specified.

3 Propagators. For each internal line, label the momentum q_i (and include the direction with an arrow), and integrate over all possible values.

Propagator	Contribution	Representation
Scalar	$\displaystyle\int \frac{d^4 q_i}{(2\pi)^4} \frac{i}{q_i^2 - m^2 + i\epsilon}$	$\bullet\text{-----}\bullet$ $q_i \rightarrow$

4 Cancel the Delta Function. Multiply the particle, vertex, and propagator contributions together, and integrate over virtual particles $\{q_i\}$. The result yields

$$-i(2\pi)^4 \mathcal{A}\delta(\Delta p)$$

for the overall process, from which we may read out the amplitude.

We repeat this process for all diagrams and simply add all amplitudes.

Example 7.4: Compute the cross section for the scalar analog to **Compton scattering**, the process by which a relatively low energy photon loses a small amount of energy and is deflected by an electron. For our scalar version of Compton scattering,

$$\eta_1 + \phi_2 \rightarrow \eta_3 + \phi_4 \,,$$

where the subscripts correspond to the labeling of the 4-momentum, and $m_\eta = 0$.

Solution: For this system, with no identical outgoing particles, $S = 1$. In the center-of-mass frame, the ϕ and η have an equal and opposite momentum p_0, giving

$$p_1 = \begin{pmatrix} p_0 \\ p_0 \\ 0 \\ 0 \end{pmatrix} ; \quad p_2 = \begin{pmatrix} E_2 = \sqrt{m_\phi^2 + p_0^2} \\ -p_0 \\ 0 \\ 0 \end{pmatrix} . \tag{7.29}$$

Supposing the outgoing η particle scatters off at a direction θ with respect to the original, conservation of energy and momentum requires that

$$p_4 = \begin{pmatrix} p_0 \\ p_0 \cos\theta \\ p_0 \sin\theta \\ 0 \end{pmatrix} . \tag{7.30}$$

We've arbitrarily chosen the outgoing particle to fly out in the x-y plane, but, clearly, it doesn't matter to the final solution. We may quickly compute the kinematic component of the cross section (good also for genuine electromagnetic Compton scattering):

$$\frac{d\sigma}{d\Omega} = |\mathcal{A}|^2 \frac{1}{(8\pi)^2} \frac{1}{(p_0 + E_2)^2} \,.$$

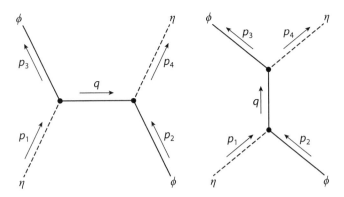

Figure 7.13. The t-channel (left) and s-channel (right) diagrams describing scalar Compton scattering.

In Figure 7.13 we drew the lowest-order Feynman diagrams for this process. We follow the Feynman rules for the t-channel:

1 *External Lines.* The external lines simply give a factor of 1.

2 *Vertex Factors.* There are two vertices, contributing a total of

$$-\lambda^2 (2\pi)^8 \delta(p_1 - q - p_3)\delta(p_2 + q - p_4).$$

3 *Propagators.* There is one propagator, giving a total contribution

$$-i\lambda^2 (2\pi)^4 \int \frac{d^4 q}{q \cdot q - m^2}\delta(p_1 - q - p_3)\delta(p_2 + q - p_4)$$

$$= -i\lambda^2 (2\pi)^4 \frac{1}{(p_1 - p_3)^2 - m^2}\delta(p_1 + p_2 - p_3 - p_4),$$

where the integral over q quickly yields $q = p_1 - p_3$ over the first δ-function.

4 *Cancel the Delta Function.*

$$A_t = \frac{\lambda^2}{(p_1 - p_3)^2 - m^2}$$

$$= \frac{\lambda^2}{2 p_0 (p_0 \cos\theta - E_2)}.$$

Similarly, the s-channel amplitude is

$$A_s = \frac{\lambda^2}{(p_1 + p_2)^2 - m^2}$$

$$= \frac{\lambda^2}{2 p_0 (p_0 + E_2)},$$

which may be combined to yield

$$A = \frac{\lambda^2}{2} \left[\frac{(1 + \cos\theta)}{(p_0 + E_2)(p_0 \cos\theta - E_2)} \right]. \tag{7.31}$$

This expression doesn't get too much simpler in generality, but in the limit of $p_0 \ll m$, we get the low-energy relation

$$A \simeq -\frac{\lambda^2}{2m^2}(1 + \cos\theta).$$

This yields a differential cross section

$$\frac{d\sigma}{d\Omega} = \frac{\lambda^4}{m^6(32\pi)^2}(1 + \cos\theta)^2. \tag{7.32}$$

By way of comparison, in electromagnetism, the low-energy Compton scattering cross section (Thomson scattering) is given by the relation

$$\frac{d\sigma}{d\Omega}\bigg|_{Th} = \frac{\alpha_e^2}{m_e^2}\left(\frac{1 + \cos^2\theta}{2}\right), \tag{7.33}$$

where α_e is known as the **fine-structure constant** and represents a good measure of the strength of the electromagnetic force

$$\alpha_e = \frac{q_e^2}{4\pi}, \tag{7.34}$$

which has a numerical value of approximately

$$\alpha_e^{-1} \simeq 137.036.$$

The fine-structure constant is small enough that for for many applications in quantum electrodynamics, higher-order diagrams will produce increasingly smaller contributions.[8] It is the first of many seemingly arbitrary parameters in the Standard Model. As Richard Feynman put it [63]:

> It is a simple number that has been experimentally determined to be close to 0.08542455. (My physicist friends won't recognize this number, because they like to remember it as the inverse of its square: about 137.03597 with about an uncertainty of about 2 in the last decimal place. It has been a mystery ever since it was discovered more than fifty years ago, and all good theoretical physicists put this number up on their wall and worry about it.) Immediately you would like to know where this number for a coupling comes from: is it related to pi or perhaps to the base of natural logarithms? Nobody knows. ...We know what kind of a dance to do experimentally to measure this number very accurately, but we don't know what kind of dance to do on the computer to make this number come out, without putting it in secretly!

---

[8] Except for the infinities. As it happens, diagrams with internal loops will tend to produce infinite contributions regardless of how small the coupling constant is. Dealing with these issues is the realm of renormalization and will be dealt with in Chapter 11.

7.4 QED

Fermions and photons add a significant additional layer of complexity to QFT calculations, since both spin *and* 4-momentum needs to be exactly conserved at every vertex in a Feynman diagram. We can see this explicitly by considering the electromagnetic interaction term (equation 6.10):

$$\mathcal{L}_{\text{Int}} = -q_e \, A^\mu \overline{\psi} \gamma^\mu \psi \, .$$

Both the electron and the photon fields need to be quantized (equations 5.55, 6.31):

$$\hat{\psi}(x) = \int \frac{d^3 p}{(2\pi)^3} \frac{1}{\sqrt{2E_p}} \left[\hat{b}_{\vec{p},r} u_r(p) e^{-ip\cdot x} + \hat{c}^\dagger_{\vec{p},r} v_r(p) e^{ip\cdot x} \right];$$

$$\hat{\overline{\psi}}(x) = \int \frac{d^3 p}{(2\pi)^3} \frac{1}{\sqrt{2E_p}} \left[\hat{b}^\dagger_{\vec{p},r} \overline{u}^{(r)}(p) e^{ip\cdot x} + \hat{c}_{\vec{p},r} \overline{v}^{(r)}(p) e^{-ip\cdot x} \right];$$

$$\hat{A}_\mu(x) = \int \frac{d^3 p}{(2\pi)^3} \frac{1}{\sqrt{2E_p}} \varepsilon^{(r)}_\mu(\vec{p}) \left[\hat{a}_{\vec{p},r} e^{-ip\cdot x} + \hat{a}^\dagger_{\vec{p},r} e^{ip\cdot x} \right],$$

where $\hat{b}^\dagger$ creates an electron, $\hat{c}^\dagger$ creates a positron, and $\hat{a}^\dagger$ creates a photon. Putting this all together produces quantum electrodynamics (commonly known as **QED**), which includes an interaction Hamiltonian with terms like

$$\mathcal{H}_{\text{Int}} \propto q_e \hat{a}^\dagger \hat{b}^\dagger \hat{b} \varepsilon_\mu v \gamma^\mu \overline{v},$$

along with seven other combinations of operators. This particular variant annihilates an electron and creates an electron and a photon, but as with the scalar theory, any twist and turn of the fundamental vertex will do.

7.4.1 Feynman Rules for Quantum Electrodynamics

Having done a similar exercise with the toy scalar model, we're now prepared to write down the Feynman rules for quantum electrodynamics (and subsequently, electroweak, weak, and strong interactions). *Unlike* with the scalar model, we will not generally derive these rules for electromagnetism. Our collected Feynman rules are given in Appendix C.

Preliminaries

In addition to the preliminaries for the scalar theory, the spin for each external particle needs to be labeled with a spin $s_1 \ldots s_n$, and each internal line with a spin $r_1 \ldots r_n$. Further, we label each vertex with a different index, μ, ν, ... to correspond with the 4-component of the photon and the related γ-matrices.

The Amplitude

1 External Lines.

Particle	Contribution	Representation
Outgoing electron	$\bar{u}_s(p)$	
Incoming electron	$u_{(s)}(p)$	
Outgoing positron	$v_s(p)$	
Incoming positron	$\bar{v}_s(p)$	
Outgoing photon	$\epsilon^{(r)\mu}*$	
Incoming photon	$\epsilon^{(r)\mu}$	

where the index on the external photons will be coupled to the corresponding index at the related vertex.

2 Vertex Factors.

Vertex	Contribution $\times \left[(2\pi)^4 \delta(\Delta p)\right]$	Representation
Electromagnetism	$iq_e \gamma^\mu$	

where the index μ relates to the photon coming out of it, and the sum is done in the usual way.

3 Propagators.

Propagator	Contribution	Representation
Electrons and positrons	$i \int \dfrac{d^4 q_i}{(2\pi)^4} \dfrac{(\slashed{q}_i + m_e \mathbf{I})}{q_i^2 - m_e^2}$	$q_i \rightarrow$
Photons	$-i \int \dfrac{d^4 q_i}{(2\pi)^4} \dfrac{g_{\mu\nu}}{q_i^2}$	$q_i \rightarrow$

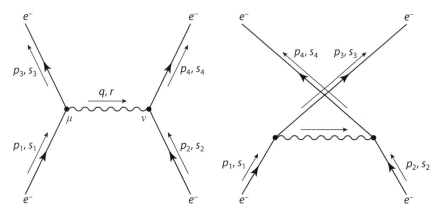

Figure 7.14. The t-channel (left) and u-channel (right) diagrams for $e^- - e^-$ interactions.

These factors look fairly complex, but an easy way to keep all the terms contracted with their appropriate partners is to follow each fermionic line backward and write the terms from left to right. Doing so will produce scalar combinations of the form

$$[\overline{u}\gamma^{\mu}u] .$$

4 Cancel the Delta Function. We remove the factor of $-i(2\pi)^4\delta(\sum p)$, and we're left with $\mathcal{A}$.

5 Antisymmetrization. We include a minus sign between diagrams that differ only in the interchange of two incoming or outgoing electrons or positrons, or of an incoming electron with an outgoing positron (or vice-versa).

7.4.2 Electron-Electron Interactions at Low Energy

The Feynman rules are relatively easy to write but somewhat more complicated to apply. To give a taste of both the power and the complications for even simple QED calculations, we look at the most natural interaction, between two electrons. In QED, electron-electron interactions are known as **Møller scattering**. The preliminary step for any QED calculation requires us to draw and label the relevant diagrams, which we do in Figure 7.14.

We then follow the Feynman rules in the t-channel.

1 External Lines. There are two incoming electrons and two outgoing ones. Writing out their contribution gives

$$\overline{u}(3)u(1)\overline{u}(4)u(2),$$

where we have used a shorthand of the form

$$u(1) = u_{s_1}(p_1)$$

to simplify further discussions.

2 Vertex Factors. There are two vertices, which we'll label as μ and ν, respectively. Inserting the γ-matrices between the bispinor terms (as required) yields

$$-q_e^2 (2\pi)^8 \overline{u}(3)\gamma^\mu u(1)\overline{u}(4)\gamma^\nu u(2)\delta(p_1 - q - p_3)\delta(p_2 + q - p_4).$$

3 Propagators. There is only one propagator, the internal photon, over which we'll integrate.

$$iq_e^2 (2\pi)^4 \int d^4 q \frac{g_{\mu\nu}}{q \cdot q} \overline{u}(3)\gamma^\mu u(1)\overline{u}(4)\gamma^\nu u(2)\delta(p_1 - q - p_3)\delta(p_2 + q - p_4)$$

$$= iq_e^2 (2\pi)^4 \frac{g_{\mu\nu}}{(p_1 - p_3)^2} [\overline{u}(3)\gamma^\mu u(1)][\overline{u}(4)\gamma^\nu u(2)]\delta(p_1 + p_2 - p_3 - p_4),$$

where we've taken $q = p_1 - p_3$.

4 Cancel the Delta Functions. Thus, we get an amplitude for the diagram of

$$A_t = -q_e^2 \frac{g_{\mu\nu}}{(p_1 - p_3)^2} [\overline{u}(3)\gamma^\mu u(1)][\overline{u}(4)\gamma^\nu u(2)]. \tag{7.35}$$

The amplitude of the u-channel may be found similarly, in which the labels of the outgoing electrons are switched:

$$A_u = q_e^2 \frac{g_{\mu\nu}}{(p_1 - p_4)^2} [\overline{u}(4)\gamma^\mu u(1)][\overline{u}(3)\gamma^\nu u(2)]. \tag{7.36}$$

Note the difference in sign between the two components due to the fermionic swap between the two diagrams (Feynman rule 5).

The t-channel is useful for another reason—it can be used to calculate the Yukawa potential directly. Recall that the scattering amplitude is intimately related to the interaction energy, such that

$$A = \int \langle f| \hat{H}_{\text{Int}}(t')|i\rangle dt', \tag{7.37}$$

which is simply the quantity computed in the penultimate step in the Feynman rules. For a stationary (or highly nonrelativistic) electron $[\overline{u}(3)\gamma^0 u(1)] = 2m$, a positive constant. This interaction energy is in momentum space,

$$V(q) = A = -\frac{q_e^2}{q \cdot q}$$

where $q = p_1 - p_3$. Since the initial and final states of the electrons are unchanged, to lowest order, $q_0 = 0$, so $q \cdot q = -|\vec{q}|^2$.

The physical space potential (the Yukawa potential) may be computed via the Fourier transform (following equation 6.4),

$$V(\vec{r}) = q_e^2 \int \frac{d^3 k}{(2\pi)^3} \frac{1}{|\vec{k}|^2} e^{i\vec{k}\cdot\vec{x}} = \frac{q_e^2}{4\pi r}, \tag{7.38}$$

as expected. We now have a generalized approach for estimating the semiclassical interaction between two particles: simply compute the Feynman amplitude and take the Fourier transform.

7.4.3 Casimir's Trick

We already knew the low-energy solution for electron-electron interaction, but QED allows us to examine scatters at higher energy. Consider two beams in the center-of-mass frame with momentum p_0. We've computed the kinematic portion of the equal-mass collisional cross section in the center-of-mass frame before (equation 7.21):

$$\frac{d\sigma}{d\Omega} = \frac{|\mathcal{A}|^2}{2(16\pi)^2 E_0^2}.$$

The amplitudes for Møller scattering (equations 7.35, 7.36) look straightforward enough, but the contractions over the spinors are going to present some problems.

Nothing prevents us from computing the amplitude for two spin-up electrons going in and two spin-up electrons going out, or any other particular combination. We simply plug in the various values for the bispinors and be done with it. Indeed, this is exactly what we asked you to do in problem 5.1. However, this approach is not only tedious, it produces somewhat horrific algebra. It's also fairly unnecessary.

In most experimental conditions, the incoming beams of electrons will be oriented randomly, and outgoing spins won't be measured. In other words, we'd like some sort of *average* over $\langle |\mathcal{A}|^2 \rangle$. To do this, we need to note the troublesome terms in our calculation. Multiplying out the amplitude squared gives

$$|\mathcal{A}_t|^2 = q_e^4 \frac{\left[\overline{u}(3)\gamma^\mu u(1)\right]\left[\overline{u}(3)\gamma^\nu u(1)\right]^* \left[\overline{u}(4)\gamma_\mu u(2)\right]\left[\overline{u}(4)\gamma_\nu u(2)\right]^*}{\left[(p_1 - p_3)\cdot(p_1 - p_3)\right]^2} + 3 \text{ similar terms.} \tag{7.39}$$

This is, to put it bluntly, a nightmare.

Fortunately, the invariance properties of the γ-matrices can lend a hand. "Casimir's trick" as David Griffiths dubbed it [80], involves summing over all input and output spins and produces the odd-looking result

$$\sum_{\text{all spins}} \left[\overline{u}(a)\gamma^\mu u(b)\right]\left[\overline{u}(b)\gamma^\nu u(a)\right]^* = Tr\left[\gamma^\mu(\not{p}_b + m_b\mathbf{I})\gamma^\nu(\not{p}_a + m_a\mathbf{I})\right], \tag{7.40}$$

where $Tr(\mathbf{M})$ takes the trace (diagonal sum) of a vector.[9] As a reminder, the slashed terms are a contraction over a vector and the γ-matrices,

$$\not{p} = p_\mu \gamma^\mu,$$

which is represented by a 4×4 matrix. On the face of it, Casimir's trick hasn't gotten us much. However, spinors have been eliminated entirely. All that remains are various multiplications and contractions over γ-matrices given in Appendix A.

Expanding the first term in the square of the amplitude (equation 7.39) yields

$$\langle |\mathcal{A}|^2 \rangle = \frac{q_e^4}{4} \frac{Tr\left[\gamma^\mu(\not{p}_3 + m_e\mathbf{I})\gamma^\nu(\not{p}_1 + m_e\mathbf{I})\right] Tr\left[\gamma_\mu(\not{p}_4 + m_e\mathbf{I})\gamma_\nu(\not{p}_2 + m_e\mathbf{I})\right]}{(p_1 - p_3)^4}.$$

---

[9] We won't *prove* Casimir's trick, but suffice it to say that it's a straightforward, but tedious, application of the orthogonality properties of the spinors.

Even this "simplification" looks grotesque. The only comfort is that the first trace and the second are presumably identical by the symmetry of the scatter.

We may distribute the multiplicative terms and get, for instance,

$$Tr[\gamma^\mu \not{p}_3 \gamma^\nu \not{p}_1] = p_{3,\alpha} p_{1,\beta} Tr[\gamma^\mu \gamma^\alpha \gamma^\nu \gamma^\beta]$$
$$= 4 p_{3,\alpha} p_{1,\beta} \left[g^{\mu\alpha} g^{\nu\beta} - g^{\mu\nu} g^{\alpha\beta} + g^{\mu\beta} g^{\alpha\nu} \right]$$
$$= 4 \left[p_3^\mu p_1^\nu - g^{\mu\nu} p_1 \cdot p_3 + p_3^\nu p_1^\mu \right],$$

in which some terms are simpler than others. For instance,

$$Tr(\gamma^\mu \not{p}_3 \gamma^\nu m_e) = 0,$$

since it involves three γ-matrices. Recognizing that

$$p_1 \cdot p_3 = p_2 \cdot p_4 = E_0^2 - p_0^2 \cos\theta,$$

we may write:

$$\langle |\mathcal{A}_t|^2 \rangle = \frac{q_e^4}{4} \frac{8(p_1 \cdot p_3)^2 + 8(p_1 \cdot p_4)^2 + 16m^2(p_1 \cdot p_3) + 16m^4}{-(p_1 - p_3)^4},$$

with similarly complex terms for the uu and tu components. Writing it all out we obtain the relation

$$\frac{d\sigma}{d\Omega} = \left(\frac{q_e^2}{4\pi} \right)^2 \frac{1}{E_0^2} \left(\frac{(4p_0^2 + 2m_e^2)^2}{4p_0^4 \sin^4\theta} - \frac{8p_0^4 + 12m_e^2 p_0^2 + 3m_e^4}{4p_0^4 \sin^2\theta} + \frac{1}{4} \right), \tag{7.41}$$

which we've plotted in Figure 7.15 along with the classical result.

The term in parentheses, as we've noted, is the fine-structure constant

$$\alpha_e = \frac{q_e^2}{4\pi}. \tag{7.42}$$

We're in the process of describing perhaps *the* most classical electromagnetic interaction, so it's not surprising that the fine-structure constant shows up front and center.

Møller scattering (unlike the classical Rutherford result) produces a maximum both at $\theta = 0$ (low scatter) and $\theta = \pi$, full reflection. This strange result is nothing more than a reminder that all electrons are identical. The scattering process makes no distinction between whether the outgoing electron is the "same" as the incoming one, or whether it's the "other" incoming electron.

7.4.4 Electron-Positron Annihilation

We don't want to overwhelm you with too many examples, but it may be prudent to see at least one more calculation, one which has no classical analog: electron-positron annihilation:

$$e^- + e^+ \to \gamma + \gamma.$$

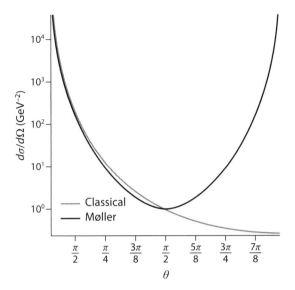

Figure 7.15. A comparison of the classical and quantum cross section for Møller scattering.

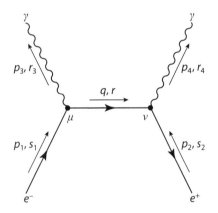

Figure 7.16. The t-channel for the electron-positron annihilation. The u-channel is produced by a twisting of the outgoing photons.

The t-channel for this process is shown in Figure 7.16, and the u-channel can be constructed similarly.

A few observations are in order. First, the down-going line p_2 corresponds to a positron with positive momentum and energy. Thus, at the right vertex, $p_2 + q = p_4$. Second, while there are two identical outgoing particles ($S = 1/2$), they are bosons, which means that the two amplitudes are added rather than subtracted. Third, since the energy of the outgoing photon is conserved, the t-term is

$$(p_1 - p_3)^2 = -m_e^2 \,,$$

which simplifies things a lot.

We take the electron-positron pair initially nearly at rest. Following the symmetric scattering relation (equation 7.22), we may write the cross section as

$$\frac{d\sigma}{d\Omega} = |\mathcal{A}|^2 \frac{1}{2(16\pi)^2 m_e \, p_0},$$

where for the remainder of the calculation, we may take $p_0 \ll m_e$. Following the Feynman rules, we may evaluate the amplitude,

$$\mathcal{A}_t = q_e^2 \frac{\bar{v}(2)\not{\epsilon}(4)(\not{q}_i + m_e)\not{\epsilon}(3)u(1)}{-2m_e^2}.$$

The expression is similar for the u-channel, but with $3 \leftrightarrow 4$.

There is a crucial difference between this example and the previous one. We need to be much more cautious about our spin states. For concreteness, we may consider the case of the spin-0 state, which is an antisymmetrized superposition:

$$|i\rangle = \frac{1}{\sqrt{2}} \left(u_+ v_- - u_- v_+ \right).$$

The energy of the outgoing photons is easily determined, $E = m_e$, 511 keV, but the direction is unknown. However, from first principles, this particular problem is isotropic. The electron-positron pair carries no momentum or net spin, which means that there is no preferred direction in the system.

Considering a photon pair emitted in the $\pm z$-direction, we get

$$\varepsilon(3, 4) = \begin{pmatrix} 0 \\ \frac{1}{\sqrt{2}} \\ \frac{1}{\sqrt{2}} \\ 0 \end{pmatrix} \; ; \; \begin{pmatrix} 0 \\ \frac{1}{\sqrt{2}} \\ -\frac{1}{\sqrt{2}} \\ 0 \end{pmatrix}.$$

Since the net angular momentum is zero, the angular momentum of the photons will likewise vanish. From here, we may compute the differential cross-section:

$$\frac{d\sigma}{d\Omega} = \left(\frac{q_e^2}{4\pi} \right)^2 \frac{1}{2m_e \, p_0}.$$

This expressions may simply be integrated to yield a total annihilation cross section of

$$\sigma = 2\pi \frac{\alpha_e^2}{m_e \, p_0}. \tag{7.43}$$

This result seems almost anticlimactic. The energy scaling relation came from the kinematic calculation of the cross section, and the contribution from second-order diagrams must produce a factor of α_e^2.

Problems

7.1 A particle has two decay channels in timescales τ_1 and τ_2. What fraction decays into channel 1? This result is known as the **branching ratio**. What's the total timescale?

7.2 Theories can be falsified by looking for a particular decay channel. If N particles are observed for a time T, what is the probability that a true decay time τ will produce no events within that time?

   If we set a threshold of 0.05 to falsify the theory, how long does the experiment need to be run to rule out the theory?

7.3 We have frequently exploited the connection between the delta function and finite volume integrals in three- and four-dimensions. Perform the integral over a finite length

$$f(k) = \int_{-L}^{L} dx\, e^{ikx}$$

explicitly, and show by integrating over all values of k that the integral converges to a delta function for large values of L. Show that for $k = 0$, the delta-function may be directly related to a total length.

7.4 Compute the integral

$$\int dx\, \delta\left(2x + \sqrt{x^2 + 1}\right).$$

7.5 A stationary particle decays into three massless, identical daughter particles. The scattering amplitude $\mathcal{A}$ for this particular process is constant as a function of energy and angle of emission. What is the decay rate?

7.6 Show that

$$v E_1 E_2 = \sqrt{(p_1 \cdot p_2)^2 - (m_1 m_2)^2}\,,$$

where p_1 and p_2 are the 4-momenta of two particles in any frame, E_1 and E_2 are the energies of the particles, and v is the relative speed between them.

7.7 Consider a strange scattering process in the toy scalar theory

$$\phi + \phi \to \eta.$$

where for concreteness, $m_\phi = 0$, and $m_\eta > 0$. It is also recommended that you do this calculation in the center-of-mass frame.
(a) Draw the lowest-order Feynman diagram describing the process.
(b) Compute the total cross section. What are the implications of a 2-1 scattering process?

7.8 Consider the $\phi - \eta$ scalar model for which ϕ is a massive particle (and subsequently, its own antiparticle), and η is massless. Consider the reaction

$$\phi + \phi \to \eta + \eta,$$

designed to strongly resemble an analog from QED.
(a) Draw all two-vertex Feynman diagrams, labeling (if possible), s-channel, t-channel, or u-channel. Be sure to draw all momentum arrows.
(b) Calculate the differential cross section $d\sigma/d\Omega$ in as reduced a form as possible while leaving p_0 (the momenta of the incoming ϕ's) unconstrained. Do not compute the amplitude.
(c) Compute the amplitudes (separately) for all diagrams.
(d) Compute the differential cross section.
(e) In the limit of $p_0 \ll m_\phi$, compute the total cross section for annihilation.

7.9 To understand the power of Casimir's trick (and convince you of its validity), we will compute the contribution for a scattered electron

$$\sum_{\text{all spins}} [\bar{u}(1)\gamma^\mu u(3)]^* [\bar{3}\gamma^\nu u(1)]$$

for $u(1)$ corresponding to an electron moving with a momentum p in the $+z$-direction, and $u(3)$ is an electron moving with an equal and opposite momentum. The spins are unspecified. The sum will be a rank-2 tensor.

(a) Compute the sum over spins using equation (7.40) and the various spin relationships for the given spin states

(b) Compute the set of vectors $[\bar{u}(3)\gamma^\mu u(1)]$ for all possible spin combinations: up-up, up-down, down-up, and down-down.

(c) Using the previous results, compute the explicit sum of spinor contractions over all possible spins. Compare the result with your answer in part (a).

7.10 Consider a scatter between an electron and a photon in the center-of-mass frame:

$$e^- + \gamma \rightarrow e^- + \gamma.$$

This is known as Compton scattering.

(a) Draw all lowest-order Feynman diagrams.

(b) Compute the amplitudes of the diagrams, simplifying as much as possible. Do not assume anything about the spin states.

7.11 Show that

$$(\not{p}_1 + \not{p}_2 + m)\not{\epsilon}_2 u(1) = -\not{\epsilon}_2 \not{p}_2 u(1)$$

if particle 2 is assumed to be a photon with polarization ϵ_2, and particle 1 is stationary. The results from problem 5.2 may be helpful.

7.12 Consider a scatter between an electron and positron:

$$e^+ + e^- \rightarrow e^+ + e^-.$$

This is known in QED as **Bhabha scattering**.

(a) Draw the lowest-order Feynman diagrams describing this process. There should be two of them.

(b) In the center-of-mass frame, use the Feynman rules to compute the scattering amplitudes for the two diagrams. Do not simplify to the point of assuming anything about spin.

(c) Compute the following quantities in the center-of-mass frame:

    i. $p_1 \cdot p_2$

    ii. $p_1 \cdot p_3$

    iii. $p_1 \cdot p_4$

    iv. $s = (p_1 + p_2)^2$

    v. $t = (p_3 - p_1)^2$

You should label the incoming (and outgoing) energies as E, the incoming and outgoing momenta as p, and the angle of deflection θ.

(d) Compute the spin-averaged amplitude squared for Bhabha scattering, $|\mathcal{A}|^2$, for the tt channel only. Simplify until only 4-vectors and masses appear in your result.

(e) Approximate the differential cross section under the assumption that the incoming momentum is small compared with the electron mass. Again, use only the $\mathcal{A}_t$ component of the amplitude. Compare this with the classical result.

7.13 Consider electron-positron pair production in the early universe:

$$\gamma + \gamma \rightarrow e^- + e^+.$$

(a) Draw all two-vertex Feynman diagrams describing the process.
(b) Compute the scattering amplitude(s) for these diagrams. Simplify as much as possible without assuming anything about the polarization or spin states. You may find the results from problem 7.11 useful.
(c) Evaluate the expected amplitude squared $\langle |\mathcal{A}|^2 \rangle$ using Casimir's trick in the limit in which the outgoing particle momentum is much less than the mass of the electron.
(d) Compute the differential scattering cross section, to first order in p_F.

Further Readings

- Feynman, Richard. *QED: The Strange Theory of Light and Matter.* Princeton, NJ: Princeton University Press, 1985. While this was intended as a popular science book, it presents an excellent semiconceptual, semimathematical introduction to the interested novice.
- Griffiths, David. *Introduction to Elementary Particles*, 2nd ed. Hoboken, NJ: Wiley-VCH, 2008. Interested readers will find chapters 6 and 7, on the Feynman calculus and QED, very readable but still rigorous enough to prove satisfying.
- Gross, Franz. *Relativistic Quantum Mechanics and Field Theory.* Hoboken, NJ: Wiley-VCH, 1993.
- Peskin, Michael E., and Dan V. Schroeder. *An Introduction to Quantum Field Theory.* Boulder, CO: Westview Press, 1995. One of the classics in quantum field theory.
- Tong, David. "Lectures on Quantum Field Theory."
  http://www.damtp.cam.ac.uk/user/tong/qft.html. Cambridge University, 2006.

8 | The Weak Interaction

Figure 8.1. Chien-Shiung Wu (1912–1997) at Columbia University (1958). Wu and her team found parity violations in the weak interaction, paving the way to electroweak unification. Credit: Acc. 90–105 – Science Service, Records, 1920s–1970s, Smithsonian Institution Archives.

The Standard Model is so much more than a list of particles and interactions, esthetically arranged. Rather, it's a process by which those particles and interactions are shown to be intricately linked to one another through a series of symmetries.

Pedagogically, we presupposed the existence of an electron, treating it, in effect, as "most fundamental." In truth, it's really just most *familiar*. Electrons are charged, plentiful, and extremely stable. Our particle zoo then doubled through Dirac's work and the prediction of the positron. The photon was the first big surprise, as it arose from local U(1) gauge invariance in the underlying Lagrangian, and the development of the electromagnetic interaction.

Thus far, more than halfway through this book, we have predicted only a tiny fraction of known particles. But we've developed a method for doing so, one that will yield, in short order, all the fundamental particles, their interaction Lagrangians, and no leftovers.

We started with the familiar just so we would be on a solid foundation when we started into unfamiliar territory and less pedestrian particles. Few have proven as wily and confounding as the neutrino.

Table 8.1. The Three Generations of Leptons, with the Corresponding Mass of the Charged Leptons.

l^-	ν_l	m_l (MeV)
e^-	ν_e	0.511
μ^-	ν_μ	105.7
τ^-	ν_τ	1776.8

Note: As of this writing, the masses of the neutrinos are not known, but the existence of mass (as well as their mass differences) is constrained by neutrino oscillation. We shall discuss this topic further in Chapter 10.

8.1 Leptons

8.1.1 Neutrinos

Though neutrinos are painfully difficult to detect, the energetics of neutron decay give us some key insights into their properties:

$$n \rightarrow p + e^- + \bar{\nu}_e. \tag{8.1}$$

Conservation of charge guarantees that the unseen particle is electrically neutral, and conservation of spin requires that it be a fermion. If not massless, the antineutrino must be very, very light. Finally, the staggering length of the decay time demands that the mediating interaction be very, very weak [163]. Indeed, neutrinos are an almost omnipresent indicator of any weak process.

Neutrinos are part of a family of fermions known as **leptons**,[1] which include both the neutrinos and a trio corresponding of charged particles: the electron, the muon, and the tau particle (Table 8.1). The antileptons have the same mass and opposite charge to that of their leptonic counterparts.

The most obvious feature of the lepton seen in the table is that they come in three generations or **flavors** of successively greater mass. Or, rather, the *charged* lepton has successively greater mass. While we're relatively certain that at least some flavors of neutrino have mass, we don't yet know how much.

This multigenerational structure is puzzling. Isidor Rabi, when informed about the discovery of the muon in 1936 famously complained, "Who ordered *that*?" [138]. For the moment, we'll make note of the three generations of leptons and quarks, and deal with the fallout later.

Neutrinos, in all their guises, are among the most common particles in the universe, with a mean density of approximately 400 million primordial particles per cubic meter [56,130]. And yet, neutrinos are nearly undetectable; a few trillion pass through

---

[1] Contra quarks, which have their own branch of the family tree.

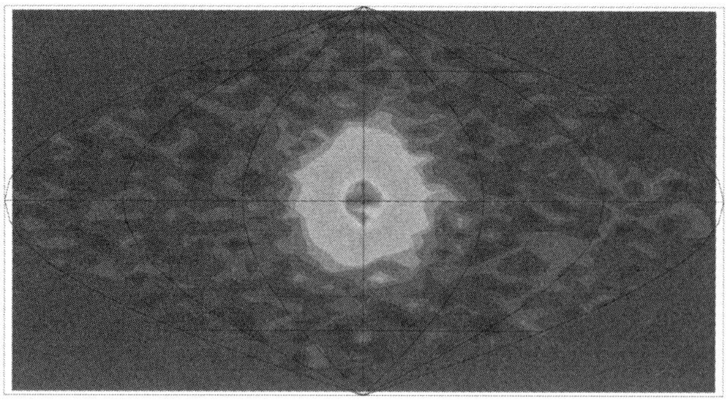

Figure 8.2. The sun, as seen in neutrinos. Credit: Kamioka Observatory, ICRR (Institute for Cosmic Ray Research), The University of Tokyo.

your thumbnail unimpeded every second. They are ubiquitous, produced in the atmosphere [10, 137], in nuclear reactors, and, significantly, in the sun (Figure 8.2), which we'll consider as a worked example.

The fusion of hydrogen into helium involves a number of weak reactions, the net result of which takes the form

$$4H \rightarrow {}^4He + 2e^+ + 2\nu_e,$$

first suggested by Hans Bethe [29] less than a decade after neutrinos were proposed, long before they were discovered experimentally. His foresight was rewarded with the 1967 Nobel Prize in Physics.

Fusion produces roughly as many neutrinos (depending on the cycle) as it does photons, and with energies near the mega-electron-volt range. In 1964 Ray Davis and John Bahcall proposed a neutrino detector in the Homestake Mine in South Dakota [23, 24, 42]. Despite a similar luminosity in photons and neutrinos, and a 380 m^3 tank, only a few neutrinos were detected per day. The weak interaction is *very* weak.

8.1.2 The Fermi Model of the Weak Interaction

All fermions participate in the weak interaction, but many weak interactions have a strong or, more commonly, electromagnetic version which dwarfs and thus masks the effect. Electrons repel one another weakly, just as they do electrically, but unless the impact parameter is on the scale of an atomic nucleus, the contribution of the weak interaction is effectively zero for any realistic system.

However, processes involving neutrinos have no electromagnetic pathway, as is the case for neutron decay (equation 8.1). While the ultimate description of the weak interaction will involve a mediator, as a first stab Enrico Fermi [163] proposed an interaction of the form

$$\mathcal{L} = 2G_F(\overline{\psi}_p \gamma^\lambda \psi_n)(\overline{\psi}_e \gamma_\lambda \psi_\nu), \tag{8.2}$$

where the subscript for each spinor refers to the particle type, and where G_F is known as the **Fermi constant**.

We *know* that this Lagrangian is incorrect in a number of ways. For one thing, protons and neutrons are composite particles. A more accurate calculation would involve a down quark decaying into an up quark, and as there are two of the former in a neutron and two of the latter in a proton, we would need to compute the total amplitude of *either* down decaying into an up. Likewise, as we've alluded, the weak interaction violates P symmetry, an effect not incorporated into the Lagrangian. Most significantly, though, the Fermi theory doesn't incorporate a mediator particle.

As this discussion is meant to simply give a taste of how the weak interaction works, we won't concern ourselves overly with constants of order unity. Rather, we'll focus on energy scales. Dimensionally, the Fermi constant must have units of $[E]^{-2}$, since Dirac spinors have units of $[E]^{3/2}$, and the Lagrangian density of $[E]^4$. The value of the constant is approximately [119]

$$G_F = (292.8 \, \text{GeV})^{-2}. \tag{8.3}$$

These hundred-giga-electron-volt energy scales serve both as a guideline for the masses of the as-yet-underdeveloped mediators for the weak interaction and the Higgs boson, as well as the scale on which the electromagnetic and weak interactions were unified.

Example 8.1: What is the decay rate of a stationary neutron in the Fermi model?

Solution: We explored the kinematics of a three-body decay with two light daughters in the previous chapter (equation 7.17):

$$\Gamma = \iint \frac{|\mathcal{A}|^2}{(4\pi)^3} \frac{\Delta E - E_e}{m_n^2} \sqrt{E_e^2 - m_e^2} \, d E_e d(\cos \theta),$$

where E_e is the energy of the electron, and θ is the relative angle between the emitted electron and neutrino. We need compute only the amplitude of this process. Applying the equivalent of the Feynman rules (that is, a one-vertex calculation) for the Fermi model, we get

$$\mathcal{A} = 2G_F[\overline{u}(p)\gamma^\mu u(n)][\overline{u}(e)\gamma_\mu v(v)]. \tag{8.4}$$

While we are free to consider each individual spin state, it's far easier to average over all spin states, by using Casimir's trick (equation 7.39):

$$\sum_{\text{All spins}} |\mathcal{A}|^2 = 4G_F^2[\overline{u}(p)\gamma^\mu u(n)][\overline{u}(n)\gamma^\nu u(p)]^*[\overline{u}(e)\gamma_\mu v(v)][\overline{v}(v)\gamma_\nu u(e)]^*.$$

The amplitude becomes much simplified because the proton will very nearly be at rest, owing to the relatively small amount of energy liberated in the decay. As the neutron may be in either spin state (doubling the rate), the overall amplitude is

$$\langle|\mathcal{A}|^2\rangle = 32G_F^2 m_n m_p E_v \left[E_e + \sqrt{E_e^2 - m_e^2} \cos \theta \right],$$

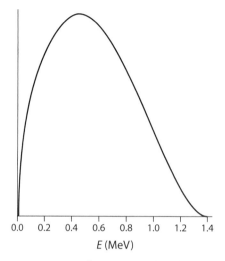

Figure 8.3. The energy spectrum for Fermi's model of beta decay. This has the same functional form (but a few corrections in overall rate) as realistic beta decays. See for example, Figure 7.4.

which, after integration over $\cos\theta$ gives a differential decay rate of

$$\frac{d\Gamma}{d E_e} = \frac{2 G_F^2}{\pi^3} (\Delta E - E_e)^2 E_e \sqrt{E_e^2 - m_e^2}.$$

Our form produces the energy spectrum in Figure 8.3 (to be compared with actual beta decay spectra, as in Figure 7.4). Integrating over all possible energies (from m_e to ΔE) and converting to physical units yields a characteristic decay time of

$$\tau = \frac{1}{\Gamma} \simeq 1300\,\text{s},$$

which, given the grossly unphysical assumptions that we made about the nature of protons, neutrons, and a lack of a mediator, is comfortably close to the true neutron decay time of 882 s (Table 7.1).

8.2 Massive Mediators

If our framework for gauge symmetries is worth anything, there must be a mediator for the weak interaction, contrary to the Fermi model. Indeed, there are *three*: the W^+ and the W^- (which are antiparticles of one another) and the neutral[2] Z^0 (which, like the photon, is its own antiparticle).[3] The masses of the weak mediators are *huge* [28]:

- $M_W = 80.385 \pm 0.015\,\text{GeV}$
- $M_Z = 91.1876 \pm 0.0021\,\text{GeV}$

---

[2] At high energies, **neutral current interactions** mediated by the Z^0 become important, especially in elastic scattering between neutrinos. We will not focus on neutral currents at this time.

[3] Unlike other mediators which are given names like "photon" or "gluon," the weak mediators are simply called "W particles," and "Z particles." Steven Weinberg named them [156] for Weak and Zero charge, respectively.

For interactions significantly less energetic than the mediator masses, the virtual mediators have a very short range,

$$\frac{1}{M_W} \simeq 2.5 \times 10^{-18} \text{ m},$$

well confined to the interior of an atomic nucleus.[4] For massive mediators, the propagator
is somewhat different than for the photon, taking the form

$$-i\frac{\left(g_{\mu\nu} - \frac{q_\mu q_\nu}{M_W^2}\right)}{q \cdot q - M_W^2} \simeq i\frac{g_{\mu\nu}}{M_W^2} \tag{8.5}$$

at low-energy conditions.

We'll hold off a bit on developing the weak Lagrangian from gauge symmetry relations,
but we may make several assumptions now:

1. Electrical charge will be conserved for every interaction and thus at the corresponding
   Feynman vertex.
2. As spin will be conserved, there must be two fermions at each Feynman vertex.
3. Experimentally, neutrinos and electrons seem coupled (conservation of **lepton number**),
   and protons and neutrons seem coupled (conservation of baryon number).

Based on nothing more than these guesses we might suppose that weak interactions take
the form

$$\mathcal{L}_{\text{Int}} \sim g_W W_\mu^- \bar{e} \gamma^\mu \nu_e, \tag{8.6}$$

with additional terms for proton-neutron (or, more precisely, up quark–down quark)
coupling and various complex conjugates. We haven't *derived* this interaction Lagrangian
but simply guessed at it from analogy. From these interactions we may construct a
somewhat more sophisticated neutrino decay model, the lowest order of which is shown
in Figure 8.4.

Application of the Feynman rules reveals a simple scaling relation for this diagram at
low energies:

$$\mathcal{A} \sim G_F = \frac{\sqrt{2}}{8}\frac{g_W^2}{M_W^2}.$$

Solving for g_W from the experimentally obtained values of G_F and M_W produces a weak
coupling constant:

$$g_W \simeq 0.6530. \tag{8.7}$$

This value can be put into context and compared to electromagnetism "on equal footing"
by computing the weak fine-structure constant,

$$\alpha_W = \frac{g_W^2}{4\pi} \simeq \frac{1}{29.473}, \tag{8.8}$$

---

[4] This is why we almost always refer to it as the weak *interaction*, rather than the weak *force*. It manifests
itself almost exclusively in changes in particle identity.

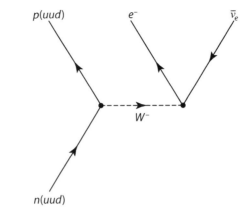

Figure 8.4. A simplified model of neutron decay employing a massive mediator (but ignoring a few other complications).

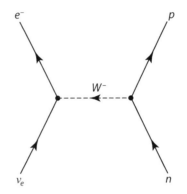

Figure 8.5. Neutrino scattering off a neutron.

to be contrasted with $\alpha_e \simeq 1/137$. Were it not for the massive mediators, the weak interaction would be *stronger* than electromagnetism. In practice, however, the weak interaction lives up to its name.

Example 8.2: As a worked example of the weakness of the weak interaction, we shall consider the oft-cited example of firing a beam of neutrinos through solid lead.

Solution: We focus on the scattering process

$$\nu_e + n \rightarrow p + e^-,$$

which can be represented in a Feynman diagram (Figure 8.5).

The neutrino is assumed to be relativistic, and the neutron can be considered at rest. We further suppose that the neutrino energy is far lower than the mass of the proton.[5] Under

---

[5] Contrary to some very exciting astrophysical cases [3] in which neutrinos have been detected with energies in excess of 100 TeV.

those circumstances, the kinematic component of the cross section may be written as

$$\sigma = \frac{\Delta E + E_\nu}{32\pi m_N m_p E_\nu} \int |\mathcal{A}|^2 d(\cos\theta),$$

where $\Delta E = m_n - m_p \simeq 1.29\,\mathrm{MeV}$. We may, in computing the amplitude, once more use Casimir's trick to average over spins and obtain the much simplified

$$\langle|\mathcal{A}|^2\rangle = 2\left(\frac{g_W}{M_W}\right)^4 E_\nu m_p(\Delta E + E_\nu)m_N.$$

The total cross section for the collision is thus

$$\sigma = \frac{1}{8\pi}(\Delta E + E_\nu)^2 \left(\frac{g_W}{M_W}\right)^4, \tag{8.9}$$

valid at all energies in the range where the neutrino energy is less than the mass energy of a proton or neutron. At very low energies ($\lesssim$ MeV), the cross section in physical units becomes

$$\sigma \simeq 1.2 \times 10^{-47}\,\mathrm{m}^2, \tag{8.10}$$

above which the cross section rises quadratically with energy. As lead has a neutron density of approximately $n = 2.7 \times 10^{30}\,\mathrm{m}^{-3}$, the mean free path is

$$\lambda_{free} = \frac{1}{n\sigma} = 3.1 \times 10^{16}\,\mathrm{m}$$
$$\simeq 3.3\ \text{light-year}.$$

Neutrinos (low-energy neutrinos, at least) really *can* go through a light-year's worth of lead unimpeded.

We know (or at least can guess) that massive mediators present a problem in gauge theories. We learned in our discussion of the electromagnetic force that local gauge invariance produces (*requires*, in fact) massless mediators, and that is also true for gauge theories in general. Explaining the mechanism for giving mass to the weak mediators was a big deal and resulted in the 2013 Nobel Prize in Physics for François Englert and Robert Brout [55] and Peter Higgs [88]. But we're getting ahead of ourselves. We still need to develop the weak interaction as a gauge theory to demonstrate why the unification is necessary in the first place.

8.3 SU(2)

The gauge theory approach was such a success in electromagnetism that we'll want to try it with the weak field, but without the advantage of history. While Maxwell's equations were discovered before Weyl worked out the consequences of U(1) symmetry, for subsequent theories we need to begin with an educated guess about the appropriate symmetry group.

In 1954, Chen-Ning Yang and Robert Mills [167] developed what has come to be known as the first **Yang-Mills theory**, on the formulation we just described, but applied to non-Abelian groups. This will introduce a few kinks into the system.

8.3.1 The Fermionic Doublet

There is a straightforward recipe for developing a gauge theory:

1. Construct a globally invariant Lagrangian and find the conserved Noether current.
2. Assume a *local* gauge invariance.
3. Find the interaction Lagrangian which cancels any explicit gauge dependence.
4. Add the free-field Lagrangian for the vector mediator.

We'll try out this recipe using what will turn out to be the $SU(2)$ group. We start by postulating the existence of a doublet of bispinors:

$$\Psi = \begin{pmatrix} \psi_u \\ \psi_d \end{pmatrix}. \tag{8.11}$$

Conceptually, these two components are as similar to one another as a spin-up electron is to a spin-down. Within the context of the weak interaction, the particle doublet might be

$$\Psi = \begin{pmatrix} \nu_e \\ e^- \end{pmatrix} \quad \text{or} \quad \begin{pmatrix} u \\ d \end{pmatrix},$$

though we will focus on the lepton doublet for the time being. At first glance, the components of these doublets seem far more different from one another than do two electrons separated by a spin flip. For one thing, they have a different electric charge. But fear not, while electrons and neutrinos are clearly not symmetric under electromagnetism, for the moment, we're considering symmetry only under $SU(2)$. We'll see how the two may be combined in the next chapter.

We may also define an adjoint:

$$\overline{\Psi} \equiv (\overline{\nu}_e \quad \overline{e}).$$

Just as we will use e to represent the bispinor of an electron, $\overline{e}$ will be the adjoint of an electron, *not* the antiparticle. Knowing nothing else, we might expect the two components to evolve independently. That is, we'd expect a Lagrangian of the form

$$\mathcal{L} = \overline{\nu}_e[i\gamma^\mu \partial_\mu - m_{\nu_e}]\nu_e + \overline{e}[i\gamma^\mu \partial_\mu - m_e]e.$$

We've already hit a snag. Whatever Lagrangian we come up with must be invariant under exchange of electrons for neutrinos, and vice-versa, but this is manifestly *not* true unless they have the same mass—ideally zero. The observed masses of electrons and neutrinos are quite different. Indeed, until 1998 [70], experimental evidence pointed to neutrinos having no mass at all. Thus, the existence of particle masses should be thought of as a problem to be dealt with, and is one that we'll address once we dive headfirst into electroweak unification.

Under the (clearly incorrect) assumption that all fermions are massless, the free-field doublet Lagrangian may be written as

$$\mathcal{L} = i\overline{\Psi}\gamma^\mu \partial_\mu \Psi. \tag{8.12}$$

We may now explore the implications of demanding a symmetry.

8.3.2 The Conserved Current

Having guessed at an invariant Lagrangian, we then apply a global gauge transformation to the fermionic doublet of the form

$$\Psi \rightarrow e^{-ig w\vec{\theta}\cdot\vec{X}}\Psi; \tag{8.13}$$

$$\overline{\Psi} \rightarrow \overline{\Psi}e^{ig w\vec{\theta}\cdot\vec{X}}.$$

As a word of caution, since we're dealing with row-and-column vectors, and 2×2 operators, the ordering of the doublet, the conjugate, and the operator clearly matters. Further, transformations must leave

$$\overline{\Psi}\Psi = \overline{\nu}_e \nu_e + \overline{e}e$$

invariant, justifying our requirement that both components of the doublet have the same mass.

The notation is a bit tricky, since we combine terms in two different ways: as the components of our doublets, and as the components of each individual bispinor. This represents a conservation of **lepton number** (and specifically **electron number**). Note further that because of this notation and the normalization of the antiparticle modes, antiparticles have a lepton number of -1:

- Electron, neutrino: $L_e = 1$
- Positron, antineutrino: $L_e = -1$

Given the form, the symmetry group is simply $SU(2)$, for which the generators are simply the Pauli matrices

$$\sigma_1 = \begin{pmatrix} 0 & 1 \\ 1 & 0 \end{pmatrix}; \quad \sigma_2 = \begin{pmatrix} 0 & -i \\ i & 0 \end{pmatrix}; \quad \sigma_3 = \begin{pmatrix} 1 & 0 \\ 0 & -1 \end{pmatrix} \tag{8.14}$$

since

$$\sigma_i \sigma_i = \mathbf{I}; \quad \sigma_i = \sigma_i^\dagger$$

for all three Pauli matrices. This is the *same* symmetry used for describing the spin of Dirac particles, as we already noted. The relation of ν_e to e is the same as the relation of spin up to spin down, and accordingly, we refer to the quantity as **weak isospin**, normally labeled T_3. For the upper component, $T_3 = 1/2$. For the lower component, $T_3 = -1/2$.

Applying a global SU(2) transform to the doublet yields

$$\frac{\partial \Psi}{\partial \theta^i} = -i g_W \sigma_i \Psi.$$

Noether's theorem quickly yields *three* conserved currents:

$$J^{(i)\mu} = g_W \frac{\partial \mathcal{L}}{\partial(\partial_\mu \phi_a)} \frac{\partial \phi_a}{\partial \theta^i}$$

$$= g_W \overline{\Psi} \gamma^\mu \sigma_i \Psi. \tag{8.15}$$

These currents are non intuitive, so it's worth multiplying them out explicitly:

$$J^{(1)\mu} = g_W \left(\overline{\nu}_e \gamma^\mu e + \overline{e} \gamma^\mu \nu_e \right); \tag{8.16}$$

$$J^{(2)\mu} = -i g_W \left(\overline{\nu}_e \gamma^\mu e - \overline{e} \gamma^\mu \nu_e \right); \tag{8.17}$$

$$J^{(3)\mu} = g_W \left(\overline{\nu}_e \gamma^\mu \nu_e - \overline{e} \gamma^\mu e \right). \tag{8.18}$$

The first two currents represent a coupling between neutrinos and electrons. The third looks an awful lot like electromagnetic current, but with neutrinos participating somehow, and a minus sign between the electron and neutrino contributions. We'll return to the relation between $J^{(3)}$ and the electromagnetic current in the next chapter.

8.3.3 Local Gauge Invariance

Step 2 in a Yang-Mills theory involves assuming a *local* gauge invariance. The underlying idea (as with electromagnetism) is that the gauge parameters $\vec{\theta}$ can vary over spacetime:

$$\Psi \to e^{-i g_W \vec{\sigma} \cdot \vec{\theta}(x)} \Psi,$$

where we again use $\vec{\sigma}$ to represent a 3-D Hilbert space vector composed of the Pauli matrices. As with U(1), the spacetime derivative in the Lagrangian produces a troublesome term under the gauge transformation,

$$\partial_\mu \Psi \to e^{-i g_W \vec{\sigma} \cdot \vec{\theta}} \partial_\mu \Psi - i g_W (\vec{\sigma} \cdot \partial_\mu \vec{\theta}) e^{-i g_W \vec{\sigma} \cdot \vec{\theta}} \Psi,$$

and thus the Lagrangian transforms as

$$\mathcal{L} \to i \overline{\Psi} \gamma^\mu \partial_\mu \Psi + g_W \overline{\Psi} \gamma^\mu (\vec{\sigma} \cdot \partial_\mu \vec{\theta}) \Psi.$$

Following our work with the electromagnetic force, we need to get rid of the erroneous gauge term, and so the covariant derivative of the weak theory will take the form

$$\partial_\mu \to D_\mu = \partial_\mu + i g_W \vec{\sigma} \cdot \vec{W}_\mu, \tag{8.19}$$

where $\vec{W}_\mu$ is a set of three vector fields, $W_\mu^{(i)}$. Multiplying out the covariant derivative produces an interaction term in the Lagrangian of the form

$$\mathcal{L}_{\text{Int}} = -\vec{W} \cdot \vec{J}, \tag{8.20}$$

where the vector of 4-currents $\vec{J}$ is defined by equation (8.15).

We might further (incorrectly) suppose that the individual components of $\vec{W}$ transform under the gauge symmetry as

$$\vec{W}_\mu \rightarrow \vec{W}_\mu + \partial_\mu \vec{\theta}.$$

You can get a good sense of why this transformation is incorrect by expanding the transformed Lagrangian:

$$\mathcal{L}' = i\overline{\Psi}e^{ig_W\vec{\sigma}\cdot\vec{\theta}}\gamma^\mu\left[\partial_\mu + ig_W\vec{\sigma}\cdot\left(\vec{W}_\mu + \partial_\mu\vec{\theta}\right)\right]e^{-ig_W\vec{\sigma}\cdot\vec{\theta}}\Psi$$

$$= i\overline{\Psi}\gamma^\mu\partial_\mu\Psi - g_W\overline{\Psi}e^{i\vec{\sigma}\cdot\vec{\theta}}\gamma^\mu\vec{\sigma}\cdot\left(\vec{W}_\mu + \partial_\mu\vec{\theta}\right)e^{-i\vec{\sigma}\cdot\vec{\theta}}\Psi,$$

where the $\partial_\mu\vec{\theta}$ term is "swallowed" by the gauge transformation. However, there remains a problematic term in the form of the σ in front of the parentheses. Invoking the Baker-Campbell-Hausdorff relation gives

$$e^{i\vec{\sigma}\cdot\vec{\theta}}\sigma_j e^{i\vec{\sigma}\cdot\vec{\theta}} = \sigma_j + i\sum_i \theta^i[\sigma_i, \sigma_j].$$

Thus,

$$\mathcal{L}' = i\overline{\Psi}\gamma^\mu D_\mu\Psi + 2i\overline{\Psi}\gamma^\mu\sum_{i,j,k}\theta^i W_\mu^{(j)}\sigma_k\epsilon_{ijk}\Psi,$$

where we explicitly utilized the structure constants for the Pauli matrices, and where ϵ_{ijk} is the Levi-Civita tensor. To get rid of this term *entirely*, we need a slightly more complicated gauge transformation for the vector fields:

$$W_\mu^{(i)} \rightarrow W_\mu^{(i)} + \partial_\mu\theta^i - 2g_W\sum_{j,k}\epsilon_{ijk}W_\mu^{(j)}\theta^k. \tag{8.21}$$

At first glance, this latter term seems unnecessary, and even seems to introduce a new complication; however, it *is* gauge invariant. What's more, we can (following electromagnetism) quickly find the weak field Lagrangian as

$$\mathcal{L}_W = -\sum_i \frac{1}{4}F_{\mu\nu}^{(i)}F^{\mu\nu(i)} \tag{8.22}$$

but with the crucial difference

$$F_{\mu\nu}^{(i)} = D_\mu W_\nu^{(i)} - D_\nu W_\mu^{(i)}$$

$$= \partial_\mu W_\nu^{(i)} - \partial_\nu W_\mu^{(i)} + 2g_W\sum_{jk}\epsilon_{jkl}W_\mu^{(k)}W_\nu^{(l)}. \tag{8.23}$$

This doesn't look too bad until you realize that the free Lagrangian has fourth-order terms in $W^{(i)}$, which means that it's nontrivial to create a "free" solution for $W^{(i)}$ particles. Combining all this results in an SU(2) invariant Lagrangian:

$$\mathcal{L} = \overline{e}\left[i\gamma^\mu\partial_\mu\right]e + \overline{v}_e\left[i\gamma^\mu\partial_\mu\right]v_e - \vec{J}\cdot\vec{W} - \frac{1}{4}\vec{F}_{\mu\nu}\cdot\vec{F}^{\mu\nu}. \tag{8.24}$$

Notably, terms like $M^2\vec{W}^\mu\cdot\vec{W}_\mu$ are absent, in direct contradiction to the large experimental masses of the weak mediators. For now, we'll simply assert that electrons and weak mediators have mass (as do neutrinos, but a mass so slight that it functionally doesn't enter our calculations), and we'll justify those masses in due course.

8.3.4 The Charged Current

In weak isospin, as with ordinary quantum spin, operators may be grouped in linear combinations. In particular, the spin operators may be related to the so-called **ladder operators**, which simply transform a (weak iso-) spin-down state to a (weak iso-) spin-up or vice-versa:

$$J_\mu^+ = \frac{1}{2}\left(J_\mu^{(1)} + i J_\mu^{(2)}\right) = g_W \bar{\nu}_e \gamma_\mu e,$$ (8.25)

and

$$J_\mu^- = \frac{1}{2}\left(J_\mu^{(1)} - i J_\mu^{(2)}\right) = g_W \bar{e} \gamma_\mu \nu_e,$$ (8.26)

which, combined, describe a flip from an electron to a neutrino, or vice-versa. For charge to be conserved, the vector mediator must itself be charged. Fortunately,

$$W_\mu^\pm = \frac{1}{\sqrt{2}}\left(W_\mu^{(1)} \mp i W_\mu^{(2)}\right)$$ (8.27)

produces the desired particles. Inspection reveals that W^+ and W^- are simply complex conjugates (and thus antiparticles) of one another. The remaining boson $W^{(3)}$ will roughly correspond to the Z^0, but it will take a little work to figure out the exact mass and coupling-constant relation.[6]

In this form, equation (8.20) may be rewritten to produce the interaction Lagrangian,

$$\mathcal{L}_{int} = -\frac{1}{\sqrt{2}} J^+ \cdot W^+ - \frac{1}{\sqrt{2}} J^- \cdot W^- - J^{(3)} \cdot W^{(3)},$$ (8.28)

and the Faraday tensors may be similarly combined to produce

$$F_{\mu\nu}^\pm = F_{\mu\nu}^{(1)} \mp i F_{\mu\nu}^{(2)},$$

where the weak-field Faraday tensor elements include some troubling cubic and quartic terms that will make some of our calculations a bit tricky (See Appendix C for the amplitudes of vertices with three and four mediators at a vertex).

The free-field Lagrangian for $W^\pm$ and $W^{(3)}$ particles is

$$\mathcal{L}_{W, Free} = -\frac{1}{4} F_{\mu\nu}^{(3)} F^{(3)\mu\nu} - \frac{1}{2} F_{\mu\nu}^- F^{+\mu\nu},$$ (8.29)

describing a neutral vector field and two charged vector fields as antiparticles of one another.

---

[6] We will deal with the fallout of electroweak unification in the next chapter, but the important point is that while $SU(2)$ gives a lot of insight into the weak interaction, the true symmetry is $SU(2)_L \otimes U(1)$, for all of electroweak.

8.4 Helicity

Our recipe for the weak interaction is still missing two key ingredients:

1. Handedness.

   As Wu [166] demonstrated in 1957, the weak interaction is not ambidextrous, but the SU(2) Lagrangian, as written, is. In experimental weak decay reactions, **all** neutrinos are created so that they are left-handed, with antineutrinos produced as right-handed.

2. Mass.

   From simple gauge arguments, both the $W^{\pm}$ and the Z^0 particles are predicated to be massless—and they clearly aren't. We may insert those effects into our Lagrangian with terms like

   $$\mathcal{L}_{Mass} = M_W^2 W_{\mu}^+ W^{-\mu},$$

   but we cannot avoid the fact that these terms are gauge dependent, and thus the Yang-Mills theory seems not to describe nature as it stands.

As it happens, these questions are not independent; handedness is closely connected to mass.

8.4.1 Pion Decay

We say that "neutrinos are left-handed," but in fact, it's nearly impossible to directly measure the spin of a neutrino. We have to infer it from a neutrino's partners in a reaction. As neutrinos are essentially massless, they travel essentially at the speed of light. Since lepton number is conserved, electrons and other charged leptons are often produced in weak reactions as well, and *they* are nonrelativistic.

Consider a spin-up fermionic particle propagating in the $+z$-direction:

$$u_+(p) \propto \begin{pmatrix} \sqrt{\frac{m}{E+p}} \\ 0 \\ \sqrt{\frac{E+p}{m}} \\ 0 \end{pmatrix},$$

where we've omitted the normalization for clarity. In Weyl basis, the upper two components correspond to a left-handed **helicity**, while the bottom two correspond to a right-handed one. For massless particles, there is no distinction between the helicity—that is, the dot product of the spin and the momentum—and the handedness. If a massless neutrino is left-handed in one Lorentz frame, it will be left-handed in *all* Lorentz frames.

For massive particles, this is not the case. The probability of measuring a particular spin given a helicity state is simply proportional to the square of the amplitude. For a particle with a helicity of $+1/2$, the probability of a left-handed state is

$$P_L = \frac{\frac{m}{E+p}}{\frac{m}{E+p} + \frac{E+p}{m}}$$

$$= \frac{1}{2}(1-v), \tag{8.30}$$

and the right-handed probability is simply the complement.

Example 8.3: Pions are spinless composite particles made of a quark and an antiquark. The negatively charged π^- is composed of $(\bar{u}d)$, has a mass of 139.6 MeV, and can decay via the weak channels

$$\pi^- \to e^- + \bar{v}_e$$

or

$$\pi^- \to \mu^- + \bar{v}_\mu.$$

Assuming that the weak interaction is ambidextrous, what is the branching ratio of electrons versus muons?

Solution: Two-body decays depend only on the momentum of the outgoing particles and the amplitude of the decay (equation B.2). While the form of the amplitude should be the same for all generations of leptons (a principle known as **weak universality**), from kinematic considerations

$$\Gamma \propto E_l p_F^2.$$

We simply need to compute the outgoing momenta of the outgoing particles in each of the allowed leptonic decay channels. Pion decay into an electron produces an ultrarelativistic daughter, with a momentum and speed of

$$p_e = 69.8\,\text{MeV} \qquad v_e = 0.999973,$$

respectively. The muon, with a mass of $m_\mu = 105.7$ MeV, yields a momentum and velocity of:[7]

$$p_\mu = 29.8\,\text{MeV} \qquad v_\mu = 0.271,$$

respectively. If the weak force were ambidextrous, we might suppose

$$\frac{\Gamma_e}{\Gamma_\mu} \overset{?}{=} \frac{p_e^2 E_e}{p_\mu^2 E_\mu} \simeq 3.49,$$

which suggests that charged pions decay preferentially into electrons as opposed to muons.

---

[7] τ^- particles aren't produced at all in this process, as they are energetically forbidden.

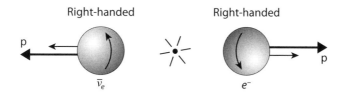

Right-handed   Right-handed

p $\bar{\nu}_e$ e^- p

Figure 8.6. The decay of a spin-0 pion should produce two left-handed particles or two right-handed particles.

Contrary to the results in example 8.3, the decay into electrons occurs on a timescale approximately 10,000 *longer* than the decay into muons. Pions are spin-0, which means that the spins of the outgoing particles need to cancel one another. If the electron is spin-up, then the antineutrino must be spin-down, but with opposing momenta. Thus, we'd expect that both should be left-handed or both right-handed (Figure 8.6).

If the weak Lagrangian is intrinsically left-handed, the interaction term might *actually* take the form

$$\mathcal{L} = \frac{1}{\sqrt{2}} g_W W_\mu^+ \bar{\nu}_{e,R} \gamma^\mu e_L,$$

where the $\bar{\nu}_{e,R}$ is a shorthand for the right-handed antineutrino adjoint spinor, and e_L is the shorthand for the left-handed electron spinor.

Starting with equation (8.30), if we produce a right-handed antineutrino, the probability of measuring a *left*-handed electron or muon is

$$p_{E,L} = 1.34 \times 10^{-5},$$

and

$$p_{\mu,L} = 0.364.$$

If only left-handed electrons or muons are allowed, then the decays are suppressed by a factor

$$\frac{\Gamma_e}{\Gamma_\mu} = 3.49 \times \frac{1.34 \times 10^{-5}}{0.364} = \frac{m_e^2 (m_\pi^2 - m_e^2)^2}{m_\mu^2 (m_\pi^2 - m_\mu^2)^2} = 1.28 \times 10^{-4}.$$

This result is very similar to the experimentally measured branching ratio, 1.23×10^{-4} [119], where the small discrepancy arises from the energy dependence of the scattering amplitude.

The weak interaction must make the distinction between left- and right- handedness and so parity by itself is *not* a symmetry of the weak force.

8.4.2 Charge and Parity

We need to redesign the interaction terms in our Lagrangian to accommodate the handedness of the weak interaction. Fortunately, we have a tool for distinguishing between left-handed and right-handed spinors: γ^5 (equation 5.33). Starting with a potentially ambidextrous spinor, we may select the left- or right-handed components via

$$\psi_{L,R} = \frac{1}{2}\left(\mathbf{I} \mp \gamma^5\right)\psi \tag{8.31}$$

Thus, if a Lagrangian is to interact with only left-handed particles, we need to replace currents appropriately. For instance, the left-handed J_μ^+ current can be written as

$$J_\mu^+ = \frac{1}{2}g_w\bar{\nu}_e\gamma^\mu(1-\gamma^5)e \tag{8.32}$$

with everything else remaining the same. The symmetry for the weak force, in other words, isn't SU(2), it's SU(2)$_L$, a symmetry acting *only* on left-handed particles and doing nothing to right-handed ones.[8] The universe distinguishes between left and right, but the combination of C and P (charge and parity inversion) does seem to be conserved—at least in everything we've seen so far.[9]

We have already seen the charge conjugation operator (equation 5.38). The parity operator is even simpler:

$$\psi_L^P = \hat{\mathbf{P}}\psi_L = \psi_R$$
$$\psi_R^P = \hat{\mathbf{P}}\psi_R = \psi_L,$$

which may be combined to produce

$$\psi_L^{CP} = -i\sigma_2\psi_R^*$$
$$\psi_R^{CP} = i\sigma_2\psi_L^*.$$

The implication is that Lagrangian interaction terms like

$$\mathcal{L}_{W_-} = -g_w\frac{1}{2\sqrt{2}}W_\mu^- \cdot [\bar{e}_R\gamma^\mu\nu_{e,L}]$$

transform to

$$\hat{\mathbf{CP}}(\mathcal{L}_{W_-}) = -g_w\frac{1}{2\sqrt{2}}W_\mu^+ \cdot [\bar{\nu}_R\gamma^\mu e_L] = \mathcal{L}_{W_+}.$$

There is a similar W^+ term, as we will see, that transforms similarly, leaving the total Lagrangian unchanged.

8.5 Feynman Rules for the Weak Interaction

While we have neglected, for the time being, neutral currents and the Z^0 boson (out of necessity), we've developed a fairly comprehensive theory of charged-current interactions

---

[8] Again, this is an approximation. SU(2)$_L$ doesn't quite describe the weak interaction by itself. Rather, the *combination* of SU(2)$_L$ and U(1) will describe both the electromagnetic and the weak interactions.

[9] Disclaimer: CP is experimentally violated in some weak interactions, but we'll get to that in due course.

in the weak interaction. The theory has an interaction Lagrangian of the form

$$\mathcal{L}_{\text{Int}} = -\frac{1}{2\sqrt{2}} g_W \left[W_\mu^+ \bar{\nu}_e \gamma^\mu (1 - \gamma^5) e + W_\mu^- \bar{e} \gamma^\mu (1 - \gamma^5) \nu_e \right], \tag{8.33}$$

along with a a pair of vector mediators.[10] We can, essentially by inspection, modify the Feynman rules to accommodate these interactions, as follows.

2 Vertex Factors. Unlike with electromagnetism, we have several possible vertices in the weak interaction, but for leptonic interactions with charged currents, the contribution to the amplitudes are the same.

Vertex	Contribution $\times \left[(2\pi)^4 \delta(\Delta p) \right]$	Representation
Charged current (leptons)	$-i \dfrac{g_W}{2\sqrt{2}} \gamma^\mu (1 - \gamma^5)$	

The possibility also exists for vertices with three and four $W^\pm$ and Z^0 bosons, but we will have no need of them in our discussions.

3 Propagators. For massive vector propagators (both the $W^\pm$, as well as Z^0) add

Propagator	Contribution	Representation
$W^\pm$	$-i \int \dfrac{d^4 q_i}{(2\pi)^4} \dfrac{\left(g_{\mu\nu} - \frac{q_{i,\mu} q_{i,\nu}}{M_W^2} \right)}{q_i^2 - M_W^2}$	
Z^0	$-i \int \dfrac{d^4 q_i}{(2\pi)^4} \dfrac{\left(g_{\mu\nu} - \frac{q_{i,\mu} q_{i,\nu}}{M_Z^2} \right)}{q_i^2 - M_Z^2}$	

In the limit where the mediator is much more massive than any other energy scales in the system, the mediator (along with the charged vertices) simply reduces to the Fermi term.

While the $SU(2)_L$ Yang-Mills theory produced some great predictions for the weak interaction, we needed to simply *assert* masses to make the whole thing work. This is untenable, as we've noted frequently. Thus, the time has finally come to develop one of the most powerful parts of the Standard Model, the theory of electroweak interactions. We will do so in the next chapter.

---

[10] The introduction of γ^5 into the vertices introduces some complications when using Casimir's trick to compute the average square amplitude over spins. The basic approach remains identical, but with the γ-matrix terms in equation (7.40) replaced with terms like $\gamma^\mu (1 - \gamma^5)$ both in the initial sum over spins and in the trace. You will find the trace theorems in Appendix A useful, and problem 8.9 puts the inclusion of handedness into practice.

Problems

8.1 Assuming that the luminosity in neutrinos is the same as that for light, and that solar neutrinos are monoenergetic, with $E \simeq 0.6\,\mathrm{MeV}$, estimate the flux of neutrinos at the surface of the earth.

8.2 For each of the following weak interactions, identify the missing particles by employing conservation of charge, lepton, and baryon number. Using online data from the Particle Data Group, (pdg.lbl.gov) determine approximately, how much energy is liberated (or consumed) from each.
(a) $p + p \to d + ?$ (where d is a deuteron)
(b) $? + n \to p + e^-$
(c) $\tau^+ \to \nu_e + ?$

8.3 Proton fusion into deuterium is a vital step in the production of helium in the sun. We may think of the process semiclassically, requiring that two protons must get within a distance of $\sim 1/M_W$ to fuse (and for the weak force to kick in). Under these assumptions, what is the approximate required speed of protons to undergo fusion? To what temperature does that speed correspond? (In practice, this is a very significant overestimate, since *typical* protons must not be fusing all the time, or the sun would quickly consume its fuel).

8.4 Consider a lepton bispinor in the state $\Psi = \begin{pmatrix} 0 \\ 1 \end{pmatrix} \psi(x)$, a pure electron. Now consider a gauge transformation involving a rotation by an angle θ around the x-axis in SU(2) space.
(a) Compute the components of the bispinor in the rotated gauge.
(b) Compute the mass-like product $\overline{\Psi}\Psi$ before and after gauge transformation, and compare the results.
(c) Compute the sum of the currents $J^{(0)\mu}$ in the rotated gauge, and compare with the original.
(d) The weak Lagrangian takes the form $-\vec{J} \cdot \vec{W}$. Assume that in the original gauge, $W^{(3)}$ is the only nonzero gauge term. Using equation (8.21), explicitly calculate the vector fields in the new gauge, under a small rotation. Generalize for an arbitrarily large value of θ. Remember that this is a global gauge transformation.
(e) Compute the interaction energy in the primed and unprimed gauges.

8.5 Consider the process
$$e^- + e^+ \to W^- + W^+.$$
Draw all possible two vertex diagrams, and estimate the order-of-magnitude cross section using dimensional analysis.

8.6 Consider the collisional process
$$e^- + e^+ \to \nu_e + \overline{\nu}_e.$$
For the purposes of this problem you may assume that neutrinos are massless, and the electron and positron are moving at nonrelativistic speeds. You may also ignore spin.
(a) Draw all lowest-order Feynman diagrams describing the scattering process. Just to save you the effort, the lowest-order diagrams should have two vertices.
(b) In this problem, focus only on charged, t-channel interaction. Calculate the amplitude for this scatter.
(c) Calculate the total cross section of the scatter.
(d) This is not the only thing that can happen to an electron-positron pair. By comparison, the *electromagnetic* interaction
$$e^- + e^+ \to \gamma + \gamma$$
has a cross section of
$$\sigma = \frac{\pi \alpha^2}{2m|\vec{p}|}.$$
What is the fractional probability of producing a neutrino-antineutrino pair rather than two photons?

8.7 The γ^5 matrix is crucial to our understanding of handedness. Recall

$$\psi_{L,R} = \frac{1}{2}\left(\mathbf{I} \mp \gamma^5\right)\psi.$$

Compute the following explicitly.
(a) $\{\gamma^5, \gamma^0\}$
(b) $\gamma^5\gamma^5$
(c) $\left(\frac{1}{2}\left(\mathbf{I}+\gamma^5\right)\right)\left[\frac{1}{2}\left(\mathbf{I}-\gamma^5\right)\right)$
(d) $\overline{\psi}_L\psi_L$ (*Note:* $\overline{\psi}_L$ is the adjoint of ψ_L).
(e) $\overline{\psi}_L\gamma^0\psi_L$
(f) $\overline{\psi}_L\psi_R + \overline{\psi}_R\psi_L$

8.8 A D^- meson comprises ($\overline{c}d$), a down and an anticharmed quark. It has a mass of approximately 1.87 GeV and a spin of zero. While the following are not the dominant decay processes, what are the approximate relative rates of the three energetically allowed processes:

$$D^- \rightarrow \tau^- + \overline{\nu}_\tau$$

$$D^- \rightarrow \mu^- + \overline{\nu}_\mu$$

$$D^- \rightarrow e^- + \overline{\nu}_e$$

Be sure to include spin considerations.

8.9 One of the most important decays in the weak model is

$$\mu^- \rightarrow e^- + \overline{\nu}_e + \nu_\mu.$$

(a) Draw the lowest-order Feynman diagram(s) for the decay.
(b) What is the amplitude for the decay? Note that W bosons are much more massive than muons. Assume the muon is at rest and in a definite spin state.
(c) Use Casimir's trick to estimate the spin-averaged amplitude of the decay. You may find the trace relation

$$\mathrm{Tr}[\gamma^5\gamma^\mu\gamma^\nu\gamma^\lambda\gamma^\sigma] = 4i\epsilon^{\mu\nu\lambda\sigma}$$

helpful. The permutation matrix $\epsilon^{0123} = 1$, with even permutations also giving 1 and odd permutations giving -1.
(d) Calculate the decay time of a muon. Assume the neutrinos and electron are all massless.

Further Readings

- Cottingham, W. N., and D. A. Greenwood. *An Introduction to the Standard Model of Particle Physics.* Cambridge: Cambridge University Press, 1998. Cottingham and Greenwood have a number of excellent worked examples of weak interaction phenomenology in chapter 9.
- Griffiths, David. *Introduction to Elementary Particles, 2nd ed.* Hoboken, NJ: Wiley-VCH, 2008. Chapters 10 (on the weak interaction, and the subsequent development of the Feynman formalism) and 11.4 (a brisk but illuminating introduction to Yang-Mills theories) are particularly useful for the first-time student of the subject.
- Gross, Franz. *Relativistic Quantum Mechanics and Field Theory.* Hoboken, NJ: Wiley-VCH, 1993. Chapter 13 (on gauge theories) has a particularly nice discussion of the Yang-Mills approach.
- Peacock, John A. *Cosmological Physics.* Cambridge: Cambridge University Press, 2001. While this work provides a very good overview of many areas of particle and classical cosmology, I particularly recommend Chapters 8 and 9, which deal with cosmological implications of the Standard Model, and relics (including neutrino relics) of the early universe.

9 | Electroweak Unification

Figure 9.1. (Left–right) Sheldon Glashow (1932–), Steven Weinberg (1933–), and Abdus Salam (1926–1996), proposed the unification of the electroweak force, for which they received the Nobel Prize in Physics in 1979. Credits: Glasgow, courtesy of Luboš Motl; Weinberg, courtesy of JwH; Salaam, courtesy of Dutch National Archives/Bart Molendijk/Anefo.

Yang-Mills is beautiful, but as it was originally conceived, wrong in several particulars, mostly pertaining to the disparate masses of the electron and neutrino, and to the very *existence* of mass for the W and Z particles. In the 1960s Sheldon Glashow, Steven Weinberg, and Abdus Salam [78, 146, 156] united electromagnetism and the weak interaction into a single gauge theory.

GWS theory, as it's generally known, requires a bit more fine-tuning than the single degree of freedom in electromagnetism, wherein we needed specify only the fundamental charge of the electron. But upon its completion, we'll find a theory that can predict a remarkable range of phenomena.

9.1 Leptons and Quarks

The electroweak model is predicated on the idea that leptons and quarks are each grouped into doublets of the form

$$\Psi_{\text{Lepton}} = \begin{pmatrix} v_e \\ e \end{pmatrix}; \quad \Psi_{\text{Quark}} = \begin{pmatrix} u \\ d \end{pmatrix}. \tag{9.1}$$

We've focused on leptons so far, but to get a sense of the power of the electroweak unification theory, we need to consider the similarities and differences between these two doublets. The chief difference—and one outside of GWS theory—is that quarks participate in the strong interaction, while leptons do not.

The electric charges on the particles also differ. The up quark has a charge of $+2/3$, while the neutrino is neutral. However, there's an important similarity, one which will drop out of the theory naturally, namely, the *differences* between the charges of the doublets are the same; that is,

$$Q_u - Q_d = Q_\nu - Q_e = +1.$$

Finally, the mass of both first-generation quarks is on the scale of a mega-electron-volt, as is that of the electron. But neutrinos are so light that until neutrino oscillations were discovered in 1998 [70], neutrinos were assumed to be completely massless (Figure 1).

9.1.1 $SU(2)_L \otimes U(1)$

When generating Yang-Mills theory, we started with the supposition that $\overline{\Psi}\Psi$ is invariant under a gauge transformation, and we quickly found that the group $SU(2)$ fit the bill. But strictly speaking, $SU(2)$ isn't the *only* group that will satisfy the invariance condition. $U(1)$ works, as well.

There's a hitch, though. We suppose that only *left-handed* fermions participate in the $SU(2)$ symmetry, and thus we want to focus on an $SU(2)_L \otimes U(1)$ symmetry transformation. As a result, we'll write down a generalized form of possible gauge transformations for left-handed and right-handed fields separately:

$$\Psi_L \to e^{-\frac{i}{2}g'Y_L\theta^0} e^{-\frac{i}{2}g_W \vec{\theta}\cdot\vec{\sigma}} \Psi_L, \tag{9.2}$$

where we've assumed the existence of two different "charges," g' and g_W, the latter of which we've already seen.[1] The factor Y_L is known as the **weak hypercharge** of the associated left-handed particles. Similarly, right-handed particles are invariant under

$$\Psi_R \to e^{-\frac{i}{2}g'Y_R\theta^0} \Psi_R, \tag{9.3}$$

only the $U(1)$ transformation, and no $SU(2)$.

We've overgeneralized a bit. The transformations require only three coefficients, and yet we've introduced five: the two coupling constants g_W and g', and up to three different different weak hypercharges, one for the left-handed doublet, and one for each of two right-handed singlets. We'll tidy things up in short order.

---

[1] The sign convention and factors of 2 seem a bit arbitrary, and they are. We have selected them and others so that electric charges will have their familiar values. We'd be able to construct a perfectly sensible theory wherein the electron had a charge of $+3$, the up quark had -2 and so on. The physics would be the same. It would just be confusing to read.

We know what comes next—we can compute the conserved currents. Indeed, we've already done so. Using the U(1) gauge parameter θ^0 produces a left-handed Noether current:

$$J_\mu^{(0,L)} = \frac{g'}{2} Y_L \overline{\Psi}_L \gamma_\mu \Psi_L$$

$$= \frac{g'}{2} Y_L \left(\overline{\nu}_L \gamma_\mu \nu_L + \overline{e}_L \gamma_\mu e_L \right).$$

As current terms like this will arise often, we'll use the shorthand

$$j_\mu^{(ee,L)} = \overline{e}_L \gamma_\mu e_L \tag{9.4}$$

and similar to denote the non-normalized components of the various currents. The left-handed Noether currents are thus

$$J_\mu^{(0,L)} = \frac{g'}{2} Y_L \left(j_\mu^{(\nu\nu,L)} + j_\mu^{(ee,L)} \right)$$

$$J_\mu^{(1,L)} = \frac{g_W}{2} \left(j_\mu^{(e\nu,L)} + j_\mu^{(\nu e,L)} \right)$$

$$J_\mu^{(2,L)} = i\frac{g_W}{2} \left(j_\mu^{(e\nu,L)} - j_\mu^{(\nu e,L)} \right) \tag{9.5}$$

$$J_\mu^{(3,L)} = \frac{g_W}{2} \left(j_\mu^{(\nu\nu,L)} - j_\mu^{(ee,L)} \right).$$

The right-handed currents are much simpler:

$$J_\mu^{(0,R)} = \frac{g'}{2} Y_R \left(j_\mu^{(\nu\nu,R)} + j_\mu^{(ee,R)} \right). \tag{9.6}$$

This gives the interaction Lagrangian

$$\mathcal{L}_{\text{Int}} = -B \cdot (J^{(0,L)} + J^{(0,R)}) - \sum_{i=1}^{3} W^{(i)} \cdot J^{(i,L)},$$

where B is the vector field generated by U(1),[2] and $W^{(i)}$ are the three vector fields generated by SU(2). Following equation (8.27), we may rotate our basis to separate out the charged mediators:

$$\mathcal{L}_{\text{Int}} = -B \cdot (J^{(0,L)} + J^{(0,R)}) - W^{(3)} \cdot J^{(3,L)} - \frac{g_W}{\sqrt{2}} \left(W^+ \cdot j^{(\nu e,L)} + W^- \cdot j^{(e\nu,L)} \right) \tag{9.7}$$

if you'll pardon the mixing of conventions for describing the current terms. The neutral currents are another matter entirely. We expand *only* the left-handed components of the weak current interactions:

$$\mathcal{L}_{L,\text{Neut}} = -\frac{1}{2} \left[(g' Y_L B + g_W W^{(3)}) \cdot j^{(\nu\nu,L)} + (g' Y_L B - g_W W^{(3)}) \cdot j^{(ee,L)} \right]. \tag{9.8}$$

---

[2] You may be wondering why we don't label the U(1) field A, the electromagnetic field, as we did in our initial discussion. As we will shortly see, the U(1) field by itself *isn't* the electromagnetic field. That is the nature of electroweak unification.

This is curious. The U(1) gauge field and the third component of the SU(2) gauge field each interact with the neutral currents, but by different factors. But there's a way to make things a bit neater. Our choice of basis (the z-axis in weak isospin space) is largely arbitrary. We will rotate into a more appropriate one.

9.1.2 The Weinberg Angle

The similarity of the two neutral currents is quite suggestive, as is the gauge freedom, which allows us to rotate B and $W^{(3)}$ fields into two new fields, A (which will ultimately be our electromagnetic field), and Z (the weak Z^0 particles):

$$\begin{pmatrix} B \\ W^{(3)} \end{pmatrix} = \begin{pmatrix} \cos\theta_W & \sin\theta_W \\ -\sin\theta_W & \cos\theta_W \end{pmatrix} \begin{pmatrix} A \\ Z \end{pmatrix}.$$

(9.9)

Here, θ_W is known as the **Weinberg angle**.

Applying this rotation doesn't *really* change our Lagrangian, of course. It is a gauge choice that simply expresses two new vector fields as a linear superposition of two old ones. Substituting the rotation in equation (9.9) into equation (9.8) expands the former to

$$\mathcal{L}_{L,Neut} = -A \cdot j^{(vv,L)} \overbrace{\left(Y_L \frac{g'}{2} \cos\theta_W - \frac{g_W}{2} \sin\theta_W \right)}^{0}$$

$$-A \cdot j^{(ee,L)} \overbrace{\left(Y_L \frac{g'}{2} \cos\theta_W + \frac{g_W}{2} \sin\theta_W \right)}^{-q_e}$$

$$-Z \cdot j^{(vv,L)} \left(\frac{g'}{2} Y_L \sin\theta_W + \frac{g_W}{2} \cos\theta_W \right)$$

$$-Z \cdot j^{(ee,L)} \left(\frac{g'}{2} Y_L \sin\theta_W - \frac{g_W}{2} \cos\theta_W \right).$$

(9.10)

This is a lot to take in, but the upshot is that with an appropriate choice of rotation angle, we can make the electromagnetic interaction with the neutrinos disappear entirely.[3]

We noted earlier that we had overconstrained the parameterization of the gauge transformation. It is time to cash in that chit and define, by convention, that the weak hypercharge of the left-handed lepton doublet is

$$Y_L = -1.$$

---

[3] The Weinberg rotation may seem like cheating, but it's really quite similar to the old introductory physics chestnut of a block on a plane. The sensible young physicist will rotate his or her coordinate axes to run parallel to the plane rather than the ground. The physics is the same, but the representation is much simpler.

Under this constraint, the unfamiliar charge term may be written as

$$g' = -g_W \tan \theta_W, \tag{9.11}$$

and also,

$$q_e = g_W \sin \theta_W, \tag{9.12}$$

where we have set the definition of q_e to be positive.

In appropriate units, the Weinberg angle is

$$\sin \theta_W = \frac{q_e}{g_W} \simeq \left(\frac{0.302}{0.653} \right) \simeq 0.46. \tag{9.13}$$

Substituting equations (9.11) and (9.12) into the interaction Lagrangian in generality yields some interesting results.

9.1.3 Leptonic Charges

After specifying both the Weinberg angle and the weak hypercharge of the left-handed lepton doublet, $Y_L = -1$, equation (9.10) simplifies dramatically:

$$\mathcal{L}_{L,Neut} = q_e A_\mu [\bar{e}_L \gamma^\mu e_L] - g_W \frac{Z_\mu}{2 \cos \theta_W} \left([\bar{v}_L \gamma^\mu v_L] - \cos(2\theta_W) [\bar{e}_L \gamma^\mu e_L] \right).$$

Neutrinos and electrons respond very differently, not only to the electromagnetic field but also to the weak field. That difference arises from a difference in the weak isospin T_3: $+1/2$ for the upper component (neutrinos, up quarks), and $-1/2$ for the lower (electrons, down quarks). Generalization of the Lagrangian (equation 9.10) yields a relation between electrical charge (in electron units) and the weak isospin and hypercharge:[4]

$$Q = \frac{Y_W}{2} + T_3. \tag{9.14}$$

We may write a simplified interaction term with the neutral weak field:

$$\mathcal{L}_{Int,Z} = -g_W \frac{Z_\mu}{\cos \theta_W} \left[T_3 - Q \sin^2 \theta_W \right] [\bar{\psi} \gamma^\mu \psi]. \tag{9.15}$$

We're not quite done. We still have to figure out the Lagrangian for the right-handed fermions. The gauge freedom allows us to assign a different phase transformation to every component of the spinor, so the weak hypercharge of the right-handed neutrino and right-handed electron need not have the same hypercharge at all.

Put another way, each generation of leptons and quarks has a left-handed doublet and two ostensibly independent right-handed singlets (with no weak isospin). The hypercharge for each of the singlets, however, is selected so that the electric charge is the same as for the left-handed particle.

---

[4] You will have the opportunity to develop this in problem 9.2.

This step is rather straightforward, since right-handed fermions have a weak isospin of zero. If we demand that the electric charge of left- and right-handed fermions be identical, we quickly get

$$Y_R = -2 \text{ (right-handed electrons)}; \quad Y_R = 0 \text{ (right-handed neutrinos)}, \tag{9.16}$$

from which the neutral form of the interaction can be computed very simply from equation (9.15), but with a weak isospin of $T_3 = 0$. This means that right-handed electrons couple to the weak field, unlike right-handed neutrinos (which may, in fact, not exist at all).

Since the left- and right-handed components respond differently, the general form of the neutral interaction becomes[5]

$$\mathcal{L}_Z = -g_W \frac{Z_\mu}{\cos \theta_W} \overline{\psi} \gamma_\mu (c_V - c_A \gamma^5) \psi, \tag{9.17}$$

where

$$c_V = T_{3,L} - Q \sin^2 \theta_W; \quad c_A = T_{3,L}. \tag{9.18}$$

In this expression, $T_{3,L}$ refers to the weak isospin of the left-handed version of the particle only. This can be integrated into the Feynman rules in a very straightforward way (Appendix C).

Finally, we must say a few words about right-handed neutrinos. Right-handed neutrinos *essentially* don't exist. If neutrinos were massless (and they're not, quite), then there would be no way to generate a right-handed neutrino at all, as—unlike the right-handed electrons or quarks—they don't couple to either the $W^\pm$ or Z^0 fields.

We can now write the full leptonic Lagrangian:

$$
\begin{aligned}
\mathcal{L} = &\ \overline{v}_e (i \gamma^\mu \partial_\mu) v_e + \overline{e} (i \gamma^\mu \partial_\mu) e + q_e A_\mu \overline{e} \gamma^\mu e \\
&- \frac{g_W}{2\sqrt{2}} \left(W_\mu^+ \overline{v}_e \gamma^\mu \left[(1 - \gamma^5) \right] e + W_\mu^- \overline{v}_e \gamma^\mu \left[(1 - \gamma^5) \right] e \right) \\
&- g_W Z_\mu \left(\overline{v}_e \gamma^\mu \left[\frac{1}{2} (1 - \gamma^5) \right] v_e + \overline{e} \gamma^\mu \left[c_V - c_A \gamma^5 \right] e \right) \\
&+ \mathcal{L}_{A,Z,W}.
\end{aligned}
\tag{9.19}
$$

The first line is the free-field solution of the neutrinos and electrons, along with the ambidextrous electrical force. The second line is the charged-current weak-field interaction. The third and fourth times are the neutral interactions that we've worked so hard on, along with the free-field Lagrangian for the photons and weak mediators (equation 8.29). This Lagrangian holds not only for electrons (and their corresponding neutrinos) but for muons and taus as well. Most important, *all* these terms have been confirmed experimentally!

---

[5] The V refers to the vector portion of the interaction, which is ambidextrous. The A refers to the axial vector, or pseudovector. See Table 5.1 for a reminder about their transformation properties.

Table 9.1. The First Generation of Particles in the Standard Model.

Symbol	Handedness	Weak Isospin (T_3)	Weak Hypercharge (Y_W)	Electric Charge (Q)
ν_L	Left	1/2	−1	0
ν_R	Right	0	0	0
e_L	Left	−1/2	−1	−1
e_R	Right	0	−2	−1
u_L	Left	1/2	1/3	2/3
u_R	Right	0	4/3	2/3
d_L	Left	−1/2	1/3	−1/3
d_R	Right	0	−2/3	−1/3

Note: The right-handed neutrino may not, in fact, exist.

9.1.4 Quarks

Most of our work thus far has focused exclusively on leptons. Quarks also interact weakly. The only real distinction between quarks and leptons within the context of the electroweak interaction is that quarks have a different weak hypercharge, $Y_L = \frac{1}{3}$ for the left-handed doublet.

It is a fact of nature, and not yet well understood, that the charges of quarks and electrons are simple ratios of one another. The strong interaction requires that quarks are grouped in three, and thus it is interesting that any group of three quarks will produce an integer charge. But we don't have a fundamental basis for saying why some particles get the weak hypercharge that they do. As the Nobel laureate Frank Wilczek [162] has noted:

> While little doubt can remain that the Standard Model is essentially correct …it is not a complete or final theory. The fermions fall apart into five lopsided pieces with peculiar hypercharge assignments; this pattern needs to be explained. Also the separate gauge theories, which are quite mathematically and conceptually similar, are fairly begging to be unified.

This is simply another way of noting that while the *form* of the electroweak Lagrangian is quite simple, it still hides a multitude of sins—sins that we will, for now, simply ignore. That we have the freedom to set the weak hypercharge for each singlet and doublet doesn't necessarily justify that they seem to be arbitrary, just neat and orderly (Figure 9.2). By asserting this property of nature, along with the constraint that left- and right-handed particles have the same electric charge, we are able to quickly fill in the rest of the hypercharges for the quarks. We've done so in Table 9.1.

Having found a generalization of the weak interaction, we can introduce a new Feynman rule for neutral currents.

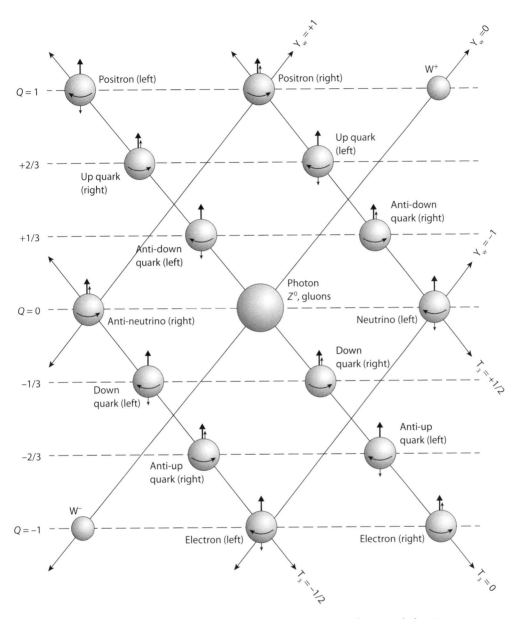

Figure 9.2. A visualization of the weak isospin, weak hypercharge, and electric charge of all the particles in the Standard Model.

2 Vertex Factors

Vertex	Contribution $\times \left[(2\pi)^4\delta(\Delta p)\right]$	Representation
Neutral current	$-ig_W\gamma^\mu(c_V - c_A\gamma^5)$	

where the vector and axial coefficients, c_V and c_A, respectively, are defined in equation (9.18).

We're still missing quite a few important pieces of the puzzle: the masses of the W's, Z's, electrons, and quarks, along with a reasonable explanation for *why* the neutrinos are essentially massless.

9.2 Spontaneous Symmetry Breaking

Gauge theories predict massless mediator particles, and yet, observationally, the weak interaction includes mediators with *extremely* high masses. The Higgs mechanism (as it is now generally known) was developed in the early 1960s by several groups, beginning with the work of Phil Anderson [16] and Julian Schwinger [149], and culminating in 1964 with nearly simultaneous publication by Robert Brout and François Englert [55], Peter Higgs [88], and Gerald Guralnik, Richard Hagen, and Tom Kibble [84].

The Higgs mechanism is based on the idea of **spontaneous symmetry breaking.** We have already seen a hint of symmetry breaking in the GWS $SU(2)_L \otimes U(1)$ model. *Spontaneous* symmetry breaking is slightly different. An inverted paraboloid, for instance, clearly has a rotational symmetry, and in principle, a marble might precariously balance at the top. But once it falls in one direction or another, though the paraboloid itself remains symmetric, the marble will respond to a decidedly nonsymmetric local force.

9.2.1 Higgs-like Particles

As a simplified introduction to the Higgs mechanism, consider a scalar field with the Lagrangian

$$\mathcal{L} = \partial_\mu \phi \partial^\mu \phi^* - V(\phi); \tag{9.20}$$

$$V(\phi) = V_0 - \mu^2 \phi \phi^* + \lambda (\phi \phi^*)^2. \tag{9.21}$$

The Higgs potential is a bit unfamiliar because the "masslike" term (the one with the quadratic part of the Lagrangian) has the wrong sign (Figure 9.3).[6] Thus, near $\phi \simeq 0$, the particles appear massless.

However, there is a ground state, the **vacuum expectation value** (or VEV) at

$$|\phi|_g = \frac{v}{\sqrt{2}} \equiv \frac{\mu}{\sqrt{2\lambda}}. \tag{9.22}$$

We can select $V_0 = \mu^4/4\lambda$ to put the ground state at zero energy. We should say "ground states," since there are an infinite array of states at $|\phi| = v/\sqrt{2}$. As the scalar field may have any complex phase at all, each one is part of the ground state.

---

[6] The shape of the Higgs potential, on the other hand, should look familiar. We encountered it in Figure 3.3 when discussing the false vacuum that drove the dynamics of the inflaton field.

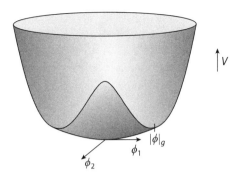

Figure 9.3. The Higgs potential in the complex plane.

How does the potential look to a field trapped near the ground state? We can see by simply expanding the potential as a Taylor series:

$$\phi(x) = \frac{v + H(x)}{\sqrt{2}} e^{i\vartheta(x)},$$

where we have swapped two dynamic fields, ϕ_1 and ϕ_2, for two *new* fields, $H(x)$ and $\vartheta(x)$. If both the notation and the form of ϑ is reminiscent of a gauge transformation, that's good. It was intended to be. We can, without too much difficulty, expand the scalar Lagrangian (equation 9.20) in terms of these new dynamic variables:

$$\mathcal{L} = \frac{1}{2}\partial_\mu H \partial^\mu H + \frac{1}{2}(v + H)^2 \partial_\mu \vartheta \partial^\mu \vartheta + V_0 + \mu^2 |\phi|^2 - \lambda |\phi|^4$$

$$\simeq \frac{1}{2}\partial_\mu H \partial^\mu H + \frac{1}{2}v^2 \partial_\mu \vartheta \partial^\mu \vartheta - \mu^2 H^2. \tag{9.23}$$

Though initially the ϕ and ϕ^* components are, in some sense, equivalent to one another, once the field has *relaxed* to or near a minimum energy condition, the two degrees of freedom are more easily expressed in terms of the following:

1. A real-valued (neutral) scalar field H with mass

$$M_H = v\sqrt{2\lambda}. \tag{9.24}$$

2. A massless scalar field ϑ, which is completely uncoupled from H.

We noted earlier the similarity between the Higgs and inflaton potentials. This isn't too surprising. Inflation presumably occurred around the decoupling of the strong and electroweak interactions. Likewise, the GWS model describes the decoupling of electromagnetism and the weak interaction, and so the similarity in potentials makes a sort of intuitive sense.[7]

---

[7] At the same time, it's worth considering the question of whether V_0 should be set to make the ground-state energy density exactly zero. One of the well-known issues in cosmology is that the vacuum energy density seems to be approximately 120 orders of magnitude lower than V_0, but not identically zero.

9.2.2 Interactions

The Higgs Lagrangian (9.20) has a U(1) gauge invariance, and thus we can quickly compute the interaction terms arising from a local gauge invariance,

$$\phi \to \phi e^{-iq\theta(x)},$$

where the gauge field θ is distinct from (but clearly closely related to) the massless scalar field, ϑ. This gauge transformation quickly gives rise to a covariant derivative,

$$D_\mu = \partial_\mu + iq\, A_\mu.$$

where A_μ is the free-field solution for a massless vector field, identical, in this case, to a photon. The full Lagrangian is thus

$$\mathcal{L} = D_\mu \phi\, D^\mu \phi^* - V(|\phi|) - \frac{1}{4} F_{\mu\nu} F^{\mu\nu}.$$

Expanding the covariant derivative yields an interaction term:

$$\mathcal{L}_{\text{Int}} = -q^2 \phi \phi^* A_\mu A^\mu. \tag{9.25}$$

This is a bit unusual (four particles at a vertex) but not unprecedented. We may expand ϕ near the minimum, and something interesting happens:

$$\mathcal{L}_{\text{Int}} = -\frac{1}{2} q^2 (v + H)^2 A \cdot A$$

$$\simeq -\frac{M_H^2}{4\lambda} q^2 A \cdot A - \frac{1}{\sqrt{2\lambda}} q^2 M_H H A \cdot A - \frac{1}{2} q^2 H^2 A \cdot A, \tag{9.26}$$

where these terms, respectively, correspond to

1. A massive A field with

$$m_A = \frac{q}{\sqrt{2\lambda}} M_H. \tag{9.27}$$

2. A three-particle interaction term with the newly massive H particle, and two A particles.
3. A four-particle interaction term with two H particles, and two A particles.

The symmetry breaking has a characteristic energy scale. For instance, the mass of the vector boson is roughly of the same order as the mass of the H particle, provided that the coupling constant q and the dimensionless parameter λ are both of order unity.

The full development of the Higgs theory isn't conceptually much different from what we've outlined here, but it's worth noting that we have both generated a coupling to the vacuum state (the massive vector fields) as well as predicted a new *particle* which can, in principle, be produced in high enough energy interactions.

It is worth spending a moment considering the θ field in all this. At the end of the symmetry breaking, ϑ is left with no mass and no interaction terms. This is hardly surprising given the underlying U(1) symmetry of the Lagrangian (both before and after symmetry breaking). Symmetry breaking generically produces terms like this, known as **Goldstone bosons** [79, 113], which don't couple to anything.

9.3 The Higgs Mechanism

All the pieces are in place for us to complete the GWS model. We've seen how spontaneous symmetry breaking can turn a massless field into a massive one. All that remains is to apply these principles to the electroweak interaction. And to do that, we need to go back to basics.

9.3.1 The Ground State

Our understanding of the Standard Model has been predicated on the idea that there is a *fundamental* particle, the fermion, from which gauge invariances required the existence of vector (spin-1) bosons. Within the electroweak model, there is another possibility: there is a scalar field which is just as fundamental as the fermionic ones—and *not* generated by a gauge symmetry. The Higgs doublet takes the form

$$\Phi = \begin{pmatrix} \phi_1 \\ \phi_2 \end{pmatrix}, \tag{9.28}$$

which is simply inserted into a fundamental Lagrangian of the universe:

$$\mathcal{L} = \partial_\mu \Phi^\dagger \partial^\mu \Phi - V(\Phi) + i \overline{\Psi}^{(l)} \gamma^\mu \partial_\mu \Psi^{(l)} + i \overline{\Psi}^{(q)} \gamma^\mu \partial_\mu \Psi^{(q)} + \dots + \mathcal{L}_{\Phi,\Psi}.$$

This Lagrangian is Lorentz invariant and predicts massless particles of every type. The leptons and quarks need to be put in by hand – the ellipsis is a reminder that we have *six* fermionic doublets, two families, and three generations. We accept them simply as facts of nature.

We simply *assert* that the weak hypercharge of the scalar doublet is $Y_H = 1$ and transforms under $SU(2)_L \otimes U(1)$ as

$$\Phi \rightarrow \exp\left[-i\left(\frac{g'}{2}\theta^0\right)\right] \exp\left[-i\frac{g_W}{2}\vec{\theta} \cdot \vec{\sigma}\right]\Phi. \tag{9.29}$$

The scalar Lagrangian was *designed* to produce a gauge freedom. As with our singlet example in the previous section, we assume the potential (also invariant under $SU(2)_L \otimes U(1)$) takes the parameterized form

$$V(\Phi) = V_0 - \mu^2 \Phi^\dagger \Phi + \lambda(\Phi^\dagger \Phi)^2.$$

This is about the simplest potential possible which both obeys the required symmetry and contains a false vacuum. The form is, in some sense, parametric. We don't have any motivation for either μ or the dimensionless λ except that they describe a relatively simple curve.

As a gauge choice, we are free to select a vacuum state for Φ at

$$\Phi_g = \begin{pmatrix} 0 \\ \frac{v}{\sqrt{2}} \end{pmatrix}.$$

We have chosen both this particular ground state and the corresponding hypercharge with an eye toward its interaction with the vector fields. Working from equation (9.14), and noting that the doublet has a weak hypercharge of $Y_H = 1$, and the downstairs component of the scalar field has a weak isospin, $T_3 = -1/2$, we find the ground state is electrically neutral.

This is a *very* good gauge choice for the theory. When we talk about the spontaneous symmetry breaking of the Higgs doublet, we're saying a number of different things at once. The Weinberg angle rotation creates a distinction between the photon and the Z^0. In essence, our choice of basis simply says that the massless mediator of the electroweak theory (photon) is *also* the particle that doesn't couple to the Higgs particle (electrically neutral).

Near the ground state, the scalar doublet assumes a form

$$\Phi = \begin{pmatrix} \phi_1 \\ \frac{v+H}{\sqrt{2}} e^{i\vartheta} \end{pmatrix} \tag{9.30}$$

where H will be the Higgs field. We may expand out the scalar Lagrangian near this new basis as

$$\mathcal{L}_\Phi = \left[\partial_\mu \phi_1^* \partial^\mu \phi_1 \right] + \frac{1}{2} \left[(v+H)^2 \partial_\mu \vartheta \, \partial^\mu \vartheta \right] + \frac{1}{2} \partial_\mu H \partial^\mu H - \mu^2 H^2.$$

The first term corresponds to the uncoupled Goldstone bosons, which we won't worry about again. The second term produces another Goldstone term, but this one couples only to the Higgs field. The latter two terms represent the free-field Lagrangian for the Higgs itself, including an expected mass term.

In 2012, the ATLAS [1] and CMS [37] collaborations at the Large Hadron Collider (LHC) experimentally detected the Higgs boson with a mass [28] of

$$M_H = 125.6 \pm 0.3 \, \text{GeV}, \tag{9.31}$$

giving a value of:

$$\mu \simeq 177.6 \, \text{GeV}$$

and a subsequent vacuum expectation value of approximately 144 GeV.

9.3.2 Vector Fields and the Higgs

The main triumph of the Higgs model is not in predicting its *own* mass (or even its own existence) but, rather, in predicting the mass of the weak mediators.

The $SU(2) \otimes U(1)$ symmetry of the scalar doublet generates the same four vector fields as does the fermionic doublet. The local gauge transformation can be used to quickly develop a covariant derivative which gives the coupling between the scalar and vector fields,

$$\partial_\mu \Phi \rightarrow D_\mu \Phi = \left[\partial_\mu + i \frac{g'}{2} B_\mu + i \frac{g_W}{2} \vec{\sigma} \cdot \vec{W}_\mu \right] \Phi, \tag{9.32}$$

modifying the scalar contributions to the electroweak Lagrangian as

$$\mathcal{L}_{(\Phi)} = D_\mu \Phi^\dagger D^\mu \Phi - V(\Phi). \tag{9.33}$$

It is significant that the scalar Lagrangian is quadratic in the covariant derivative, unlike the Dirac Lagrangian which is only linear. The quadratic terms are important because they will ultimately produce a mass for the vector fields.

While it's possible to expand equation (9.33) into a dozen or so terms, it isn't necessary to do so. We've seen, for instance, that the scalar-only terms simplify once the doublet has relaxed into the ground state. The second-order interaction terms are far more interesting:

$$\mathcal{L}_2 = \Phi^\dagger \left[\left(\frac{g'}{2} B + \frac{g_W}{2} \vec{\sigma} \cdot \vec{W} \right)^2 \right] \Phi.$$

As there is only one nonzero term in the ground state doublet, all off-diagonal terms (terms linear in σ_1 and σ_2) can be ignored. Recalling $\sigma_i^2 = \mathbf{I}$, we find that the second-order interaction Lagrangian becomes

$$\mathcal{L}_2 = \frac{1}{8} (v + H)^2 \left[\left(g' B - g_W W^{(3)} \right)^2 + g_W^2 (W^{(1)})^2 + g_W^2 (W^{(2)})^2 \right]. \tag{9.34}$$

This simplifies dramatically when we incorporate the Weinberg rotation (equations 9.9 and 9.11), which can be combined to give

$$g' B - g_W W^{(3)} = -g_W Z \sec \theta_W,$$

eliminating the coupling with the photon completely, as expected. Likewise, the off-diagonal vector fields may once again be rewritten as a superposition of the charged vector fields (equation 8.27). Combining all these effects, we obtain the electroweak Lagrangian

$$\mathcal{L} = \frac{1}{2} \partial_\mu H \partial^\mu H - \frac{1}{2} M_H^2 H^2 + \frac{1}{8} (v + H)^2 g_W^2 \left[2 W^+ \cdot W^- + \frac{Z \cdot Z}{\cos^2 \theta_W} \right] + \mathcal{L}_{A,W,Z} + \mathcal{L}_\Psi, \tag{9.35}$$

where $\mathcal{L}_\Psi$ is the leptonic contribution (equation 9.20), and $\mathcal{L}_{A,Z,W}$ is the free-field Lagrangian for the massless vector fields (equation 8.29). Given the form of the Lagrangian, the mass terms can be found via

$$\mathcal{L}_{Mass,Z,W} = \frac{1}{2} M_Z^2 Z^2 + M_W^2 W^+ \cdot W^-,$$

but there is no comparable term for photons. By inspection of equation (9.35), the masses are thus

$$M_\gamma = 0; \tag{9.36}$$

$$M_W = \frac{v g_W}{2} = \frac{M_H}{\sqrt{8\lambda}}; \tag{9.37}$$

$$M_Z = \frac{v g_w}{2 \cos \theta_W} = \frac{M_W}{\cos \theta_W}. \tag{9.38}$$

At long last, we have a concrete prediction from the Higgs model. The Weinberg angle can be computed from both the relative strength of the electromagnetic and weak interactions *and* the ratio of masses of the W and Z. While the specific value of the Weinberg angle isn't a prediction of the theory, we should get the *same* value whether we estimate from the ratio of electric to weak charges or from the ratio of W and Z masses.

Experimentally [28], the W's and Z's have masses of

$$M_W = 80.385 \pm 0.015 \, \text{GeV} \quad M_Z = 91.1876 \pm 0.002 \, \text{GeV},$$

yielding a Weinberg angle of

$$\theta_W = 28.17 \pm 0.02°.$$

We may independently estimate the Weinberg angle from the ratio of the electric charges and weak charges [28]:

$$\sin \theta_W = \frac{q_e}{g_W} \rightarrow \theta_W \simeq 28.77°.$$

These results aren't quite identical; as we will see in Chapter 11, effective interaction energies are a function of energy. However, even without the detailed calculation (which does, indeed, reconcile the two estimates), the two independent approaches to θ_W justify our faith in the electroweak model.

9.3.3 The Higgs Boson

The Higgs mechanism motivates the existence of massive vector fields, but there is also a *particle* associated with the theory, one capable of being detected. We already have the machinery at our disposal to determine the interactions, cross sections, and decay rates associated with the Higgs boson.

The scalar-vector Lagrangian (equation 9.35) hides a number of direct coupling terms between the Higgs boson and the massive vector fields, namely,

$$\mathcal{L}_{\text{Int}} = g_W H \left[M_W W^+ \cdot W^- + \frac{M_Z}{2 \cos \theta_W} Z \cdot Z \right] + \frac{1}{4} g_W^2 H^2 \left[W^+ \cdot W^- + \frac{1}{2 \cos^2 \theta_W} Z \cdot Z \right], \tag{9.39}$$

where the first set of interactions is typically much more significant at high energy. Our Feynman rules, of course, must now include these additional vertices from which the vertex amplitudes may be computed directly from equation (9.39). Expanding our Feynman rules still further, we have

2 Vertex Factors

Vertex	Contribution $\times \left[(2\pi)^4 \delta(\Delta p)\right]$	Representation
Higgs boson (charged)	$i g_W M_W g_{\mu\nu}$	
Higgs boson (neutral)	$i g_W \dfrac{M_Z}{\cos\theta_W} g_{\mu\nu}$	
Two Higgs bosons (charged)	$i \dfrac{g_W^2}{2} g_{\mu\nu}$	
Two Higgs bosons (neutral)	$i \dfrac{g_W^2}{2\cos^2\theta_W} g_{\mu\nu}$	

We've truncated our list somewhat with an eye toward including only those diagrams relevant to calculations discussed in this book.[8] The complete set of Feynman rules appears in Appendix C.

9.4 Higgs-Fermion Interactions

The fermions still haven't acquired a mass in the theory, and this must be reckoned with experimentally, The electron mass is significant, $m_e \simeq 0.511\,\text{MeV}$, and the first-generation quark masses are in the range of a few mega-electron-volts. The charged leptons and quarks in the second and third generations are more massive still.

Meanwhile, the neutrino's mass is so close to zero that until the detection of neutrino oscillations in 1998 [70] (more on this in the next chapter!) there was the tantalizing possibility that the neutrino was entirely massless.

9.4.1 Fermion Mass Terms

Much like the weak mediators and the Higgs boson itself, fermions acquire mass through the spontaneous symmetry breaking of the Higgs field. The electroweak model produces

---

[8] In particular, we've included a single two-Higgs, two-boson vertex. Interactions of this sort will be very important when we consider renormalization in Chapter 11 and supersymmetry in Chapter 12.

a bewildering array of ostensibly independent fields. For a given family and generation, there is a left-handed Dirac doublet and two singlets, along with the Higgs doublet. Our choice of gauge sets the weak hypercharge of the Higgs doublet to

$$Y_H = 1,$$

and nature sets the weak hypercharge of the left-handed pair

$$Y_L = -1 \quad \text{leptons;}$$

$$Y_L = 1/3 \quad \text{quarks.}$$

Further, consistency between the electrical charge of right- and left-handed particles requires that

$$Y_{u,R} = Y_L + 1;$$

$$Y_{d,R} = Y_L - 1.$$

For a Lagrangian with two fields, additional gauge-invariant terms are possible. For instance, consider a generalized Yukawa interaction,

$$\mathcal{L}_{\Phi,\Psi} = -\lambda_i \overline{\psi}_i \phi_i \psi_i + \text{H.C.,} \tag{9.40}$$

with a different coupling constant for each fermion. Anticipating that handedness will be important, we expand the quadratic spinor term into a left-right pair:

$$\overline{\psi}\psi = \overline{\psi}_L \psi_R + \overline{\psi}_R \psi_L.$$

Not all terms in equation (9.40) are gauge invariant, but by a careful use of our charges, we can quickly select the few that are:

$$\mathcal{L}_{\Phi,\Psi} = -\lambda \left[\overline{\Psi}_L \Phi \psi_{R,2} + \overline{\psi}_{R,2} \Phi^\dagger \Psi_L \right] \tag{9.41}$$

produces terms that are manifestly gauge invariant, where the weak isospin and hypercharge of the Higgs doublet explicitly cancel the corresponding weak isospin and hypercharge of the left-handed spinor doublet.

In our preferred basis, ϕ_2 becomes real and nonzero after spontaneous symmetry breaking, and thus equation (9.41) simply reduces to

$$\mathcal{L}_m = -\lambda \frac{(v+H)}{\sqrt{2}} \overline{\psi}_2 \psi_2 . \tag{9.42}$$

Particle 2, the downstairs fermion, acquires a mass of

$$m_2 = \lambda_f \frac{v}{\sqrt{2}}.$$

The masses of fermions are simply the product of the Higgs vacuum expectation value and a dimensionless coupling constant. For an electron, the coupling constant is the rather meager

$$\lambda_e \simeq 2.9 \times 10^{-6}.$$

In addition to mass, this product also adds a possible interaction in our Feynman rules, a vertex between fermions and Higgs particles.

2 Vertex Factors

Vertex	Contribution $\times \left[(2\pi)^4 \delta(\Delta p)\right]$	representation
Higgs-fermion	$-\frac{i}{2}\dfrac{m_f}{M_W} g_W$	

9.4.2 Masses and Couplings

The Yukawa interaction produces some nice results. For one thing, it naturally yields a coupling between the Higgs boson and fermions.

However, our first guess at the interaction Lagrangian also suggests (erroneously) that only the downstairs components of the fermionic doublets acquire mass. While it is true that the neutrino is so light as to be nearly massless, the up quark has a measured mass [119] of

$$m_u = 2.3^{+0.7}_{-0.5} \text{ MeV}.$$

This is hardly massless, especially compared to the electron. What's going on here?

We haven't quite exhausted the space of invariant Lagrangians. Indeed, simply by making the matrix of coupling constants nondiagonal, we may introduce additional contributions to equation (9.41) with the substitution

$$\lambda_2 \Phi \rightarrow \lambda_1 \sigma_1 \Phi + \lambda_2 \Phi,$$

where σ_1 is simply the first Pauli matrix. This form allows a coupling for *all* fermions of the form given in equation (9.42).

It is sometimes said that the Standard Model does not allow for the neutrino to have mass. This isn't quite true, as we've seen. If the up quark can acquire mass through the Higgs mechanism, then neutrinos can as well. However, given that the upper limit on the sum of the neutrino masses is less than an electron-volt, this puts an upper limit on any of the neutrino coupling constants of

$$\lambda_\nu \lesssim 10^{-12}.$$

The Yukawa coupling terms are given in Table 9.2, but the key point is that the coupling for neutrinos is six orders of magnitude smaller than any other.

There are a number of other surprises. While the coupling constants, and thus the masses, of the various fermions increase in each generation,[9] there is no obvious pattern

---

[9] Some caution must be applied when talking about the mass of a particular species of quark. There is considerable mixing between generations, a topic we'll take up in detail in Chapter 10. For now, these numbers are meant to be illustrative.

Table 9.2. Approximate Values for the Coupling Constant λ_f for Each of the Charged Fermions [119].

Generation	Charged Lepton, l^-	Up-Type Quark	Down-Type Quark
1	3×10^{-6}	1×10^{-5}	3×10^{-5}
2	6×10^{-4}	7×10^{-3}	5×10^{-4}
3	1×10^{-2}	1	2×10^{-2}

to the values. Down quarks are more massive than up quarks, for instance, but top quarks are *much* more massive than bottom quarks. It is intriguing that the coupling constants span only a few orders of magnitude within any given generation. Though their values are left unconstrained by symmetry arguments, they're clearly not *that* arbitrary.

All the coupling constants are small, with the tantalizing exception of the top, which is, within experimental errors, 1. This is of more than passing interest. As we will see in Chapter 11, effective coupling terms in the Lagrangian (including values of λ_f) are generally energy dependent. Just as the Higgs field gives the top quark mass, the top quark (and others) produce corrections to the Higgs potential. At very high top quark mass, the Higgs vacuum becomes unstable, eventually decaying to a lower vacuum state [13]. As the situation stands, the Higgs potential appears to be metastable, but this suggests that there aren't hidden Standard Model particles at yet-higher masses just waiting to be discovered.

Given the smallness of the other eight coupling constants, it is simply a matter of taste whether one chooses to round down $\sim 10^{-12}$ to zero and simply to *assert* that neutrinos are massless in this model. We will discuss neutrino oscillation in the next chapter, but experimental evidence points to neutrinos having mass. What makes neutrinos special, in this context, is that they are electrically neutral. The implication is that a right-handed neutrino is a true singlet—uncoupled from the other particles in its sector. This leaves neutrinos in a unique position of acquiring their mass outside of the Higgs mechanism.

9.5 A Reflection on Free Parameters

It is not obvious why nature chooses the symmetries that it does. Certainly, the gauge symmetries in the electroweak model are some of the simpler ones, and among the simplest that introduce a parity violation. But beyond the choice of symmetry group, nature seems also to have selected three generations of both quarks and leptons, and rational-valued weak hypercharges for both leptons (-1) and quarks ($1/3$).

It is hoped that these values will be explained by a forthcoming unification scheme, and so they aren't quite so frustrating as the plethora of free parameters describing the electroweak Lagrangian: the $SU(2)$ and $U(1)$ charges, the nine coupling constants between the Higgs and various fermionic fields, and the two parameters describing the Higgs potential. Thirteen free parameters in a simplified theory seems like an awful lot.

Worse, still, is that many of these parameters seem, in a sense, to be unnatural. The dimensionless parameters are generally several orders of magnitude smaller than unity, for instance. Even the characteristic energy μ which describes the Higgs field is 10 orders of magnitude lower than the Planck scale. This discrepancy is a manifestation of the **hierarchy problem**, in which measured masses are seemingly unnaturally small.

Our ad hoc list of parameters seems a little long at first and, frankly, at second glance. The presumption is that some, or perhaps most of them, will be predictable from a *more* fundamental theory. At present, we simply need to simply assert that this is how the universe *is*. Even so, starting with the given list of parameters, we can predict an awful lot, including the electron charge, the masses of the Higgs, $W^{\pm}$ and Z^0, a massless photon field, the neutrino, Maxwell's equations, and the detailed Feynman rules for the electroweak interactions. That isn't too shabby.

Problems

9.1 Consider a spinor doublet $\begin{pmatrix} \psi_u \\ \psi_d \end{pmatrix}$ in an unrotated gauge. The left- and right-handed components globally transform according to equations (9.2) and (9.3), respectively. The left-handed particle has a weak hypercharge Y_L, and the right-handed singlets have $Y_{u,R}$ and $Y_{d,R}$. Consider gauge rotations in θ^0 and θ^3 only.
   (a) What is the explicit left-handed doublet in the rotated gauge? Terms should include factors like $\psi_{u,L}$.
   (b) Using the result from problem 8.7, compute $\overline{\psi}\psi$ in the rotated gauge and compare the result with the unrotated gauge.
   (c) Suppose we constrain Y_L. What additional constraints (if any) arise if we demand that if the coupled up doublets transform with a phase of $e^{i\theta}$, that the down doublets transform with $e^{-i\theta}$ for all values of θ^0 and θ^3?

9.2 Beginning with equation (9.10) and noting that the weak isospin of the neutrino is $T_3 = +1/2$, and that of the electron is $T_3 = -1/2$, and using the definitions in equations (9.11) and (9.12):
   (a) Generalize equation (9.10) explicitly in terms of q_e (when coupling to the A field), g_W (when coupling to the Z field), Y_L, and T_3.
   (b) Knowing that right-handed particles must have the same charge as their left-handed counterparts (but have no weak isospin), compute the relationship between right-handed weak isospin and left-handed isospin in generality. Is this consistent with the result of the previous problem?

9.3 Consider an as-yet-undiscovered doublet

$$\Psi = \begin{pmatrix} \psi_u \\ \psi_d \end{pmatrix}$$

with a left-handed weak hypercharge of $Y_L = 2$. What are the electric charges of the up and down particles? What are the weak hypercharges of the up and down singlets?

9.4 In the text we developed a scalar model of broken symmetry which used a complex scalar field with a broken $U(1)$ symmetry (equation 9.26). Draw all nonvanishing vector-Higgs vertices in this model, and give their amplitudes.

9.5 Suppose there was a free-field Lagrangian for a *real*-valued scalar doublet

$$\Phi = \begin{pmatrix} \phi_1 \\ \phi_2 \end{pmatrix}$$

of the form

$$\mathcal{L} = \frac{1}{2}\partial_\mu \phi_1 \partial^\mu \phi_1 + \frac{1}{2}\partial_\mu \phi_2 \partial^\mu \phi_2 - V(\Phi),$$

where

$$V(\Phi) = (\phi_1^2 + \phi_2^2 - C)^2,$$

and where C is a positive constant. This Lagrangian is invariant under SO(2) transformations.

(a) What is the conserved current J^μ of the SO(2) symmetry?

(b) Write the vector field and interaction Lagrangian for the SO(2) symmetry. You should call the vector field A^μ and assign an arbitrary coupling constant q.

(c) What are the possible "ground states" of the Lagrangian (stable equilibria)?

(d) Call the ground-state value ϕ_0. Clearly, the generalized solution is

$$\phi_1 = \phi_0 \cos(\theta) \qquad \phi_2 = \phi_0 \sin(\theta),$$

which means, more generally, that the solutions can be written as

$$\phi_1 = (\phi_0 + R)\cos\theta; \quad \phi_2 = (\phi_0 + R)\sin\theta.$$

Write the total Lagrangian (initial Φ fields and interactions) as a function of $R(x)$ and $\theta(x)$.

(e) Based on the preceding result, what is the mass of the R field, the θ field, and the A field?

(f) Draw a Feynman diagram illustrating how the R field couples to the A field.

9.6 We would like to calculate the decay of the Higgs into various channels.

(a) Draw the lowest-order Feynman diagram of a Higgs decaying into a quark-antiquark pair. For what flavor(s) of quarks is this decay allowed?

(b) Compute the decay rate as a function of the mass of the outgoing particles.

(c) At which fermion mass will the decay rate be maximized?

9.7 Draw the lowest-order Feynman diagram of a Higgs decaying into two photons.

9.8 Show that the individual combination of Higgs and fermionic fields

$$\overline{\psi}_{d,R} \Phi^\dagger \psi_L$$

is invariant under $SU(2)_L \otimes U(1)$ transformations. You should make explicit use of equations (9.2, 9.3, and 9.29).

Further Readings

- Cottingham, W. N., and D. A. Greenwood. *An Introduction to the Standard Model of Particle Physics.* Cambridge: Cambridge University Press, 1998. Chapters 10 and 12 have a nice discussion of symmetry breaking in general, and the GWS model, respectively.
- Gross, Franz. *Relativistic Quantum Mechanics and Field Theory.* Hoboken, NJ: Wiley-VCH, 1993. The interested reader may find the discussion of quantization of the electroweak model in Chapter 15 useful.
- Quigg, Chris. *Gauge Theories of the Strong, Weak, and Electromagnetic Interactions,* 2nd ed. Princeton, NJ: Princeton University Press, 2013. Chapters 5–7 have an especially excellent discussion of the electroweak interaction, along with a vivid explanation of the motivation for the Majorana mass generation for neutrinos.
- Wilczek, F. and D. J. Gross. The Future of Particle Physics. In *Physics in the 21st Century,* 71–96. Singapore: World Scientific, 1997.

10 | Particle Mixing

Figure 10.1. Nicola Cabibbo (1935–2010), taken in 2006. Cabibbo was one of the first to recognize that mass and flavor eigenstates were not necessarily the same. Credit: © Marcella Bonna.

Though we're well aware of the sheer number of particles—three generations each of two families of fermions, each of which is itself a doublet, to say nothing of different helicity states, antiparticles, mediators, and so on—we've focused our discussion on the electron and the electron neutrino. We've chosen this path in large part because leptons are generally found in isolation, simplifying things dramatically.

With quarks, the situation gets far more complicated, as the number of bound states is enormous.[1] Enrico Fermi, when asked by a student about the name of this or that particle famously replied [34]:

> If I could remember the names of all these particles, I would have become a botanist.

We won't dwell on the discovery of every particle or every partially successful unification scheme, as doing so will distract us from our ultimate goal of *understanding* the Standard Model. But to put the overall story in some context, see the timeline in Figure 10.2, wherein we can get a sense of what we knew, and when.

We've seen many of these milestones in passing, but to study the implications of multiple generations of particles, we consider the state of affairs in the early 1960s,

---

[1] See, for instance, Figure 10.3, to get a sense of what can be created from just the three lightest quarks.

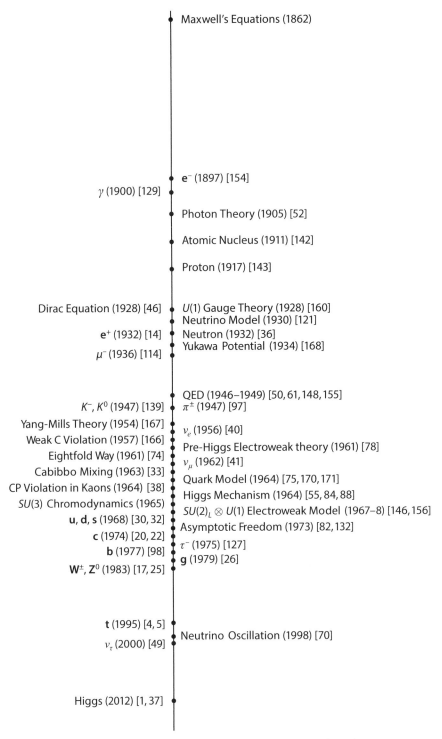

Figure 10.2. An incomplete timeline of the Standard Model, selected to advance our discussion. Discovery of fundamental particles are boldfaced.

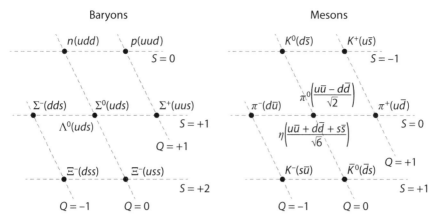

Figure 10.3. The Eightfold Way. The charges and strangeness of the lightest baryons (left) and mesons (right), including their quark configurations.

after the muon and kaons (the first of the second-generation elementary and composite particles) had been discovered.

10.1 Quarks

We noted how odd it is that there are three generations rather than the seemingly more natural *one*, but until now, the different generations of quarks and leptons have had a "live and let live" relationship to one another, and each generation seemed to differ from the others only with regard to mass. But not only do different generations interact with one another in a profound way, as we'll see, but multiple generations are *required* to produce CP violation, and thus matter-antimatter asymmetry. And to understand *that* we consider the role of the strange quark.

10.1.1 *The Strange Quark*

The ups and downs are the lightest quarks by far (Table 10.1). This is not terribly surprising, since the up and down quarks in turn make up the lightest baryons, the protons and neutrons. Until 1947, protons and neutrons were all there were, hadronically speaking.[2]

But in the late 1940s and 1950s there was an eruption of discovery of new **strange** particles, including kaons and other exotica. *Strangeness* was not intended to be pejorative but rather was a description of a new quantum number. As we now understand it, a particle with a strangeness $S = 1$ contains a strange quark, $S = -1$ has a strange antiquark, and otherwise, $S = 0$.

As with electric charge, strangeness was presumed to be a conserved quantity of physical systems. As a practical matter, strangeness often *is* conserved by energetic

---

[2] **Hadrons** are the superset of baryons (three quarks) and mesons (a quark-antiquark pair).

Table 10.1. The quark doublets.

Name	Mass (GeV)	Charge
Up	$(1.7\text{–}3.1)\times 10^{-3}$	$+2/3$
Down	$(4.1\text{–}5.7)\times 10^{-3}$	$-1/3$
Charmed	1.29 ± 0.08	$+2/3$
Strange	0.10 ± 0.03	$-1/3$
Top	172.9 ± 0.7	$+2/3$
Bottom	4.19 ± 0.1	$-1/3$

considerations, since the strange quark has a mass of 95 MeV, more than 10 times that of the *next*-lightest quark. But as we will see, this conservation is not absolute.

In 1961, Murray Gell-Mann found that categorization of the baryons and mesons then known formed simple geometric patterns [74]. The lightest groups each formed octets (Figure 10.3), in what has become known as the **Eightfold Way**. In some sense, the Eightfold Way is a mere historical curiosity, as the quarks themselves were hypothesized in 1964 [75, 170, 171] and discovered through deep inelastic scattering of very high energy ($\sim$20 GeV) electrons off protons in the Stanford Linear Accelerator four years later [30, 32]. While quarks are not detected in isolation, analysis of the outgoing electron scattering and energy distribution demonstrated that the proton has substructure consistent with Gell-Mann's and Zweig's predictions.

Though we now know that there are six flavors of quark (Table 10.1), at relatively low energies we can simply pretend that there are only three, which means that we're especially concerned about the interplay between the down and its massive doppelgänger, the strange.

10.1.2 The Cabibbo Angle

Consider the decay of a negative pion:

$$\pi^- \to e^- + \bar{\nu}_e. \tag{10.1}$$

This is about as simple as a weak reaction can get. A first-generation quark-antiquark pair ($d\bar{u}$) decays into a first-generation lepton-antilepton pair (Figure 10.4).

But pion decay isn't the only possible one of this type. The decay of a negatively charged kaon looks remarkably similar to equation (10.1):

$$K^- \to e^- + \bar{\nu}_e. \tag{10.2}$$

A kaon consists of a ($s\bar{u}$) pair. If strangeness (or generation generally) were conserved, this kaon decay should be impossible. But experimentally, not only does leptonic kaon decay happen, it does so at a rate comparable to that of pion decay!

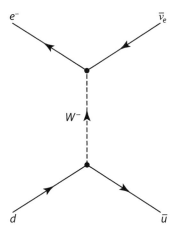

Figure 10.4. π^- decay.

Consider the weak-interaction Lagrangian that we derived in the last two chapters. In it, pairs of quarks appear in two different contexts. In the first, and arguably most fundamental, we have terms of the form

$$\mathcal{L}_W = -\frac{g_W}{2\sqrt{2}} W_\mu^+ \left[\bar{u}\gamma^\mu (1 - \gamma^5) d' + \bar{c}\gamma^\mu (1 - \gamma^5) s' + \ldots, \right] + \ldots, \tag{10.3}$$

where we've deliberately ignored a lot of very similar terms. One element might look surprising. The d' and s' particles are written with primes to indicate that they are eigenstates of the weak interaction but *not* necessarily eigenstates of mass.

On the other hand, the Higgs mechanism produces terms of the form

$$\mathcal{L}_M = m_d \bar{d} d + m_s \bar{s} s + \ldots. \tag{10.4}$$

In this case, the d and s states are eigenstates of the mass operator. There's no reason that d and s can't be linear combinations of d' and s', provided that the two sets of quarks were related by a unitary transformation. We may think of the first two generations of quark doublets under $SU(2)_L$ as

$$\begin{pmatrix} u \\ d' \end{pmatrix} ; \quad \begin{pmatrix} c \\ s' \end{pmatrix},$$

where d' represents a mixing between down and strange (but mostly down), and s' represents a mixing between strange and down (but mostly strange). The down-type quarks (down or strange) mix via

$$\begin{pmatrix} d' \\ s' \end{pmatrix} = \begin{pmatrix} \cos\theta_C & \sin\theta_C \\ -\sin\theta_C & \cos\theta_C \end{pmatrix} \begin{pmatrix} d \\ s \end{pmatrix}, \tag{10.5}$$

where $\theta_C \simeq 13.1°$ is the **Cabibbo angle**, named after Nicola Cabibbo, who proposed the relation in 1963 [33]. Cabibbo's inference predated the quark hypothesis by about a year, and the discovery of quarks by half a decade, but the basic idea—that strangeness is not a conserved quantity in the weak interaction—formed the basis of what we now understand about quark mixing.

As a general rule, we consider only the down-type quarks to be mixed. This is a matter of convention. We could, for instance, imagine that both u' and d' are eigenstates of the weak interaction. However, it's far easier to "rotate away" the flavor mixing of either the up-type or down-type quarks.[3]

A kaon, which nominally has only a strange and an antiup quark can decay as though it had a down quark. Using this mechanism, we can compute the relative amplitudes of down-to-up transition compared with a strange-to-up transition,

$$\frac{\mathcal{A}_{s \to u}}{\mathcal{A}_{d \to u}} = \frac{\sin \theta_C}{\cos \theta_C} \simeq 0.23.$$

Weak quark interactions don't *quite* conserve generation. Experimentally, we can see this in the comparison of kaon and pion decay. Computing the relative decay rates for a two-body decay (equation B.2), we get

$$\frac{\Gamma(K^- \to e^- + \overline{e}_\nu)}{\Gamma(\pi^- \to e^- + \overline{e}_\nu)} = \tan^2 \theta_C \left(\frac{m_\pi}{m_K}\right)^2 \left(\frac{m_K^2 - m_e^2}{m_\pi^2 - m_e^2}\right) \simeq 0.19.$$

Cabibbo's approach was generalized by Makoto Kobayashi and Toshihide Maskawa in 1973 to *three* generations of quarks. The **Cabibbo-Kobayashi-Maskawa** (or **CKM**) [33,92] matrix allows a transformation between the mass and weak eigenstate,

$$\begin{pmatrix} d' \\ s' \\ b' \end{pmatrix} = \begin{pmatrix} V_{ud} & V_{us} & V_{ub} \\ V_{cd} & V_{cs} & V_{cb} \\ V_{td} & V_{ts} & V_{tb} \end{pmatrix} \begin{pmatrix} d \\ s \\ b \end{pmatrix}, \tag{10.6}$$

where, as a reminder, the unprimed quarks correspond to the mass states, and the primed correspond to the weak eigenstates. The elements of the CKM matrix can be constructed from Euler angles, and represented by [119][4]

$$V_{ij} = \begin{bmatrix} 1 & 0 & 0 \\ 0 & c_{23} & s_{23} \\ 0 & -s_{23} & c_{23} \end{bmatrix} \begin{bmatrix} c_{13} & 0 & s_{13}e^{-i\delta} \\ 0 & 1 & 0 \\ -s_{13}e^{i\delta} & 0 & c_{13} \end{bmatrix} \begin{bmatrix} c_{12} & s_{12} & 0 \\ -s_{12} & c_{12} & 0 \\ 0 & 0 & 1 \end{bmatrix}$$

$$= \begin{pmatrix} c_{12}c_{13} & s_{12}c_{13} & s_{13}e^{-i\delta} \\ -s_{12}c_{23} - c_{12}s_{23}s_{13}e^{i\delta} & c_{12}c_{23} - s_{12}s_{23}s_{13}e^{i\delta} & s_{23}c_{13} \\ s_{12}s_{23} - c_{12}c_{23}s_{13}e^{i\delta} & -c_{12}s_{23} - s_{12}c_{23}s_{13}e^{i\delta} & c_{23}c_{13} \end{pmatrix}, \tag{10.7}$$

---

[3] You will have the opportunity to do so in problem 10.2.
[4] I've used a shorthand where c_{12}, for example, is the cosine of the θ_{12} mixing angle (what we were formerly calling the Cabibbo angle, before we were made aware of three additional flavors of quark).

though the exact form of this matrix is simply a convention and depends on the order in which we apply rotations in this parameter space.

In general, an $n \times n$ unitary matrix (for an n-generation model) has n^2 constraints. The individual phases of the quarks may be rotated freely, and as there are $2n$ quarks in total, this means that $2n - 1$ constraints may be "rotated away." Thus, there are

$$n^2 - (2n - 1) = (n - 1)^2$$

free parameters in the CKM matrix. This is fascinating! In a two-generation model, there's only a single parameter (the Cabibbo angle), but there are four parameters for a three-generation model. Three generations, in other words, is the smallest number that allows meaningful complex phases. The importance of this phase can't be underestimated. Kobayashi and Maskawa originally proposed a third generation of quarks, as that was the minimum that would allow CP violation [92].

Experimentally, the amplitudes of the CKM mixing angles are [119]

$$\theta_{12} = 13.04 \pm 0.05°;$$

$$\theta_{13} = 0.201 \pm 0.011°;$$

$$\theta_{23} = 2.38 \pm 0.06°.$$

These mixing angles are all quite small. In other words, the flavor states (for quarks, at least) are *very nearly* the same as the mass states, but not quite.

Given the hierarchy of angles, $\theta_{12} \gg \theta_{23} \gg \theta_{13}$, any elements in the CKM matrix that are quadratic in, for instance, $\sin\theta_{13} \sin\theta_{23}$ can be said to essentially vanish. Inspection of equation (10.7) reveals that only V_{ub}, V_{td}, and V_{ts} can potentially have a significant imaginary component.

10.1.3 *The Cronin-Fitch Experiment*

Wu's measurements of P symmetry violation in cobalt-60 decays in the 1950s [166] suggested that there is something unique about the weak force. If parity and charge symmetry can be violated individually, how about the *combination* of charge and parity? In 1964, James Cronin and Val Fitch [38] tested CP violations in neutral kaons.

Kaons, you will recall, were among the "strange" particles discovered in the late 1940s which ultimately helped kick off the quark model. Neutral kaons are particularly intriguing:

$$\left| K^0 \right\rangle = |\bar{s}d\rangle; \qquad \left| \overline{K}^0 \right\rangle = |s\bar{d}\rangle.$$

They look a lot like pions, but with a subtle difference. Pions, as we've seen, are superpositions of up and antiup quarks, and down and antidown quarks (equation 5.40), resulting in a particle which is an eigenstate of CP transformations with an eigenvalue of -1. Neutral kaons, in contrast, are *not* eigenstates of CP transformations but rather,

$$\hat{\mathbf{C}}\hat{\mathbf{P}}\big|K^0\big\rangle = -\big|\overline{K}^0\big\rangle; \qquad \hat{\mathbf{C}}\hat{\mathbf{P}}\big|\overline{K}^0\big\rangle = -\big|K^0\big\rangle.$$

Since kaons and antikaons are so similar, it is useful to construct superposition states of the two which *are* eigenstates of CP:

$$|K_1\rangle = \frac{1}{\sqrt{2}}\left(\big|K^0\big\rangle - \big|\overline{K}^0\big\rangle\right) \rightarrow CP = +1;$$

$$|K_2\rangle = \frac{1}{\sqrt{2}}\left(\big|K^0\big\rangle + \big|\overline{K}^0\big\rangle\right) \rightarrow CP = -1.$$

If CP symmetry is perfectly conserved, then K_1 and K_2 could be distinguished by their decay products

$$K_1 \rightarrow \pi^0 + \pi^0 \qquad \tau = 9 \times 10^{-11}\,s,$$

$$K_2 \rightarrow \pi^0 + \pi^0 + \pi^0 \qquad \tau = 5 \times 10^{-8}\,s,$$

since an even number of pions will have a CP value of $+1$, and an odd number will have a CP of -1. Given the difference in decay times, a beam of K_2 particles can be created simply by sending a beam of mixed kaons and observing a suitable distance away. Even at relativistic speeds, virtually all the K_1 particles will decay after a few meters, and a beam of pure K_2 particles will remain.[5] The outgoing kaons *should* decay exclusively into three pions. But this is *not* what Cronin and Fitch observed. About 0.2% of the decays were to two pions, a direct measure of CP violation in the weak interaction.

Another way of looking at the CP violation is that the reaction

$$K^0 \leftrightarrow \overline{K}^0 \tag{10.8}$$

is not symmetric. It appears that nature really does distinguish between matter and antimatter.

10.1.4 CP Violation and the CKM Matrix

Since the electroweak Lagrangian is invariant under CP transformations, the symmetry violation must be in the form of the quark multiplets themselves, and in particular, in the CKM matrix. This is most clearly seen by looking at a piece of the classical Lagrangian coupling the up and down quarks (though, of course, there are many additional terms):

$$\mathcal{L}_{u,d} = \frac{g_W}{\sqrt{2}}\left[\overline{u}_L \gamma^\mu d_L\, W_\mu^- V_{ud} + \overline{d}_L \gamma^\mu u_L\, W_\mu^+ V_{ud}^*\right], \tag{10.9}$$

where the complex conjugate appears in the second term because the CKM matrix is required to be unitary. We've also simply identified the left-handed up and down fields rather than using the $(1 - \gamma^5)$ notation.

---

[5] You will test this with realistic numbers in problem 10.5.

Examination of equation (10.9) allows us to adapt our Feynman rules to include charged quark interactions.

2 Vertex Factors.

Vertex	Contribution $\times \left[(2\pi)^4 \delta(\Delta p) \right]$	Representation
Charged current (quarks)	$-i\dfrac{g_W}{2\sqrt{2}}\gamma^\mu(1-\gamma^5)V_{ij}$	
Charged current (quarks)	$-i\dfrac{g_W}{2\sqrt{2}}\gamma^\mu(1-\gamma^5)V_{ij}^*$	

All up-type and down-type quarks mix in an identical way.

The Lagrangian transforms under CP to become

$$\mathcal{L}_{u,d}^{CP} = \frac{g_W}{\sqrt{2}}\left[\bar{d}_L \gamma^\mu u_L W_\mu^+ V_{ud} + \bar{u}_L \gamma^\mu d_L W_\mu^- V_{ud}^*\right],$$

which is identical to the original only if V_{ud} (and ultimately, all other components of the CKM matrix) is real. As a general rule, a complex phase produces a CP violation.

10.1.5 The Unitarity Triangle

Some of the best estimates of CP violation in the quark sector can be found by examining the decays of **B-mesons**, B^0 ($d\bar{b}$) [6, 100] . Like their lower-mass counterparts, B-mesons oscillate freely (and slightly asymmetrically) with their charge-conjugated partners. However, for now we'll focus specifically on one decay channel and its CP conjugated partner,

$$B^0 \to J/\psi + K_S \qquad \overline{B^0} \to J/\psi + K_S,$$

where the J/ψ particle, the **psion**,[6] comprises $c\bar{c}$, and the K_S particle is the short-lived neutral kaon discussed earlier.[7] Both are their own antiparticles. We can get a taste of the CP violation in B-meson decay by considering the lowest-order diagrams describing the processes (Figure 10.5).[8]

---

[6] The psion, with a mass of $\sim$3.1 GeV, has the distinction of two names because it was discovered simultaneously and independently by two groups, one at the Stanford Linear Accelerator [22] and one at Brookhaven [20], and announced on the same day, November 11, 1974. This heralded the so-called November Revolution, which led to the discovery of a whole host of high-mass composite particles. The psion caused such a stir because it lived about a thousand times longer than comparable particles of similar mass, suggesting that there was something fundamentally different about it from previously known particles. Indeed, there was; that difference was the charmed quark.

[7] We called it K_1 then, but with the caveat that the neutral kaons themselves violate CP symmetry, so K_1 is similar to K_S, but not exactly.

[8] You may note that one diagram produces a K^0, while the other produces a $\overline{K^0}$, even though they ostensibly describe the same decays. As the two neutral kaons oscillate into one another, we have omitted those internal processes for the sake of clarity.

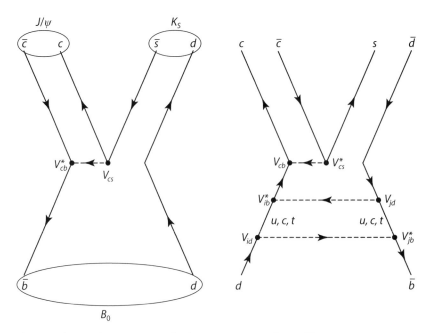

Figure 10.5. Two important contributions to B-meson decay. The two-vertex amplitude $\mathcal{A}_1$ (left) is real-valued, while the four-vertex amplitude $\mathcal{A}_2$ has a complex phase.

You might well wonder why we have included two diagrams, especially since they differ in order. A minimum of two amplitudes are required for CP violation to create an interference. Suppose the process has two amplitudes $\mathcal{A}_1$ and $\mathcal{A}_2$, where the former is real, and the latter has two complex phases:

$$\mathcal{A}_2 = |\mathcal{A}_2|\, e^{i\delta} e^{i\phi}.$$

If we suppose ϕ is invariant under CP transformations, but δ is not,

$$\mathcal{A}_2^{CP} = |\mathcal{A}_2|\, e^{-i\delta} e^{i\phi},$$

and thus

$$|\mathcal{A}|^2 = |\mathcal{A}_1|^2 + |\mathcal{A}_2|^2 + 2|\mathcal{A}_1||\mathcal{A}_2|\cos(\phi + \delta)$$

$$|\mathcal{A}^{CP}|^2 = |\mathcal{A}_1|^2 + |\mathcal{A}_2|^2 + 2|\mathcal{A}_1||\mathcal{A}_2|\cos(\phi - \delta)$$

$$\frac{|\mathcal{A}|^2 - |\mathcal{A}^{CP}|^2}{|\mathcal{A}|^2 + |\mathcal{A}^{CP}|^2} = -\frac{2\sin(\phi)\sin(\delta)|\mathcal{A}_1||\mathcal{A}_2|}{|\mathcal{A}_1|^2 + |\mathcal{A}_2|^2 + 2\cos(\phi)\cos(\delta)|\mathcal{A}_1||\mathcal{A}_2|},$$

which manifestly produces a different decay rate if both the symmetry-preserving and symmetry-violating phases are nonzero.

We can get a sense of the asymmetry of the B-meson by considering the complex phase of the amplitudes in Figure 10.5. The scattering amplitude may be simplified significantly by recalling that the CKM matrix was required to be unitary, which produces

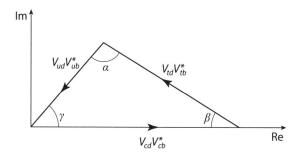

Figure 10.6. The unitarity triangle.

nine equations of the form

$$V_{ud} V_{ub}^* + V_{cd} V_{cb}^* + V_{td} V_{tb}^* = 0. \tag{10.10}$$

Because of the importance of B-meson systems in detecting CP asymmetry, equation (10.10) is generally referred to as "the" unitarity triangle, as it can be visualized as a triangle in the complex plane (Figure 10.6).[9]

If all components of the CKM matrix were real (or even if they simply had the same phase), the unitarity triangle would simply be a line segment. Its triangle-ness is a testament to the fundamental CP asymmetry of the system. In particular, we may say

$$\beta = \arg\left[-\frac{V_{cd} V_{cb}^*}{V_{td} V_{tb}^*}\right], \tag{10.11}$$

where "arg" refers to the complex phase.

The geometry of the unitarity triangle is a direct probe of the CP symmetry violation of the B-meson system:

$$\mathcal{A}_2 \propto \left[1 + \frac{V_{cd} V_{cb}^*}{V_{td} V_{tb}^*} \times (\text{kinematic contributions})\right].$$

Comparison of this last form with equation (10.11) allows us to more or less make a direct measurement of the β angle from measurements of the difference in decay rates of B-mesons and their antiparticles.

B-meson decay is not the only possible constraint on the unitarity triangle and the elements of the CKM matrix. An up-to-date list of experimental results is compiled by the Particle Data Group,[10] shown in Figure 10.7. Putting current best estimates in terms of the CP-violating term in the CKM matrix [119], we obtain

$$\delta = 68.8 \pm 4.6°.$$

CP violation in the quark family is nearly maximal! Alas, there is no comparably clean approach for measuring CP violation for leptons.

---

[9] There are five other similar triangles, each with equal area, but this is the one with the easiest angles to measure experimentally.

[10] http://pdg.lbl.gov/.

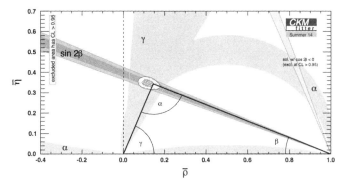

Figure 10.7. Current constraints on the unitarity triangle from CP symmetry violations in B-meson decay and other reactions. Credit: CKMfitter Group (J. Charles et al.), *Eur. Phys. J.* C41, 1–131 (2005) [hep-ph/0406184], updated results and plots are available at http://ckmfitter.in2p3.fr.

10.2 Neutrinos

Neutrinos are a nearly ubiquitous signature of weak interactions. Electron and muon neutrinos are created in the upper atmosphere from the decay of pions and muons. Radioactive decay products of unstable nuclei inside the earth shower us from below. Particle accelerators and nuclear reactors produce high-energy neutrinos that bombard us on every side.

All these sources are significant, but historically, solar neutrinos—or a deficit of them—served as the earliest and most fundamental constraints on the question of mixing in the lepton sector.

10.2.1 *The Solar Neutrino Problem*

Radiation from the sun is easy to see, but the internal mechanics are obscured by the opacity of the stellar atmosphere. The specific fusion chains that release the energy—and thus the emission spectrum of neutrinos—need to be inferred from a detailed model. But for simplicity, we simply abstract the hydrogen-to-helium fusion process as

$$4p \rightarrow \text{He} + 2e^+ + 2\nu_e,$$

suggesting that a comparable number of neutrinos and photons are produced, and at similar energies (generally in the 1 MeV range). As a rough approximation, this yields a solar neutrino flux on the surface of the earth of about 10^{15}–10^{16} m^{-2} s^{-1}. This may seem enormous, but even so, the detection rates in neutrino detectors are extremely, low since at the mega-electron-volt scale, cross sections are only of order 10^{-47} m^2 (equation 8.10).

In the 1960s, Ray Davis [42] (who led the experimental effort) and John Bahcall [23] (who did the theoretical work) built a $380\,\text{m}^3$ neutrino detector in the Homestake Mine in South Dakota—deep enough underground that background signals could be largely eliminated. The tank was filled with cleaning fluid to detect the neutrino capture reaction:

$$\nu_e + {}^{37}\text{Cl} \rightarrow {}^{37}\text{Ar} + e^-.$$

The argon was then collected to serve as a proxy for the neutrino events.

Bahcall and his collaborators made several different models for the fusion chains in the sun, but even with relatively generous assumptions, Davis found that the actual neutrino counts yielded only about a third of the predicted rate. This discrepancy persisted for three decades [24] and withstood countless efforts to refine either the solar model, the detection rate, or the effects of neutrinos propagating through the sun. None seemed to adequately explain the **solar neutrino problem**.

10.2.2 Neutrino Oscillation

The solution to the solar neutrino problem was proposed before the problem itself was even discovered. The Italian-Soviet Scientist Bruno Pontecorvo [133] suggested that neutrinos might oscillate between flavors, just as quarks form from an admixture of states. As the Homestake detector was sensitive only to electron neutrinos, they must have turned into something else en route, something not visible to the detector:

$$\nu_e \leftrightarrow \nu_\mu.$$

This transformation is known as **neutrino oscillation** and provided the first evidence for neutrino mass.

As with quarks, the flavor states and weak interaction states of neutrinos are superpositions of one another. However, because quarks interact via the strong force and leptons don't, the fundamental definition of which lepton is which is a little different than it is for quarks.

With quarks, the flavors are primarily associated with mass states. That is, a proton is "really" made of two u quarks and a d, but the weak interaction couples the u to the d' (a superposition of all "down-type" quarks).

The flavors of neutrinos, however, are defined in terms of which charged lepton they couple to under the weak force. We begin, as we did with quarks, considering a two-state model for oscillations. Supposing neutrinos come in two flavors ν_e and ν_μ and two masses m_1 and m_2, the descriptions will be superpositions of one another,

$$\begin{pmatrix} \nu_e \\ \nu_\mu \end{pmatrix} = \begin{pmatrix} \cos\theta_{12} & \sin\theta_{12} \\ -\sin\theta_{12} & \cos\theta_{12} \end{pmatrix} \begin{pmatrix} \nu_1 \\ \nu_2 \end{pmatrix}, \tag{10.12}$$

where θ_{12} is an as-yet-unknown parameter of the system. We hope it will not escape your notice that the mass mixing in the two-state model is simply a representation of SO(2) and is structurally identical to Cabibbo's work with quarks.

An electron neutrino is created in a reaction with an energy much, much larger than any of the masses,

$$E_i = \sqrt{p^2 + m_i^2}$$

$$\simeq p + \frac{m_i^2}{2E},$$

where $E \simeq p$ is the energy of the neutrinos', outgoing beam. The latter term, $m_i^2/2E$ represents the energy of a particle in a fixed mass eigenstate. Thus, if we had a particle in a pure mass eigenstate, it would evolve as

$$|\psi(t)\rangle = |i\rangle \exp\left(-i\frac{m_i^2}{2E}t\right). \tag{10.13}$$

But electron neutrinos aren't pure mass eigenstates. They can be written as superpositions:

$$|\nu_e\rangle = \cos\theta_{12}|1\rangle + \sin\theta_{12}|2\rangle.$$

At some later time, the wavefunction is

$$|\psi(t)\rangle = \exp\left(-i\frac{m_1^2}{2E}t\right)\cos\theta_{12}|1\rangle + \exp\left(-i\frac{m_2^2}{2E}t\right)\sin\theta_{12}|2\rangle,$$

and thus the probability of turning an electron neutrino into a mu neutrino is

$$P_{\nu_e \to \nu_\mu} = \left|\langle\nu_\mu|\psi(t)\rangle\right|^2 = \sin^2(2\theta_{12})\sin^2\left(\frac{\Delta m^2 L}{4E}\right) \tag{10.14}$$

$$= \sin^2(2\theta_{12})\sin^2\left(\frac{1.267(\Delta m^2/\,\mathrm{eV}^2)\,(L/\,\mathrm{km})}{(E/\,\mathrm{GeV})}\right),$$

where we've exploited the fact that neutrinos are traveling at essentially the speed of light (hence $L = t$ is the distance traveled), and we've defined Δm^2 as the differences of the masses, squared.

As the equation is written, we can't say for certain whether m_1 or m_2 is the larger mass, but oscillations in the sun help us out. As neutrinos propagate outward, there is an amplitude for interaction with electrons, as shown in Figure 10.8, introducing an additional interaction energy proportional to the density of electrons.

In the 1970s and 80s a Soviet team, Stanislav Mikheyev and Alexei Smirnov [110], and American physicist Lincoln Wolfenstein [165] independently found that the MSW effect (as it's become known) can enhance the oscillation rate of neutrinos at high density. However, the MSW effect is asymmetric, because only electron neutrinos will interact with the ambient electrons (and there are no ambient muons). As a result, the solar neutrino oscillation fraction may be used to directly constrain $m_2 > m_1$.

However, we have no such constraint relating m_2 and m_3. The hierarchy could easily be inverted with no physical or mathematical complications.

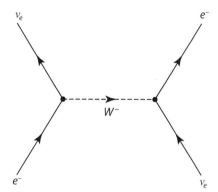

Figure 10.8. Charged boson exchange in the MSW effect.

10.2.3 Oscillation Parameters

The generalization into three generations is known as the as the Pontecorvo-Maki-Najagawa-Sakata, or PMNS, matrix [133, 108, 134])

$$|l\rangle = U_{li}|i\rangle,$$

where l is the weak lepton state (ν_e, ν_μ or ν_τ), and i is the mass eigenstate (1 through 3). The PMNS matrix takes the form identical to the CKM matrix, CP-violating complex phase and all (equation 10.7), but as suggested by the amplitude of solar neutrino oscillations, there is a lot more mixing in the neutrino sector than in the quark sector.

Estimating the oscillation parameters with any degree of certainty requires two key components: (1) a source of known flavor and energy spectrum and (2) a series of (large and heavily shielded) detectors at different distances from the source.

Atmospheric neutrinos are produced in both electron and muon versions. As high-energy ($\gg 100\,\mathrm{MeV}$) cosmic rays strike the upper atmosphere, they form pions, which can then subsequently decay into

$$\pi^- \to \mu^- + \bar{\nu}_\mu; \qquad \pi^- \to e^- + \bar{\nu}_e.$$

The ratio of primary muon to electron antineutrinos is readily predictable from the masses (equation B.3):

$$\frac{n_{\nu_\mu}}{n_{\nu_e}} = \frac{m_\pi^2 - m_\mu^2}{m_\pi^2 - m_e^2} \simeq 0.43.$$

The first definitive test of neutrino oscillation came from the Super-Kamiokande experiment [8, 18, 70]. Using a 50,000-ton tank of ultrapure water, the detector was able to measure the disappearance of atmospheric muon neutrinos via Cerenkov radiation.

At extremely high energies (and over short distances), the probability of oscillating to an electron neutrino is negligibly small, but the probability of oscillating to a ν_τ is much higher, since the mass differences are greater. So atmospheric neutrinos give us a good handle on θ_{23} and Δm_{23}^2.

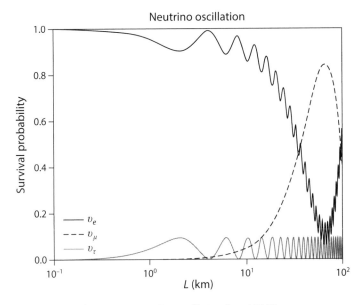

Figure 10.9. Neutrino oscillation for a 1 MeV v_e.

Table 10.2. Current Constraints on Neutrino Mixing Angles and Masses for Normal Ordered Mass Hierarchy.

	$1 \rightarrow 2$	$2 \rightarrow 3$	$1 \rightarrow 3$
$\sin^2(\theta_{ij})$	0.308 ± 0.017	$0.437^{+0.033}_{-0.023}$	$0.0234^{+0.0020}_{-0.0019}$
$\Delta m_{ij}^2 / \text{eV}^2$	$(7.54^{+0.22}_{-0.22}) \times 10^{-5}$	$(2.43 \pm 0.06) \times 10^{-3}$	$\simeq \Delta m_{23}^2$

Note: The inverted hierarchy changes the parameters at the 1% level. Credit: The Particle Data Group [119].

Nuclear reactor [9] and accelerator neutrinos [7,8,11,12] typically emerge in energy ranges of a few mega- and giga-electron-volts, respectively. But unlike with solar neutrinos, we have the option of placing multiple detectors at different distances from the source. Likewise, neutrinos from reactor-based neutrino experiments give us a handle on θ_{13} and Δm_{13}^2.

Current estimates of the mixing angles and mass differences can be found in Table 10.2 [119]. The results are intriguing. While the CKM matrix for quarks features mixing angles of only a few degrees (the largest is 13°), neutrino mixing is extremely large, with $\theta_{12} \simeq 33°$. It is a stretch to say that v_1 and v_e are even approximately the same particle.

The "natural" interpretation of the mass hierarchy of neutrinos is that $m_3 \gg m_1$ and m_2, which are each quite low in mass. But, there could also be a mass inversion, where m_3 is the lightest (Figure 10.10). At this point, we can't say, as observational constraints yield Δm_{ij}^2. But assuming normal ordering, this puts a *minimum* total mass for the neutrinos at

$$\sum_i m_i > 0.05 \, \text{eV}.$$

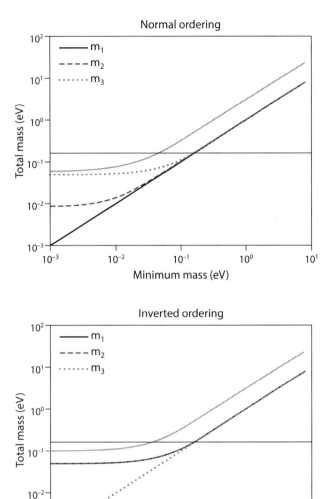

Figure 10.10. Possible mass hierarchy scenarios for neutrinos. The horizontal line indicates the mass limits suggested by cosmological data from the *Planck* satellite and baryon acoustic oscillations [130].

There are cosmological constraints on neutrino mass as well. One of the biggest mysteries in observational astrophysics and cosmology is the existence of a substance (or particle) known as dark matter. Cosmological evidence suggests that dark matter is nonbaryonic and seems to constitute about 25% of the universal energy budget [56,130]:

$$\rho_{DM} \simeq 5.2 \times 10^9 \, \text{eV/m}^3.$$

Most of the neutrinos in the universe are cosmological in origin, and at early times, they were in thermal equilibrium with photons. That blackbody neutrino background persists at a temperature of about 1.95 K [124], corresponding to a number density of

$$n_\nu \simeq 3.4 \times 10^8 \, \text{m}^{-3}$$

neutrinos and antineutrinos of all three species (in almost exactly equal numbers). At the cold energies of the cosmic background, direct detection of primordial neutrinos will likely remain impossible for the foreseeable future. But if we make the most simple-minded assumption possible—that neutrinos account for *all* dark matter—we arrive at an upper limit for mass of $\sum m_\nu \lesssim 10 \, \text{eV}$.

But there are tighter constraints than that. If the dark matter particles are sufficiently light, then they remain relativistic for a long time after the Big Bang. This is known as *hot dark matter* and is ruled out from cosmological predictions. In practice, these constraints put the limit much lower [130], at

$$\sum_i m_i < 0.16 \, \text{eV} \quad \text{(Planck)}$$

But how do we determine the masses directly? One approach is to look for masses kinematically. For instance, several experiments aim to measure the recoil of beta decays, as in the decay of tritium:

$$T \to {}^3_2 He^+ + e^- + \bar{\nu}_e.$$

In principle, a precise measurement of the recoil of the helium and the electron should yield an absolute mass estimate of the electron antineutrino mass (and thus, presumably, the neutrino mass as well) to somewhere within 0.2 to 2 eV, which is outside the range currently suggested by cosmological observations.

10.3 Neutrino Masses

Neutrinos have mass, but, compared with other fermions, just barely. If the same mechanism gives neutrinos mass as it does other particles, the coupling constant needs to be six to eight orders of magnitudes smaller for neutrinos than for the corresponding quarks or charged leptons (Table 9.2). Neutrinos are generally taken to be massless in the Standard Model. However, this assumption is not strictly required. It could simply be that neutrino coupling is much, much smaller than other Yukawa couplings. If so, the smallness of those terms is certainly *unnatural*.

However, if there is no Yukawa term (λ_f) for neutrinos, we are left with the question of where neutrino masses originate—if not Yukawa masses, then we need to move to physics *beyond* the Standard Model.

10.3.1 *The Majorana Equation*

We begin with the massless neutrino hypothesis (which, recall, we *know* to be wrong). There is an elegance to massless particles. The chirality (or spin) in one Lorentz frame will be the same as in any other. In a world of entirely massless neutrinos, there would be no right-handed neutrinos and no left-handed antineutrinos; there's simply no mechanism in the weak interaction for producing them and no probability of detecting them.

We might even, under those circumstances, simplify matters by *fiat*: neutrinos and antineutrinos are the same particle; the left-handed version we call the neutrino, and the right-handed version we call the antineutrino. This was the intriguing approach suggested by the Italian physicist Ettore Majorana in 1937 [107].

Majorana's idea is more than just a bookkeeping device. If neutrinos really were massless (and again, they aren't), then the basis spinors for the Dirac equation (equations 5.18, 5.19) would produce a simple set of relationships:

$$\text{Massless neutrinos}: u_+(p) = -v_-(p); \quad u_-(p) = v_+(p).$$

Up to a simple phase difference, a left-handed neutrino bears the same relation to a right-handed neutrino as it does to a left-handed antineutrino.

A **Majorana particle** is a fermion with itself as its antiparticle. For massless particles, there's no distinction between the Dirac formulation (which we've used for spin-1/2 particles throughout this text), and the Majorana. But for massive particles, things get a bit more complicated.

As we've spent most of our time thus far exploring the Dirac model, we start from there, writing the mass component of a Dirac Lagrangian in a somewhat novel way,

$$\mathcal{L}_M = (m_D \overline{\psi}\psi + m_D \overline{\psi^c}\psi^c),$$

where m_D is the **Dirac mass** (what we've hitherto simply referred to as "the mass" in the Dirac equation), and $\psi^C = \hat{C}\psi$ (equation 5.38) is the charge conjugation partner of the wavefunction. CPT symmetry requires particles and their antiparticles to have identical mass.

However, it's not the *only* way of generating a mass. We may add an additional off-diagonal term,

$$\mathcal{L}_M = \left(\overline{\psi} \ \overline{\psi^c} \right) \begin{pmatrix} m_D & m_M \\ m_M & m_D \end{pmatrix} \begin{pmatrix} \psi \\ \psi^c \end{pmatrix}, \tag{10.15}$$

where m_M is the Majorana mass. Majorana mass terms do not arise in the electroweak model as we've seen it. The Higgs can't produce a coupling term of this sort unless it has a weak isospin of 1. In that case, we'd require an entire Higgs *sector*.

We won't dwell on mechanisms for generating a Majorana mass term, but instead we'll examine the intriguing possibility that there is a Majorana mass term in addition to a Dirac one. We may thus generate the corresponding particle masses by finding the eigenstates of the mass matrix in equation (10.15):

$$m_\pm = m_D \pm m_M.$$

Neutrinos still can be found in two states; they will just represent *different* states than if there were no mixing. The possibility of negative masses (if $m_M > m_D$) shouldn't concern you too much. We'll deal with that in short order.

10.3.2 The See-Saw Mechanism

We have constructed a model on the basis of $SU(2)_L \otimes U(1)$ invariance. As a result, we can't simply insert new terms into the Standard Model Lagrangian without consequences. Handedness has a profound impact on the Standard Model. Thus, it makes sense to break our Dirac bispinor into left- and right-handed components via the normal mechanism:

$$\psi = \psi_L + \psi_R = \frac{1 - \gamma^5}{2}\psi + \frac{1 + \gamma^5}{2}\psi.$$

Further, as we showed in Chapter 8, terms like

$$\overline{\psi}_L \psi_L = \overline{\psi}_R \psi_R = 0$$

and thus equation (10.15) may be expanded in generality,

$$\mathcal{L}_M = \begin{pmatrix} \overline{\psi}_R & (\overline{\psi}^c)_R & \overline{\psi}_L & (\overline{\psi}^c)_L \end{pmatrix} \begin{pmatrix} 0 & 0 & m_D & m_R \\ 0 & 0 & m_L & m_D \\ m_D & m_L & 0 & 0 \\ m_R & m_D & 0 & 0 \end{pmatrix} \begin{pmatrix} \psi_R \\ (\psi^c)_R \\ \psi_L \\ (\psi^c)_L \end{pmatrix}, \tag{10.16}$$

where we've broken up the Majorana mass into left- and right-handed components such that $m_M = m_R + m_L$. Further, it can be readily shown through the properties of the γ^5 matrix that

$$(\psi^c)_R = (\psi_L)^c \qquad (\psi^c)_L = (\psi_R)^c,$$

so some care must be taken when charge conjugating a left-handed particle versus taking the left-handed component of a charge-conjugated particle.

Here's where the magic of $SU(2)_L \otimes U(1)$ symmetry can be exploited. The right-handed neutrino (or the left-handed antineutrino) are singlets in the Standard Model. As a result, they may have a Majorana mass term without charged leptons having a similar term. Thus, if $m_L = 0$ and $m_R = m_M$, then we are left with a mass matrix

$$\mathbf{M} = \begin{pmatrix} 0 & 0 & m_D & m_M \\ 0 & 0 & 0 & m_D \\ m_D & 0 & 0 & 0 \\ m_M & m_D & 0 & 0 \end{pmatrix} \tag{10.17}$$

which both satisfies the underlying symmetry and allows for a Majorana particle. The eigenvalues of the mass matrix are thus

$$m_{\pm} = \frac{\sqrt{m_M^2 + 4m_D^2} \pm m_M}{2}. \tag{10.18}$$

The situation is most interesting in the limit of $m_M \gg m_D$. In that case

$$m_+ \simeq m_M; \tag{10.19}$$

$$m_- \simeq \frac{m_D^2}{m_M}. \tag{10.20}$$

The high-mass neutrino (equation 10.19) is essentially sterile. It comprises right-handed neutrino and left-handed antineutrino states and therefore does not interact under normal circumstances.

The low-mass neutrino is the one we normally encounter in weak interactions. But something remarkable has happened. Its mass is no longer the pure Yukawa mass term but, rather, is suppressed by a factor of m_D/m_M. A large Majorana mass produces a small Dirac mass, and vice-versa. This interplay has become known as the **see-saw mechanism**.

Example 10.1: Assume that the most massive neutrino has a Dirac mass equal to the mass of the top quark:

$$m_{3,D} \simeq m_t \simeq 100 \, \text{GeV}.$$

Assuming that the Majorana mass scale is at **Grand Unified Theory** (GUT) energies,[11]

$$m_M \simeq 10^{15} \, \text{GeV},$$

what is the kinematic mass of a third neutrino species?

Solution:

$$\sum_i m_i \simeq m_3 \simeq \frac{(100 \, \text{GeV})^2}{10^{15} \, \text{GeV}} = 10^{-11} \, \text{GeV} = 0.01 \, \text{eV},$$

a realistic mass scale for neutrinos based on oscillation experiments.

10.3.3 Are Neutrinos Majorana Particles?

The symmetries of the Standard Model seem to allow for the possibility of a Majorana component and as we've seen, nature seems to have conspired to make almost every free parameter that *can* be nonzero.

A Majorana neutrino is a big deal. It allows violation of lepton number just as neutrino oscillation results in violation of electron, muon, and tau number, and manifests itself via Lagrangian interaction terms of the form

$$\mathcal{L}_M = m_M \left[\overline{\psi}_R (\psi^c)_L + (\overline{\psi^c})_L \psi_R \right],$$

which can be quantized via equation (5.55) and result in interaction terms

$$\hat{\mathcal{H}}_{\text{Int}} \sim m_M \hat{b}_{\vec{p},r}^\dagger \hat{c}_{\vec{q},s} + \ldots.$$

---

[11] More on that in Chapter 12.

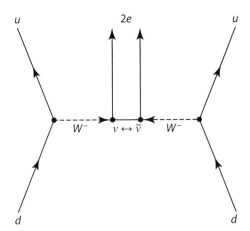

Figure 10.11. Majorana neutrinos would allow for the possibility of neutrino-less double beta decay of heavy nuclei. We show the central interaction.

This converts an antineutrino ($\hat{c}$) into a neutrino ($\hat{b}^\dagger$). Mixing terms of this sort allows reactions of the form in Figure 10.11, with an amplitude proportional to m_M^2, in which heavy nuclei can undergo double beta decay without a corresponding emission of an antineutrino.

For example, xenon-136 normally decays into

$$^{136}\text{Xe} \rightarrow {}^{136}\text{Ba} + 2e^- + 2\bar{\nu}_e$$

with an extremely long decay time ($>10^{21}$ yr). Beyond the outgoing nucleus (which will be essentially stationary), only the electrons are easily detectable, but as they are part of a four-body decay, their energy spectrum is continuous.

However, if neutrinos are Majorana particles, then a similar decay,

$$^{136}\text{Xe} \rightarrow {}^{136}\text{Ba} + 2e^-,$$

is also allowed [21]. The neutrino-less double beta decay involves only two particles, and thus the telltale signature will be a delta function of energy at half the nuclear ionization energy (Figure 10.12). The discovery of Majorana neutrinos would be an important extension to the Standard Model, but as of this writing, it remains just a tantalizing possibility.

Problems

10.1 The Λ^0 consists of (uds) and has a mass of 1.115 GeV, making it the lightest strange baryon.
  (a) How massive would the Λ^0 need to be to allow a strong (strange-preserving) decay? Give the decay explicitly. You may need to delve into the Particle Data Group online tables.
  (b) Because the strange quarks can oscillate to the down, the Λ^0 can decay, and one of the principal channels is

  $$\Lambda^0 \rightarrow p + \pi^-.$$

  Draw the lowest-order Feynman diagram for the decay. For simplicity, treat the baryons as fundamental rather than composite particles.

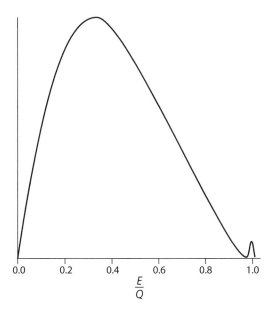

Figure 10.12. A schematic double beta decay spectrum which includes a Majorana peak at $E = Q$.

(c) Under the same assumptions, estimate (to within the correct order of magnitude) the timescale of the decay. As we have not yet discussed the strong force, take the masses of the d and $\bar{u}$ to be half the mass of the pion (This is a very gross and incorrect assumption that allows us to gloss over the strong interaction). The Cabibbo angle should appear explicitly.

10.2 Show that the decision to assign the Cabibbo angle to the lower component of the quark doublet is arbitrary.

10.3 The CKM mixing angles are hierarchical: $\theta_{12} \gg \theta_{23} \gg \theta_{13}$. To get a sense of the "size" of imaginary components, define the angles as $\theta_{12} \equiv \lambda \ll 1$, $\theta_{23} \equiv A\lambda^2$, and $\theta_{12} \equiv A\lambda^3(\rho - i\eta)$, as the last includes a complex CP-violating term. Expand out the CKM matrix to second order in λ.

This approach (though expanding to third order, rather than second) is known as the **Wolfenstein parameterization** of the CKM matrix.

10.4 One important way of describing the CP-violating terms in the CKM matrix is through the **Jarlskog invariant** J, defined as

$$\text{Im}[V_{ij} V_{kl} V_{il}^* V_{kj}^*] = J \sum_{m,n} \epsilon_{ikm} \epsilon_{jln},$$

where there is no sum on the left side. Extend the Wolfenstein parameterization to third order in λ, and compute the Jarlskog invariant to lowest nonvanishing order in λ.

10.5 Neutral kaons have a rest mass of about 0.5 GeV. They were produced in the Cronin-Fitch experiments with an energy of about 2.5 GeV, in equal amounts of the long- ($\tau = 5 \times 10^{-8}$ s) and short-lived ($\tau = 9 \times 10^{-11}$ s) particles. Assuming, wrongly, that CP symmetry is maintained in this system, what fraction of a kaon will be in the K_S state after 0.1, 10, and 20 m? This last distance is approximately the actual baseline of the Cronin-Fitch experiment.

10.6 Consider a subtly different formulation of CP asymmetry than that encountered in the text. A particular particle X_L decays with an amplitude $\mathcal{A}_1 + \mathcal{A}_2$, where both components are complex. The

CP-transformed particle $\overline{X}_R$ has a decay amplitude $\mathcal{A}_1 + \mathcal{A}_2^*$. What is the CP asymmetry

$$\frac{\Gamma_X - \Gamma_{\overline{X}}}{\Gamma_X + \Gamma_{\overline{X}}}$$

in terms of the phase angles ϕ_1 and ϕ_2 and the magnitudes of the amplitudes?

10.7 Antielectron neutrinos are produced in the sun with a typical energy of about 1 MeV. For the purpose of this problem, consider a two-state oscillation between muon and electron neutrinos. Given $\sin^2(2\theta_{12}) = 0.846$, and assuming $\Delta m_{12}^2 = 8 \times 10^{-5}$ eV, at what distance from their production site will the antineutrinos have undergone maximal oscillation (smallest fraction of electron antineutrinos)? What is the maximal fraction of muon antineutrinos? How does the oscillation scale compare with the emission region (the diameter of the sun)?

10.8 A nuclear reactor produces about 2 GW of power, approximately 2% of which is in the form of 4 MeV antineutrinos ($\overline{\nu}_e$). Assume a weak cross section of 10^{-47} m^2 and a two-neutrino model (for small distances and high energies, this is actually not a bad assumption).

    You are designing an experiment to measure neutrino oscillation by measuring the disappearance of electron antineutrinos. You place a "near" detector at a distance of 1 km and a "far" detector, to be placed at half the distance of maximum oscillation.
(a) How many antineutrinos are produced in the reactor per second?
(b) At what distance should you place the far detector? You may assume the neutrino oscillation parameters from the previous problem.
(c) Suppose you wish to design the experiment so that $\sim$100 events/year will be detected at the far detector in the absence of oscillation. How many target protons do you need?
(d) Using the parameters from Table 10.2.3, estimate the probability of oscillating to a τ neutrino in the near detector.

10.9 Let's do a quick estimate for the possibility of neutrinos as dark matter. Assume neutrinos are uniformly distributed in the galaxy and have a number density of 3.4×10^{11} m^{-3} (1000 times the universal average) and a total mass of 0.1 eV. What is the total mass (in solar units) contained within a sphere of radius 8 kpc (approximately 2.5×10^{20} m), roughly the sun's orbital radius around the center of the galaxy?

    This result can be compared with the observed stellar mass of approximately 10^{11} M$_\odot$ within the same radius.

10.10 The see-saw mechanism predicts that neutrino masses

$$m = \frac{m_D^2}{m_M},$$

where the Majorana mass m_M can be assumed to be of order the GUT energy, 10^{15} GeV.

    Assuming the Dirac masses m_D are the same as the corresponding up-type quarks in each generation, compute the masses and Δm_{ij}^2 values.

Further Readings

- Bigi, I. I., and A. I. Sanda. *CP Violation*. Cambridge: Cambridge University Press, 2000.
- Boehm, Felix, and Petr. Vogel. *Physics of Massive Neutrinos*. Cambridge: Cambridge University Press, 1992. Though this volume predates the experimental discovery of neutrino oscillation and several particle discoveries by more than half a decade, the theoretical underpinning of oscillations, Majorana masses, and the like, are given a very solid, very pedagogical presentation in this slim volume. Students may find Chapters 1 and 5 particularly relevant to this present discussion. Chapters 6 and 7 are also very helpful.
- Griffiths, David. *Introduction to Elementary Particles*, 2nd ed. Hoboken, NJ: Wiley-VCH, 2008. Readers will find Griffiths's Chapter 1 description of the history of elementary particles a good introduction.
- Kooijman, P., and N. Tuning. *CP Violation*. http://www.nikhef.nl/~h71/Lectures/2012/cp-080212.pdf. Kooijman and Tuning's lecture notes on CP symmetry violation are among the most complete and pedagogical around, with an extensive discussion of experimental confirmations.
- Nakamura, K., and S. T. Petcov. Neutrino Mass, Mixing, and Oscillations, from Chapter 11 in *The Review of Particle Physics*, K. A. Olive et al. (Particle Data Group), *Chin. Phys. C.*, 38, 090001, 2014.

11 The Strong Interaction

Figure 11.1. Murray Gell-Mann (1929–). Along with George Zweig, Gell-Mann found that SU(3) formed the basis of the strong interaction, as well as coined the word "quark." Credit: © Joi.

In *almost* every way that matters, the strong interaction is significantly cleaner than the electroweak model. The strong force is ambidextrous, unbroken, and acts upon only one family of fermions: the quarks. However, the strong interaction is significantly more complicated to probe experimentally, because quarks are never found in isolation, only in bound mesons and baryons. Unlike their counterparts in electromagnetism, the strong force carriers *themselves* have a strong charge, making many calculations possible only numerically. Even so, there's a great deal that the strong gauge symmetry can reveal about the internal working of baryons.

11.1 SU(3)

11.1.1 Color

The Eightfold Way was predicated upon an SU(3) symmetry [74]. Unfortunately, it was founded on an incorrect premise: that there were three *flavors* of quarks (up, down, and strange were all that were known at the time). Flavors matter, in that they govern the mass and electric charge of quarks, but so far as the strong force is concerned, all that matters is **color**—an analogy (and nothing more) with how humans perceive visible light [75, 170, 171]. By rather arbitrary convention, the quarks come in three colors, labeled *red*, *green*, and *blue*, which may be expressed as a column vector,

$$\Psi = \begin{pmatrix} q_r \\ q_g \\ q_b \end{pmatrix}, \tag{11.1}$$

and thus the strong force is often called **chromodynamics**. The color symmetry is unbroken. A red up quark and a blue up quark are identical in every measurable way, and a universal substitution of one for the other would leave *every* measurable quantity unchanged.

Color was introduced as a makeshift concept to make quarks of the same flavor distinguishable. The Δ^{++} baryon, for instance, comprises three up quarks, and even if two possible spin states are included, there is no way for all three quarks to satisfy the Pauli exclusion principle unless they are distinct in some way—labeled, for instance, with distinct colors.

As color is a perfect symmetry of the strong interaction, for the moment we'll take it as *axiomatic* that color can't be detected directly. Going a step further, we presuppose an important rule:

All free particles are color singlets.

For a baryon, we may suppose that our color-free rule implies that we need a red, a green, and a blue quark, while for mesons it would seem that a red-antired pair (for instance) would do nicely. But our rule about color singlets is more complex than that. Consider a meson with a color combination $|r\bar{r}\rangle$. The meson isn't a color singlet, because a rotation under SU(3) would turn the meson into $|g\bar{g}\rangle$ or some other linear combination of colors and anticolors. In fact, there are eight linear combinations of quark-antiquark colors that are simply rotations of another: the color octet. For instance,

$$\frac{1}{\sqrt{2}} \left[|r\bar{r}\rangle + |g\bar{g}\rangle \right] \quad \text{(one element of the color octet)} \tag{11.2}$$

By our ad hoc color rule, the only naturally occurring mesons—pions, for instance—are going to be color singlets:

$$|\pi\rangle = \frac{1}{\sqrt{3}} \left[|r\bar{r}\rangle + |g\bar{g}\rangle + |b\bar{b}\rangle \right] \text{(quark-antiquark singlet)} \tag{11.3}$$

Baryons are somewhat more complicated. It is not sufficient to simply say that the quarks each need to be of a different color. We know from our work on the exchange operator (equation 5.59) that swapping the label of two fermions (in which quarks are included) should yield a minus sign, so it takes a bit of thought (and application of the Levi-Civita tensor) to construct a baryonic state which is symmetric upon color swapping.

The only combinations of baryon states that will allow that are of the form

$$|p\rangle = \frac{1}{\sqrt{6}} [|rgb\rangle + |gbr\rangle + |brg\rangle - |bgr\rangle - |grb\rangle - |rbg\rangle] \quad \text{(three-quark singlet)}.$$

So far, it's just an assumption that free particles are color singlets. The explanation will have to wait until our discussion of asymptotic freedom.

11.1.2 The Gell-Mann Matrices

Colors are analogous to electric charge, but as there are three of them, the analogy clearly isn't perfect, since we're dealing with three possible charges, rather than two. To get a sense of what color charge really means, we start with the Lagrangian for a "free" quark (which we counterfactually pretend exists):

$$\mathcal{L} = i\overline{\Psi}_i \gamma^\mu \partial_\mu \Psi_i - m_i \overline{\Psi}_i \Psi_i. \tag{11.4}$$

The subscript i refers to the flavor (as opposed to the color) of the quark, and we implicitly sum over all flavors. The state Ψ refers to the three possible colors (equation 11.1).

The free-field Lagrangian is invariant under SU(3) symmetry transformation:

$$\Psi \to e^{-ig_s\vec{\theta}\cdot\vec{\lambda}}\Psi; \tag{11.5}$$

$$\overline{\Psi} \to \overline{\Psi}e^{ig_s\vec{\theta}\cdot\vec{\lambda}}. \tag{11.6}$$

We saw earlier (equation 4.22), that the SU(N) group will have $N^2 - 1$ orthogonal generators, which means that SU(3) will have eight, known as the Gell-Mann matrices:

$$\lambda_1 = \begin{pmatrix} 0 & 1 & 0 \\ 1 & 0 & 0 \\ 0 & 0 & 0 \end{pmatrix} \quad \lambda_2 = \begin{pmatrix} 0 & -i & 0 \\ i & 0 & 0 \\ 0 & 0 & 0 \end{pmatrix} \quad \lambda_3 = \begin{pmatrix} 1 & 0 & 0 \\ 0 & -1 & 0 \\ 0 & 0 & 0 \end{pmatrix}$$

$$\lambda_4 = \begin{pmatrix} 0 & 0 & 1 \\ 0 & 0 & 0 \\ 1 & 0 & 0 \end{pmatrix} \quad \lambda_5 = \begin{pmatrix} 0 & 0 & -i \\ 0 & 0 & 0 \\ i & 0 & 0 \end{pmatrix} \quad \lambda_6 = \begin{pmatrix} 0 & 0 & 0 \\ 0 & 0 & 1 \\ 0 & 1 & 0 \end{pmatrix}$$

$$\lambda_7 = \begin{pmatrix} 0 & 0 & 0 \\ 0 & 0 & -i \\ 0 & i & 0 \end{pmatrix} \quad \lambda_8 = \frac{1}{\sqrt{3}}\begin{pmatrix} 1 & 0 & 0 \\ 0 & 1 & 0 \\ 0 & 0 & -2 \end{pmatrix}.$$

The Gell-Mann matrices are very versatile. By inspection, λ_1, for instance, corresponds to the element of the quark-antiquark color octet in equation (11.2). The remainder of the elements are simply rotations of that combination in SU(3), which makes them observationally indistinguishable.

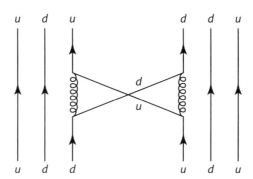

Figure 11.2. Proton-neutron interaction by an exchange of virtual pions. Don't be confused by the apparent swapping of proton (*udu*) for neutron (*udd*). The correspondence of "which particle is which" between the identities of the bottom particles and the top is simply a matter of how we've opted to twist the diagram.

The Gell-Mann matrices *also* describe the colors associated with gluons. Simply reading off from rows (anticolors) and columns (colors), we find that the gluon associated with λ_1 will be a superposition of $|r\bar{g}\rangle$ and $|g\bar{r}\rangle$ and allow a color exchange between red and green quarks.

There are eight Gell-Mann matrices, but naively we'd expect nine, since there are nine combinations of color-anticolor. The ninth is a singlet with a matrix representation of

$$\lambda_9 \overset{?}{=} \frac{1}{\sqrt{3}} \begin{pmatrix} 1 & 0 & 0 \\ 0 & 1 & 0 \\ 0 & 0 & 1 \end{pmatrix},$$

which is identical to equation (11.3), the color description of a free meson. If gluons were allowed to be found in the singlet state, then as massless particles, they would have an unlimited range and would be as omnipresent as photons. Experimentally, there are no free gluons, which means that gluons can't exist as color singlets. This raises the obvious question: if gluons are *not* color singlets, how do they communicate between protons and neutrons to hold together atomic nuclei?

The answer is, they don't. While gluons hold together protons and neutrons themselves, the internucleon strong force is effectively mediated by the pions (Figure 11.2). This was, indeed, Hideki Yukawa's original motivation in the 1930s for describing the potential that bears his name [168]. Yukawa's prediction of the pion predated its discovery by more than a decade.

In some sense, the short range of the strong interaction *between* nucleons is fundamentally different from the short range *within* them. Within nucleons, the range of the strong force is limited by **color confinement**.[1] Between nucleons, it's limited by the mass of the pion.

---

[1] A fancy name given to our rule about the colors allowed for free particles.

11.1.3 Gluon Dynamics

For the moment, we'll focus only on interquark forces and the dynamics of the gluons. We have seen this recipe before, in the discussion of the electroweak interaction. The key to determining the covariant derivative in the Lagrangian is calculation of the commutation between the generators. The SU(3) group is non-Abelian; the group members (or at least some of them) do not commute:

$$[\lambda_i, \lambda_k] = \sum_k 2i f^{ijk} \lambda_k.$$

Unlike with the Pauli matrices (for which f^{ijk} is the Levi-Civita symbol), the structure constants for the group need to be listed explicitly (problem 11.2),

$$f^{123} = 1; \quad f^{147} = f^{165} = f^{246} = f^{257} = f^{376} = \frac{1}{2}; \quad f^{458} = f^{678} = \frac{\sqrt{3}}{2},$$

and the same must be done for all cyclic permutations. Anticyclic permutations get a minus sign. That is, for instance, $f^{768} = -f^{678}$. We can quickly generate the Noether currents for the SU(3) symmetry as

$$J^{(k)\mu} = g_S \overline{\Psi} \gamma^\mu \lambda_k \Psi, \tag{11.7}$$

where g_S is the strong force coupling constant. The dynamical Lagrangian for gluons must be invariant under local SU(3) gauge transformations, but fortunately for us, we have a template for generating a gauge-invariant Lagrangian from our work with the SU(2) gauge (equation 8.21):

$$G_\mu^{(i)} \rightarrow G_\mu^{(i)} + \partial_\mu \theta^i - 2 \sum_{j,k} f^{ijk} G_\mu^{(j)} \theta^k. \tag{11.8}$$

The gluon fields can then be used to generate a covariant derivative,

$$\partial_\mu \rightarrow D_\mu = \partial_\mu + \frac{i}{2} g_S \vec{\lambda} \cdot \vec{G}_\mu, \tag{11.9}$$

where the vector refers to the eight-dimensional space spanned by the Gell-Mann matrices. And thus, the the Faraday tensor for gluons is

$$\begin{aligned} F_{\mu\nu}^{(i)} &= D_\mu G_\nu^{(i)} - D_\nu G_\mu^{(i)} \\ &= \partial_\mu G_\nu^{(i)} - \partial_\nu G_\mu^{(i)} + 2g_S \sum_{jk} f^{jkl} G_\mu^{(k)} G_\nu^{(l)}, \end{aligned} \tag{11.10}$$

which results in a chromodynamic Lagrangian,

$$\mathcal{L}_S = \overline{\Psi} \left(i\gamma^\mu \partial_\mu - m \right) \Psi - G_\mu^{(k)} J^{(k)\mu} - \frac{1}{16} F^{(k)\mu\nu} F_{(k)\mu\nu}. \tag{11.11}$$

We've achieved this result with somewhat less fanfare than with previous gauge theories.

A couple of comments are in order. As we shall see at the end of this chapter, the dynamic term in equation (11.11) isn't *really* the only possible form. It is also possible to introduce a CP-violating term, one that so far has evaded detection.

Second, following the same argument from the electromagnetic and electroweak theories, any addition of a mass term necessary spoils the gauge invariance; gluons are massless. That's okay because, as we've noted, the strong interaction uses an entirely different mechanism (color confinement, as opposed to a massive mediator) to keep the force local.

11.1.4 Feynman Rules for Quantum Chromodynamics

We have already seen how a classical Lagrangian can be turned into Feynman rules. Considering how closely related the strong interaction is to theories we've already seen, the modified Feynman rules may be read almost directly from the classical Lagrangian.

2 Vertex Factors.

Vertex	Contribution $\times \left[(2\pi)^4 \delta(\Delta p) \right]$	Representation
Strong (quark)	$-\sum\limits_{k=1}^{8} \dfrac{ig_S}{2} \lambda_k \gamma^\mu$	
Strong (three-gluon)	$-g_S f^{jkl} \left[g_{\mu\nu}(q_1 - q_2)_\lambda + g_{\nu\lambda}(q_2 - q_3)_\mu + g_{\lambda\mu}(q_3 - q_1)_\nu \right]$	
Strong (four-gluon)	$-ig_S^2 \left[f^{kli} f^{mni} (g_{\mu\lambda} g_{\nu\rho} - g_{\mu\rho} g_{\nu\lambda}) + f^{kni} f^{lmi} (g_{\mu\nu} g_{\lambda\rho} - g_{\mu\lambda} g_{\nu\rho}) + f^{kmi} f^{nli} (g_{\mu\rho} g_{\nu\lambda} - g_{\mu\nu} g_{\lambda\rho}) \right]$	

In the three-gluon contribution, all the momenta are assumed to be pointed in toward the vertex. Outgoing momenta should get a minus sign.

3 Propagators.

Propagator	Contribution	Representation
Gluons	$-i \displaystyle\int \dfrac{d^4 q_i}{(2\pi)^4} \dfrac{g_{\mu\nu} \delta^{jk}}{q_i^2}$	

In many ways, the strong force appears staggeringly similar to that of electromagnetism. The coupling constant is larger (how much larger turns out to be a function of scale), but otherwise, gluons essentially look like eight different types of photons, with the additional requirement that color must be conserved at every vertex.

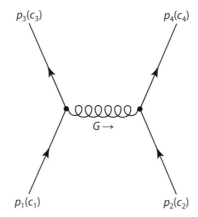

Figure 11.3. A two-vertex calculation of quark interaction.

11.1.5 *The Strong Force Yukawa Potential*

The similarities between the strong and electromagnetic interactions are obvious enough that it would be odd *not* to compute the Yukawa potential between two quarks.

As we've noted on multiple occasions, quarks don't appear in isolation, which means that we can't rightly consider "a red quark and a blue quark." Rather, we can take the nine possible combinations of two-quark colors and break them into an antisymmetric color triplet,

$$\frac{1}{\sqrt{2}}\left[|rb\rangle - |br\rangle\right]; \quad \frac{1}{\sqrt{2}}\left[|bg\rangle - |gb\rangle\right]; \quad \frac{1}{\sqrt{2}}\left[|gr\rangle - |rg\rangle\right], \tag{11.12}$$

which may be roughly interpreted as two quarks in different colors states, and a sextet,

$$|rr\rangle; \quad |gg\rangle; \quad |bb\rangle; \quad \frac{1}{\sqrt{2}}\left[|rb\rangle + |br\rangle\right]; \quad \frac{1}{\sqrt{2}}\left[|rg\rangle + |gr\rangle\right]; \quad \frac{1}{\sqrt{2}}\left[|bg\rangle + |bg\rangle\right], \tag{11.13}$$

which are monochromatic combinations. This is in direct analogy to the quark-antiquark "octet" (enumerated by the Gell-Mann matrices, representing pairs of different colors) and the color "singlet" (equation 11.3).

We'll calculate the force between two quarks of specific colors to get a sense of the interaction, but you'll have the opportunity in problem 11.3 to show that our result for quarks of different colors gives the sign (but a different coefficient) for a pair of quarks in one of the triplet states.[2]

For quarks with different flavors, the only two-vertex Feynman diagram will be the t-channel (Figure 11.3). We hold off for a moment on specifying the states but can

---

[2] This may seem like an odd choice. Why not simply compute the interaction of the triplet now? In short, because we haven't yet *proven* that quarks really exist only in color-singlet bound states. We're trying to motivate that now, which is why we start with pure-color states.

describe them more generally as

$$\Psi_1 = \begin{pmatrix} c_1^1 \\ c_1^2 \\ c_1^3 \end{pmatrix} u(1),$$

where the components of the column vector c_1 give the color or superposition color state of the quark. The Feynman rules for a quark-quark interaction thus yield the pleasing relationship

$$\mathcal{A} = -\frac{g_s^2}{4(p_1 - p_3)^2}[\overline{u}(3)\gamma^\mu u(1)][\overline{u}(4)\gamma_\mu u(2)] \sum_{k=1}^{8} (c_3^\dagger \lambda_k c_1)(c_4^\dagger \lambda_k c_2). \tag{11.14}$$

All the strong interaction information is contained in the summation over the possible mediator terms. Following equation (7.38), the amplitude more or less immediately produces a Yukawa potential,

$$V(r) = \frac{\alpha_s}{r} \frac{(c_3^\dagger \lambda_k c_1)(c_4^\dagger \lambda_k c_2)}{4}, \tag{11.15}$$

where the strong fine-structure constant α_s is defined as

$$\alpha_s = \frac{g_s^2}{4\pi}. \tag{11.16}$$

This definition is easier stated than quantified. It describes the strength of a *free* strong interaction, and no such thing exists, which is why we haven't yet given a numeric value for α_s. But we can dream, and for now, we'll assume that we may apply equation (11.15) by simply specifying the color states of the system.

Example 11.1: Using the strong Yukawa potential in equation (11.15), compute the interaction energy between two *free* quarks of differing color and flavor.

Solution: For concreteness, take c_1 (and c_3, as we are assuming no color exchange) in Figure 11.3 to be red, and c_2 (and c_4) to be green, though, clearly, the symmetry of the system suggests that this choice is somewhat arbitrary, so long as the particles differ.

The only Gell-Mann matrices which produce nonzero contributions are λ_3 and λ_8, which produce, respectively,

$$(c_3^\dagger \lambda_3 c_1)(c_4^\dagger \lambda_3 c_2) = \left[\begin{pmatrix} 1 & 0 & 0 \end{pmatrix} \begin{pmatrix} 1 & 0 & 0 \\ 0 & -1 & 0 \\ 0 & 0 & 0 \end{pmatrix} \begin{pmatrix} 1 \\ 0 \\ 0 \end{pmatrix} \right]$$

$$\times \left[\begin{pmatrix} 0 & 1 & 0 \end{pmatrix} \begin{pmatrix} 0 & 0 & 0 \\ 0 & -1 & 0 \\ 0 & 0 & 0 \end{pmatrix} \begin{pmatrix} 0 \\ 1 \\ 0 \end{pmatrix} \right] = -1$$

$$(c_3^\dagger \lambda_8 c_1)(c_4^\dagger \lambda_8 c_2) = \frac{1}{3} \left[\begin{pmatrix} 1 & 0 & 0 \end{pmatrix} \begin{pmatrix} 1 & 0 & 0 \\ 0 & 1 & 0 \\ 0 & 0 & -2 \end{pmatrix} \begin{pmatrix} 1 \\ 0 \\ 0 \end{pmatrix} \right]$$

$$\times \left[\begin{pmatrix} 0 & 1 & 0 \end{pmatrix} \begin{pmatrix} 1 & 0 & 0 \\ 0 & 1 & 0 \\ 0 & 0 & -2 \end{pmatrix} \begin{pmatrix} 0 \\ 1 \\ 0 \end{pmatrix} \right] = \frac{1}{3},$$

which sum to $-2/3$, and thus

$$V_{rg}(r) = -\left(\frac{1}{6}\right)\frac{\alpha_S}{r}.$$

Different colors attract one another, but as free quarks don't exist, and color isn't a fixed label, this result is only part of the story. In problem 11.3 you will find more generally that quark pairs in the triplet state attract with an interaction potential $V(r) = -\frac{2}{3}\frac{\alpha_S}{r}$.

Example 11.2: What is the Yukawa potential between two quarks of the same color?

Solution: The previous example may be followed directly but with the slight difference that the λ_3 contribution yields $+1$ rather than -1. Thus, $(c_3^\dagger \lambda_k c_1)(c_4^\dagger \lambda_k c_2) = \frac{4}{3}$, or

$$V_{rr}(r) = +\left(\frac{1}{3}\right)\frac{\alpha_S}{r}. \tag{11.17}$$

Quarks with like colors repel one another, just as particles with like electromagnetic charges do.

Even in the simplified Yukawa potential model, we can get a sense of why certain types of bound hadrons are allowed, but not others. Provided the quarks in a baryon are of different colors, any given one will attract the other two. However, if any two quarks are the same color, a baryon will be blown apart from pure energy considerations.

And as with electric charge, the effective strong charges are reversed for antiparticles. Mesons, which comprise quark-antiquark pairs, will preferentially be formed with the same colors. If one existed, an ambient quark of any color would find a meson to appear strong-force neutral.

While it seems energetically *reasonable* that naturally occurring composite particles are colorless (either rgb in quarks or color-anticolor combinations in mesons), we haven't yet explained why color is *never* found in isolation in nature. And to get a sense of that, we'll need to roll up our sleeves and compute Feynman diagrams with more than two vertices.

11.2 Renormalization

Every Feynman diagram that we've encountered so far has been designed to deliver the most information with the least amount of work. In electromagnetic interactions, we've gotten a lot of mileage out of simple two-vertex diagrams. Each additional vertex introduces a correction proportional to the fine-structure constant α_e ($\sim 1/137$)—suggesting that higher-order diagrams add nothing more than further precision to an already decent estimate of the scattering amplitude.

However, by the late 1940s, Julian Schwinger, Richard Feynman, Sun-Itiro Tomonaga, and Freeman Dyson realized that even a single closed loop would introduce an infinite perturbation [81]. They introduced a new approach known as **renormalization** to "sweep the infinities under the rug." We'll deal with the implications of this swindle in short order, but first, we'll need to see where the infinities come from and what renormalization is. To minimize our difficulties, we'll begin with a simple case: the mass of a scalar particle.

11.2.1 Bare and Effective Mass

In Chapter 6 we introduced a toy scalar model for two interacting fields: a massive field ϕ and a massless field η, which plays the role of an overly simplified photon (equation 6.1):

$$\mathcal{L} = \frac{1}{2}\partial_\mu \phi \partial^\mu \phi - \frac{1}{2}m^2\phi^2 + \frac{1}{2}\partial_\mu \eta \partial^\mu \eta - \lambda \phi^2 \eta.$$

While we generally group the dynamic and mass terms on one side of the ledger and interactions on the other, we've seen from our work on the Higgs field that the division is not so unambiguous between the two. Instead, we may quantize the ϕ^2 term in the Lagrangian to describe a point in spacetime wherein a ϕ gets annihilated and another gets created, with an amplitude of $\frac{1}{2}m^2$. This situation is illustrated in the left panel of Figure 11.4.

We don't normally think of mass terms as a straight-line diagram, but once we do, we open the door to introducing higher-order perturbations. For instance, we may put a loop in the middle the diagram—undetectable from the outside but presumably altering the amplitude (and hence the mass) of the ϕ particle. Computing the amplitude of the

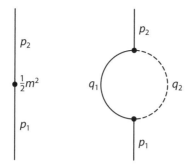

Figure 11.4. Left: The self-interaction of a ϕ particle in the toy scalar theory. Right: A second-order contribution of the ϕ self-interaction energy. Solid lines represent massive ϕ particles, while dashed lines are massless η particles.

second-order diagram from the Feynman rules, we obtain

$$A_2 = i \frac{\lambda^2}{(2\pi)^4} \int \frac{d^4q}{(q^2 - m^2)((q - p_1)^2 - m^2)}, \tag{11.18}$$

where we've substituted $q = q_2$ for tidiness.

On purely dimensional grounds, this integral is a cause for concern. We are, in principle, integrating over all momenta, which means that the integral takes the form $\int q^3 dq/q^4 = \ln(\infty)$, a most unpleasant state of affairs. This infinity arises because a propagator being created and annihilated at the same point necessarily produces a zero in the denominator. In renormalization, we imagine that a particle is created and absorbed at very *nearly* the same point to replace an infinity with something very large—of order the Planck scale. We'll soldier on and make the best of equation (11.18). We begin by neatening up a bit.

We first note the definite integral relation[3]

$$\frac{1}{ab} = \int_0^1 \frac{dx}{\left[ax + b(1 - x)\right]^2}.$$

Setting $b = (q^2 - m^2)$ and $a = [(q - p_1)^2 - m^2]$, equation (11.18) we can rewritte as

$$A_2 = i \frac{\lambda^2}{(2\pi)^4} \int_0^1 dx \int \frac{d^4q}{[q^2 - m^2 + p_1^2 x - 2q \cdot p_1 x]^2}. \tag{11.19}$$

This may not look like much of an improvement, since we've gone from a four-dimensional integral to a five-dimensional one. However, in "completing the square" the definite integral form will allow us to separate our second-order amplitude for the next stage in renormalization. We further apply the coordinate transformation

$$q \to q + x p_1,$$

---

[3] This and several other relations in this discussion are taken from Appendix E in J. J. Sakurai's *Advanced Quantum Mechanics* [145], who in turn credits Feynman. Sakurai's collection of integrals in renormalization is one of the most comprehensive and pedagogical approaches available.

which simplifies equation (11.19) to

$$\mathcal{A} = i \frac{\lambda^2}{(2\pi)^4} \int_0^1 dx \int \frac{d^4q}{[q^2 - m^2 + p_1^2 x(1-x)]^2}.$$

In principle we could solve for q^0 as a function of $\vec{q}$ and m, identify the poles, and perform a contour integral, as we did when we computed the scalar propagator (equation 7.25 and the work leading up to it).

But there's another approach. Integrating over Minkowski space is a nuisance, since space and time behave differently from one another. However, in the 1950s Gian Carlo Wick introduced an approach wherein we can treat the timelike coordinate on a par with the others [161]. Under a **Wick rotation** we perform the transformations

$$q_E^0 = iq^0; \quad q_E^i = q^i. \tag{11.20}$$

Under this transformation 4-vectors behave like ordinary four-dimensional Euclidean vectors (the subscript E is for Euclidean). This transformation yields the relations

$$d^4q = -i d^4\vec{q}_E \qquad d^4\vec{q}_E = 2\pi^2 q_E^3 dq_E \qquad q^2 = -q_E^2,$$

where the $2\pi^2$ is the solid angle over the surface of a four-dimensional unit sphere, and we've labeled $\vec{q}_E$ as a spacelike vector, as that's precisely what it is. Wick-rotating our second-order amplitude quickly gives us

$$\mathcal{A} = \frac{\lambda^2}{(2\pi)^4} \int_0^1 dx \int \frac{d^4 q_E}{[-q_E^2 - m^2 + m^2 x(1-x)]^2},$$

where we have utilized the fact that $p_1^2 = m^2$. As the amplitude is now isotropic in the Wick-rotated Euclidean space, we may directly write a perturbation to the ϕ mass as $\mathcal{A}_2 = \Delta \left(\frac{1}{2}m^2\right)$:

$$\Delta m^2 = \frac{\lambda^2}{4\pi^2} \int_0^1 dx \int_0^\infty dq_E \frac{q_E^3}{[q_E^2 + m^2(1-x+x^2)]^2}.$$

We see immediately that our worst fears are realized. If q_E is allowed to be integrated to infinity, the mass contribution will blow up. However, we can **regularize** the integral by imposing a cutoff at $q_E = M$, which we'll generally set as the Planck energy.[4] Integrating and taking only the leading terms, we get

$$\Delta m^2 \simeq \frac{\lambda^2}{8\pi^2} \ln \left(\frac{M^2}{m^2}\right) \tag{11.21}$$

in the limit of $M \gg m$.

What we measure in the lab isn't the bare mass m at all. Rather, it's the combination of

$$m_{\text{effective}} = m_{\text{bare}} + \Delta m.$$

---

[4] This is one of the crudest approaches to regularization, but by no means the only one. Others involve multiplying through by a factor of $M^2/(M^2 - q^2)$ which goes to unity at low energy and produces a cutoff term at $|q| > M$. Some involve adding additional terms that cancel the propagator at high energies. For our work, we will use a simple cutoff.

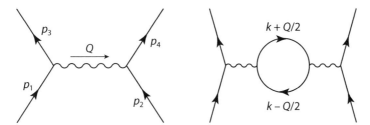

Figure 11.5. Left: The t-channel diagram for $e-e$ scattering. Right: One-loop correction to the electromagnetic force.

The effective mass of a particle is the one that enters into Newton's second law and everything else derived from it. The bare mass, on the other hand, is more or less irrelevant.

11.2.2 Effective Theories

Renormalization plays a role not only in the self-interaction of particles but also in their interactions with others. In understanding quarks—the central players in the strong interaction—it might help to first refer to their leptonic cousins. Indeed, the pioneers of renormalization did their work some two decades before quarks were even experimentally detected. I warn you in advance that renormalization of fermionic theories gets a little hairy mathematically, but the lessons learned from a familiar system (Møller scattering) will be immediately applicable to others.

The amplitude for the t-channel of Møller scattering (derived from the left panel of Figure 11.5) is

$$\mathcal{A}_1 = -4\pi\alpha_e[\overline{u}(3)\gamma^\mu u(1)]\frac{g_{\mu\nu}}{Q^2}[\overline{u}(4)\gamma^\nu u(2)], \tag{11.22}$$

where $Q \equiv p_1 - p_3$ is a measure of the overall energetics of the encounter, and where we have substituted for the fine-structure constant in anticipation of what follows.

Applying the Feynman rules for the right-hand diagram in Figure 11.5, we get

$$\mathcal{A}_2 = i\frac{\alpha_e^2}{\pi^2 Q^4}[\overline{u}(3)\gamma^\mu u(1)][\overline{u}(4)\gamma^\nu u(2)]\text{Tr}\int d^4k \frac{\gamma_\mu(\not{k}+\not{Q}/2-m)\gamma_\nu(\not{k}-\not{Q}/2-m)}{((k+Q/2)^2-m^2)((k-Q/2)^2-m^2)}. \tag{11.23}$$

There is a lot going on here, not least of which is that we've surreptitiously introduced two new rules with regard to closed fermionic loops:

1. Each closed fermionic loop multiplies the diagram by -1.
2. The contribution to the closed loop is found by following the fermions around a full loop (including the γ matrices) and taking the *trace* of the resulting terms.

Equation (11.23) looks a bit intimidating, but turning it into something more familiar is simply a matter of applying the various trace theorems in Appendix A (with especially judicious uses of equation A.16). As traces over all odd powers of γ vanish, the only nonzero contributions to the amplitude are

$$\text{Tr}\left[\gamma_\mu(\not{k} + \not{Q}/2 - m)\gamma_v(\not{k} - \not{Q}/2 - m)\right] = 4m^2 g_{\mu v} + 8k_\mu k_v - 4k^2 g_{\mu v} - 2Q_\mu Q_v + g_{\mu v}Q^2.$$

It can be shown (and we leave it as an exercise in problem 11.6) that the $Q_\mu Q_v$ terms identically vanish upon contraction with the spinors. We may further "simplify" the amplitude by following the scalar mass example to complete the square and apply a coordinate shift $k \to k + Qx$,

$$\mathcal{A}_2 = i\frac{4\alpha_e^2}{\pi^2 Q^4}[\bar{u}(3)\gamma^\mu u(1)][\bar{u}(4)\gamma^v u(2)] \int dx \int d^4k \frac{g_{\mu v}(m^2 - k^2 + Q^2 x(1-x)) + 2k_\mu k_v}{[k^2 - m^2 + Q^2 x(1-x)]^2} \tag{11.24}$$

where we have eliminated all odd powers of k and terms proportional to $Q_\mu Q_v$, which both vanish upon integration. The amplitude simplifies dramatically from the generalized relationship[5]

$$\int d^4k \frac{g_{\mu v}(\Delta - k^2) + 2k_\mu k_v}{(k^2 - \Delta)^2} = 0 \tag{11.25}$$

in four-dimensional spacetime. If $\Delta = m^2 - Q^2 x(1-x)$,

$$\mathcal{A}_2 = \frac{16\alpha_e^2}{Q^2}[\bar{u}(3)\gamma^\mu u(1)]g_{\mu v}[\bar{u}(4)\gamma^v u(2)] \int dx \int dk_E \frac{k_E^3 x(1-x)}{[k_E^2 + m^2 + |\vec{Q}|^2 x(1-x)]^2},$$

where we have performed a Wick rotation and substituted $|\vec{Q}|^2 = -Q^2$, since the scattering 4-vector has only spacelike nonzero terms.

This last form of the amplitude is remarkable. It includes exactly the $g_{\mu v}/Q^2$ contribution that mimics the first-order propagator. In other words, by comparison with equation (11.22), we may generate a perturbation to the fine-structure constant,

$$\Delta\alpha_e = -\frac{4\alpha_e^2}{\pi} \int dx \int dk_E \frac{k_E^3 x(1-x)}{[k_E^2 + m^2 + |\vec{Q}|^2 x(1-x)]^2},$$

which, to first order in $|\vec{Q}|^2$, can be written

$$\Delta\alpha_e \simeq -\frac{4\alpha_e^2}{\pi} \int dx \int dk_E \left[\frac{k_E^3 x(1-x)}{[k_E^2 + m^2]^2} - \frac{2k_E^3 x^2(1-x)^2|\vec{Q}|^2}{[k_E^2 + m^2]^3}\right], \tag{11.26}$$

Even if no momentum at all is transferred in the scatter, there is an enormous—infinite, in fact—correction to the fine-structure constant in the form of the first term on the right. If we regularize the integral over k at $k_E = M$, we get a *renormalized* fine-structure constant:

$$\alpha_R(0) = \alpha_e\left[1 - \frac{\alpha_e}{3\pi}\ln\left(\frac{M^2}{m^2}\right)\right]. \tag{11.27}$$

---

[5] This is one of many incredibly useful integrals which may be found in Appendix E of Peskin and Schroeder's *Introduction to Field Theory* [128]. For now, you may take this relation as fundamental.

Even if the limiting energy is at the Planck scale, this represents only about an 8% correction, and in reality, it doesn't matter anyway, since we measure the renormalized fine-structure constant in the lab, as opposed to the bare one. The energy-dependent term in equation (11.26) is perfectly convergent and yields a **running coupling constant**,

$$\alpha_R(Q^2) = \alpha_R(0) \left[1 + \frac{\alpha_R(0)}{3\pi} f\left(\frac{|\vec{Q}|^2}{m^2} \right) \right], \tag{11.28}$$

where, in generality,

$$f(u) = 6 \int_0^1 x(1-x) \ln(1 + xu(1-x)) dx.$$

This function produces $u \simeq u/5$ for $u = |\vec{Q}|^2/m^2 \ll 1$. At relativistic energies

$$f\left(\frac{|\vec{Q}|^2}{m^2} \right) \simeq \ln\left(\frac{|\vec{Q}|^2}{m^2} \right),$$

a limit we'll generally use when considering quark interactions.

From the sign of the results, it appears that electromagnetism grows *stronger* at higher energies or, equivalently at small scales.[6] This should not come as a surprise. The vacuum, as we've seen, consists of a bubbling sea of electron-positron pairs (and others) constantly popping into and out of existence. As the virtual positrons are attracted to the real electron and the electrons are repeled, the polarization **screens** the central electric charge much as charges are screened in a conducting material. What we observe in the lab, therefore, isn't the true electric charge but the screened version:

$$q_e = q_{bare} + \Delta q.$$

This is an example of an **effective field theory** in so far as we can essentially ignore the effects on small scales by swallowing them into the coupling constant. We can't measure directly what portion of the measured electric charge is the bare value and what is subtracted by polarization. Even at relativistic speeds, the fractional correction to the fine-structure constant is only $\sim 10^{-3}$, meaning that under many circumstances, it can be ignored.

11.2.3 *Implications of Renormalization*

If a single loop can contribute an infinity, then we can presumably exacerbate the problem by adding more loops (for instance, as illustrated in Figure 11.6).

Fortunately, we don't need to integrate every possible diagram of this sort. Instead, we may treat the third-order diagram as a perturbation on a perturbation, which in turn may be extended to fourth order, and so on,

$$1 + x + x^2 + \ldots = \frac{1}{1-x},$$

---

[6] The crossover point is roughly $l \sim 1/m_e$.

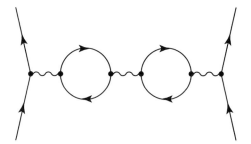

Figure 11.6. A third order $e-e$ scattering diagram.

and thus we can generalize the energy dependence on coupling constants as

$$\alpha_R(Q^2) = \frac{\alpha_R(0)}{1 - \frac{\alpha_R(0)}{3\pi} f(|\vec{Q}|^2/m^2)}. \tag{11.29}$$

Renormalization can be used to patch over myriad numerical sins with relatively little effort. But there is a cost. As Feynman described it [63]:

> The shell game that we play ... is technically called 'renormalization.' But no matter how clever the word, it is still what I would call a dippy process! Having to resort to such hocus-pocus has prevented us from proving that the theory of quantum electrodynamics is mathematically self-consistent. It's surprising that the theory still hasn't been proved self-consistent one way or the other by now; I suspect that renormalization is not mathematically legitimate.

Feynman's concerns have been echoed by countless others, but there's a sense in which renormalization must be valid—higher-order diagrams produce very precise measurements of the electron magnetic moment, for instance. But still, we should be cautious. As with the gauge-dependent terms in the electromagnetic stress-energy tensor, we recognize the problem of simply accepting a mathematical model because it produces sensible results that seem to accord with reality.

But *does* renormalization give sensible results? In the 1950s there was a great deal of concern that the charge of the electron could be perturbed in this way. It was known as the "zero charge problem" at the time and was premised on the idea that the bare charge of the electron depended, somehow, on where we chose our ultraviolet cutoff [81].

However, there's a sense in which these concerns can be a bit esoteric. We're quite used to abstracting away details on small scales. When chemists work with molecules, they generally may ignore the atomic nature of matter. When atomic physicists experiment on atoms, they may generally ignore interactions of quarks, and so on. At present, we will simply acknowledge the criticism as valid and move on.

How do we figure out whether a theory will lend itself to an infinity that we can subsequently "sweep under the rug"? Not every theory is as obliging as our toy scalar theory or even electromagnetism. The simplest test for renormalizability requires us to return to our constraints on the Lagrangian density, but in spacetime dimensionality d

(as opposed to simply assuming $d = 4$). A Lagrangian and the corresponding fields have units of

$$[\mathcal{L}] = [E]^d \qquad [\phi] = [E]^1 \qquad [A^\mu] = [E]^1 \qquad [\psi] = [E]^{3/2},$$

as we've seen. Thus, a coupling constant between n_B boson fields (scalar or vector) and n_F fermionic fields will have dimensionality

$$[\lambda] = [E]^{d - n_B - \frac{3}{2}n_F}.$$

In general, a theory is renormalizable if the coupling constant is a positive (or zero) function of energy. Our scalar theory has a constant with units of E^1, while the electromagnetic, weak, and strong coupling constants are all dimensionless (at least in 3+1 dimensions). Thus, the groups in the Standard Model are all renormalizable.

As a counterexample, consider general relativity with a coupling constant of $G = E_{Pl}^{-2}$. Gravity, it seems, is *not* a renormalizable theory—which has greatly held up its quantization. It is interesting to note that gravity would be normalizable in higher dimensions, and the Standard Model not normalizable in fewer than 3+1.

11.3 Asymptotic Freedom

Renormalization presents a number of conceptual problems in quantum electrodynamics, but when quarks were discovered in the late 1960s, a whole new set of complications was introduced. As David Gross described it [81]:

> [T]he paradigm of quantum field theory—Quantum Electrodynamics (QED)—was infrared stable; in other words, the effective charge grew larger at short distances and no one had ever constructed a theory in which the opposite occurred.

The strong interaction is just such a theory: the effective charges appear to grow *weaker* at short distances. In 1973, Gross and Frank Wilczek [82] simultaneously with H. David Politzer [132] derived a renormalization scheme for quantum chromodynamics (QCD) which explains (among many other things) the color confinement that we've been treating as axiomatic. Because gluons themselves have color, they don't just couple to quarks, they couple to each other (e.g., Figure 11.7). Quark loops and gluon loops produce very different signatures on quark-quark interactions.

11.3.1 *Quark Loops*

Consider the quark-loop contribution of quark-quark scattering. We've essentially already done this, as the calculation is nearly identical to electromagnetic loop electrons. There is a bit of a complication. In electromagnetism, electrons run free, and thus it's meaningful to talk about the coupling constant at zero energy. With quark interactions, the zero point will present a problem, and so we'll introduce a characteristic energy μ (typically in the

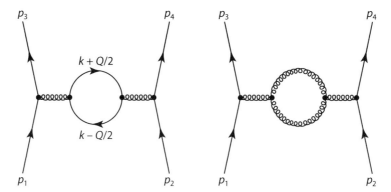

Figure 11.7. Left: A renormalizable quark loop. Right: A renormalizable gluon loop.

GeV-TeV range, depending on the context) such that the running coupling constant can be written as:

$$\alpha_S(|\vec{Q}|^2) \simeq \alpha_S(\mu^2)\left[1 + \frac{\alpha_S(\mu^2)}{3\pi}n_f \ln\left(\frac{|\vec{Q}|^2}{\mu^2}\right)\right]. \tag{11.30}$$

The factor n_f refers to the number of quark species (six, in the Standard Model), since any one of them can produce an internal loop.[7] As with electromagnetism, the creation of virtual quark loops does, indeed, increase the effective coupling constant at small scales. In contrast, the gluon loops have exactly the opposite effect, decreasing the effective coupling constant.

11.3.2 Gluon Loops

Application of the Feynman rules to the right-hand panel of Figure 11.7 produces (after a bit of work)

$$A_2 = i\frac{g_s^4}{4(2\pi)^4}\sum_{k,l,m,n=1}^{8}\int \frac{d^4k}{Q^4\left(\frac{Q}{2}-k\right)^2\left(\frac{Q}{2}+k\right)^2}[\overline{u}(3)\gamma^\mu u(1)][\overline{u}(4)\gamma^\nu u(2)]$$

$$\times f_{lmn}f_{kmn}[c_3^\dagger\lambda_k c_1][c_4^\dagger\lambda_l c_2]\left(g_{\mu\alpha}(k+3Q/2)_\beta - g_{\alpha\beta}(2k)_\mu + g_{\beta\mu}(k-3Q/2)_\alpha\right)$$

$$\times \left(g_\mu^\alpha(k+3Q/2)^\beta - g^{\alpha\beta}(2k)_\mu + g_\mu^\beta(k-3Q/2)^\alpha\right),$$

where $Q = p_3 - p_1$ is the 4-momentum of the scatter, and $k = q_3 - Q$. The various momenta simplify dramatically, as do the structure constants from the Gell-Mann matrices

$$\sum_{c,d=1}^{8} f_{acd}f_{bcd} = n_c\delta ab, \tag{11.31}$$

---

[7] Even this factor isn't *quite* right, since loops upon loops add correction factors of order unity. At the moment, we're most concerned with the sign of the effect and the approximate scaling.

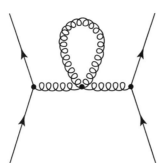

Figure 11.8. The gluon "tadpole."

where n_c is the number of colors (three in QCD). Applying the contraction yields

$$\mathcal{A}_2 = i\frac{\alpha_s^2}{4\pi^2}N_c[\overline{u}(3)\gamma^\mu u(1)][\overline{u}(4)\gamma^\nu u(2)][c_3^\dagger\lambda_k c_1][c_4^\dagger\lambda_k c_2]$$
$$\times\int\frac{d^4k}{Q^2\left(\frac{Q^2}{4}-k^2\right)^2}\left[g_{\mu\nu}\left(k^2+\frac{9}{4}Q^2\right)+5k_\mu k_\nu-\frac{9}{4}Q_\mu Q_\nu\right].$$

There are four energy-dependent terms in this perturbation:

- Terms proportional to $Q_\mu Q_\nu$ will vanish upon contraction with the spinor terms (as you will demonstrate in problem 11.6).
- Terms proportional to $g_{\mu\nu}Q^2$ will remain.
- Terms that include $g_{\mu\nu}k^2$ and $k_\mu k_\nu$ do not vanish by direct integration. Rather, there are several more diagrams, including Figure 11.8, that exactly cancel the contribution, as well as slightly modify the Q^2 term ($9/4 \to 11/4$).[8]

The nonvanishing contributions look very familiar. Simply following the example of the running coupling constant for electromagnetism allows us to quickly write a contribution to the running strong constant of

$$\Delta\alpha_S(Q^2) \simeq -11N_c\frac{\alpha_S(\mu^2)}{12\pi}\ln\left(\frac{|\vec{Q}|^2}{\mu^2}\right).$$

This relationship is only approximate, since we've neglected a lot of high-order diagrams. In the strong force, these contributions can be quite important.[9] At the energy

---

[8] We'd be remiss if we didn't point out a couple of complications. In 1967, Ludvig Faddeev and Victor Popov recognized that gauge symmetries in field theories will often introduce additional possible terms into the Lagrangian—and thus into the amplitude [57]. We saw something of this sort when we discussed the electromagnetic stress-energy tensor. Faddeev and Popov's **ghosts** are required to produce a single answer, regardless of gauge choice. Introduction of the ghosts is required to get all terms $\propto k^2$ to cancel. A deeper discussion of ghosts would take us too far afield.

[9] We have noted multiple times that our calculations are merely approximate and have hand-waved factor-of-2 corrections. Strong interactions typically don't converge at high energy, and quark confinement prohibits low-energy phenomena. As a result, computers need to be employed to effectively make strong force predictions.

The brute-force approaches typically involve creating a lattice of discrete points in spacetime and propagating gluons and quarks between them, rather than continuously. A lattice separated by spacing a cuts off energy at $\sim 1/a$, and as there are 19 orders of magnitude between the lightest baryons and the Planck scale, it would require $\sim 10^{76}$ lattice points to effectively model a proton.

scales of M_Z [119]

$$\alpha_S(M_Z^2) \simeq 0.118,$$

which then gets a multiplier for each gluon color (8). Thus, each higher-order correction in the strong interaction is of order unity. Strong calculations do not converge quickly.

11.3.3 Quark Confinement

Fermions screen, and gluons antiscreen under the strong force; the combined effect is

$$\alpha_S(Q^2) = \frac{\alpha_S(\mu^2)}{1 + \frac{\alpha_S(\mu^2)}{12\pi}(11n_c - 2n_f)\ln\left(\frac{|\vec{Q}|^2}{\mu^2}\right)}. \tag{11.32}$$

The dominant sign of the screening depends entirely upon the number of colors in the theory and the number of fermions. In this case

$$11 \times n_c - 2 \times n_f = 21,$$

which means that the gluons dominate. At low energies (large separations), $\ln(Q^2/\mu^2)$ becomes highly negative, and the running coupling constant goes to infinity once,

$$|\vec{Q}|^2 \lesssim \mu^2 \exp\left(-\frac{1}{\alpha_S(\mu^2)}\right),$$

which corresponds to roughly the few hundred mega- to giga-electron-volt range, or a confinement length of roughly

$$r \sim 10^{-16} \text{ m.}$$

Quarks flying around inside a neutron or proton experience a relatively weak interaction, but beyond a characteristic distance, forces becomes infinitely large. Thus, to liberate a quark, it is energetically favorable to create a quark-antiquark pair that leaves the remaining mesons and baryons colorless.

The opposite effect can be seen in electromagnetism, but the breakdown is at high energies (equation 11.28):

$$|Q| \simeq m_e \exp\left(\frac{3\pi}{2\alpha_E}\right) \simeq 10^{280} \text{ MeV.}$$

This point is known as a **Landau pole** after Lev Landau, who noted the constraint in the 1950s [96]. The very fact that such a pole can exist was one of the chief causes of the crisis of faith surrounding the entire endeavor of field theory. However, in the typical range of electromagnetic interactions, this pole never provides a practical barrier to analysis.

Higher-order quantum field theory calculations reveal a remarkable insight: the strength of fundamental interactions is a function of energy. The effective fine-structure constant, for instance, gets larger at higher energies as interacting particles get increasingly closer to the bare electric charge. There seems (see the cartoon in Figure 11.9) to be an approximate intersection of all three strengths at an energy scale of approximately

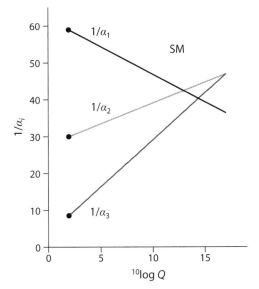

Figure 11.9. From the Particle Data Group [119], Fig 16.1. The Standard Model coupling constants nearly converge at high energies.

10^{15} GeV. This is the **Grand Unification** Theory (GUT) scale, and we will return to it in the next chapter.[10] By contrast, electroweak unification occurs on energy scales roughly 12 orders of magnitude lower, and there seems to be very little physics in between (Figure 1.5). This *desert* requires some sort of explanation, as do a number of other complications that we've so far politely ignored. The time has finally come to discuss the constraints on our ignorance.

Problems

11.1 Consider the meson color singlet

$$|\psi\rangle = \frac{1}{\sqrt{3}}\left[|r\bar{r}\rangle + |g\bar{g}\rangle + |b\bar{b}\rangle\right].$$

Apply a rotation by an angle θ using the λ_1 generator, and find the resulting color.

11.2 Compute the following commutators directly. Verify that they reproduce the structure constants given in the text.
(a) $[\lambda_1, \lambda_2]$
(b) $[\lambda_2, \lambda_8]$
(c) $[\lambda_7, \lambda_3]$

11.3 In example 11.1, we computed the interaction energy between two free quarks of differing flavors and colors and found that they are attractive. As there are no free quarks, compute, instead, the Yukawa

---

[10] It is telling that the three coupling constants do not precisely intersect one another, even once high-order corrections are included.

potential (equation 11.15) for two quarks in the antisymmetric quark triplet

$$\psi = \frac{1}{\sqrt{2}} = [|rb\rangle - |br\rangle].$$

11.4 Show that Wick rotation and contour integrals produce the same result for

$$\int \frac{d^4 k}{k^2 - m^2}$$

in the cutoff limit of high energy. Are they identical for finite cutoff? If not, why not?

11.5 Consider the mass correction for a "proton-like" scalar particle with a bare mass of 1 GeV, a dimensional coupling constant $\lambda = 0.1$ GeV, and an energy cutoff at the Planck scale. Use equation (11.21) to estimate the renormalized mass correction.

11.6 In the derivation of equation (11.24) we argued that

$$Q_\mu Q_\nu [\bar{u}(3)\gamma^\mu u(1)][\bar{u}(4)\gamma^\nu u(2)]$$

vanishes identically, where $Q = p_q - p_3$. Show that this is the case.

11.7 Consider the second-order correction to the t-channel for ϕ–ϕ scattering in our toy scalar theory,

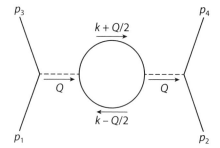

with the interaction Lagrangian,

$$\mathcal{L} = \lambda \phi^2 \eta,$$

where $m_\eta = 0$, $m_\phi = m$, and the ϕ particles are assumed to be distinguishable. In equation (7.26) we found a lowest-order amplitude:

$$\mathcal{A}_1 = -\frac{\lambda^2}{2 p_0^2 (1 - \cos\theta)}.$$

(a) Apply the Feynman rules and simplify, but do not renormalize, the second-order diagram.
(b) Use a Wick rotation to turn your amplitude into a one-dimensional integral over k_E.
(c) Evaluate the perturbation in the low-energy limit ($m \gg |Q|$).

11.8 For two massive scalar fields ϕ_A, ϕ_B and a massless scalar mediator η in 1+1 dimensional space, determine, and show, whether the interaction Lagrangian

$$\mathcal{L} = \lambda \phi_A \phi_B \eta$$

is renormalizable.

11.9 An electron propagator or mass diagram has a second-order correction,

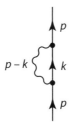

where we may consider the internal correction $\Sigma_2(p)$ as

$$\mathcal{A}_2 = \bar{u}(p)\Sigma_2(p)u(p)$$

for a mass correction, and similarly for a propagator. The self-energy Σ is what we'd get from the Feynman rules in the absence of the external spinors. It is a 4×4 matrix.

(a) Use the Feynman rules to compute Σ_2. Simplify as much as possible, but leave the term as an explicit integral over k.

(b) "Complete the square," shift $k \to k + px$, perform a Wick rotation, and evaluate the self-energy term with a regularization cutoff M. You may ignore finite terms.

(c) What is the contribution if this serves as a mass term correction?

11.10 In the calculation of bosonic loops, we used the relationship

$$\sum_{c,d} f_{acd}\, f_{bcd} = n\delta_{ab},$$

where n is the number of "colors" in an SU(n) gauge theory, f_{acd} is the structure constant for the group, and the sum is over all members of SU(n). Does this relationship hold for SU(2)? For $f^{abc} = \epsilon^{abc}$ compute the explicit structure constant contraction for $n = 2$.

11.11 Compute the explicit contraction of the Gell-Mann matrix structure constants

$$\sum_{b,c=1}^{8} f_{3bc}\, f_{8bc}$$

and

$$\sum_{b,c=1}^{8} f_{3bc}\, f_{3bc}.$$

Further Readings

- Cottingham, W. N., and D. A. Greenwood. *An Introduction to the Standard Model of Particle Physics.* Cambridge: Cambridge University Press, 1998. Though fairly brief, Chapters 16 and 17 on the strong interaction are a good introduction for the novice.
- Griffiths, David. *Introduction to Elementary Particles*, 2nd ed. Hoboken, NJ: Wiley-VCH, 2008. While Griffiths glosses over some of the more gruesome details, the discussion of asymptotic freedom and loop diagrams in Chapter 9 is instructive.
- Gross, Franz. *Relativistic Quantum Mechanics and Field Theory.* Hoboken, NJ: Wiley-VCH, 1993. Chapter 16 has an excellent discussion of renormalization schemes.
- Peskin, Michael E., and Dan V. Schroeder. *An Introduction to Quantum Field Theory.* Boulder, CO: Westview Press, 1995. The ambitious student will find the entire book worth a read, but in particular,

emphasis should be given to Chapters 11 and 12, which deal with renormalization, and to Appendix A, which has many invaluable relations.

- Quigg, Chris. *Gauge Theories of the Strong, Weak, and Electromagnetic Interactions*, 2nd ed. princeton, NJ: Princeton University Press, 2013. Chapter 8, on the strong interaction, is definitely worth a read.
- Sakurai, J. J. *Advanced Quantum Mechanics*. Boston, MA: Addison-Wesley, 1967. Sakurai's book is a classic, and Appendix E, in particular, has many gems for students wishing to perform practical renormalization calculations.

12 | Beyond the Standard Model

Figure 12.1. Howard Georgi (1947–). Georgi, along with Sheldon Glashow, was among the first to construct gauge Grand Unified Theories, including those based on SU(5) and SO(10).

12.1 Free Parameters

It is remarkable how wide a range of phenomena can be predicted from a relatively simple set of assumptions. But all the same, it's hard to shake the impression that there must be something deeper. Even in the "traditional" Standard Model, there are 19 free parameters, with another 7 arising from the observational fact of neutrino mass[1] (Figure 12.2).

Knowing nothing else, we might guess that every mass or energy should be at the Planck scale, and every dimensionless quantity should be 0, 1, 2π, or some other simple quantity. That none of the Standard Model parameters can be predicted from simple arguments is cause for at least some consternation. Some are so far from unity as to

---

[1] We are playing fast and loose with the limits of the Standard Model here. By convention, neutrino masses are generally considered outside the scope of the model, but at this point in history, I find this position difficult to defend.

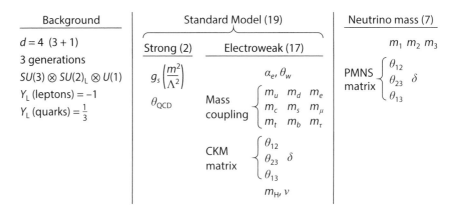

Figure 12.2. The tunable parameters of the Standard Model.

appear unnatural. By way of example, we've seen that while neutrino masses are naively expected to be either similar to the other fermions or zero—they're neither.[2]

We don't know why the parameters appear in nature as they do, but there are two broad categories of explanation:

1. There is an underlying and as-yet-undiscovered principle which will be able to predict all the seemingly arbitrary parameters.
2. The parameters (or some subset of them) are simply accidents of either history or the laws of nature.

If the parameters are accidental, there are enormous consequences for those accidents. If the up quark, for instance, were much more massive than the down, then neutrons would be the lightest baryon—a disaster for nuclear physics and chemistry as we know it, but that's not how nature played out.

There are many ways in which we have gotten lucky with regard to the parameters of the Standard Model. As Freeman Dyson argued in 1971 [51], even an increase of a few percent in α_S would mean that the strong force would become even more dominant in the interior of atoms, and diprotons would become energetically favorable to deuterium.

Even the number of fermion generations has proved fortuitous. Sakharov's criteria required a CP violation for a matter-antimatter asymmetry [144]. However, CP-violating terms enter the CKM matrix only if there are at least three generations of quarks. That there are exactly three generations seems to be a stroke of good luck.

The dimensionality of spacetime seems likewise to be unusually suited to complex life [136, 153]. Fewer than three macroscopic spatial dimensions produces too simple a physics (eliminating, for instance, cross products from physical theories), while more than three produces unstable orbits, even quantum mechanically.[3]

---

[2] This motivated our discussion of the Majorana mechanism, on which the jury is still out.
[3] This refers only to macroscopic dimensions. String theories generally rely on many more microscopic dimensions, but that discussion is well outside the scope of this book.

In 1973, Brandon Carter coined the term **anthropic principle** to describe the suitableness of the universe (both in general and at this particular cosmological epoch) to the sort of life capable of understanding it. The anthropic principle was refined further by John Barrow and Frank Tipler [27], who described the "weak" version:

> The observed values of all physical and cosmological quantities are not equally probable but they take on values restricted by the requirement that there exist sites where carbon-based life can evolve[4] and by the requirements that the universe be old enough for it to have already done so.

Merely giving a phenomenon a name doesn't really solve the problem, and we hasten to point out that this sort of speculation takes us dangerously far from what is normally considered predictive science. As with a number of similar issues discussed earlier in the book, we simply note an unresolved issue and move on.

A deeper and, we hope, resolvable issue is the the question, why these symmetries and not others? While translational and Lorentz invariance seem, in some sense, *cooked* into the equation of spacetime, the same cannot be said for $SU(3) \otimes SU(2)_L \otimes U(1)$. Could this be a reflection of something simpler?

12.2 Grand Unified Theories

At this remove in history, we simply accept electricity and magnetism as being two aspects of the same force, but it was the first great unification, begun by Maxwell and completed by Einstein. Likewise, one of the key insights into the development of the Standard Model centered around the electroweak unification, resulting in predictions of the Higgs boson, massive weak mediators, and much else.

Unification is natural. The laws of nature arise from a common source, so there's a sense in which the fundamental interactions shouldn't merely resemble one another but, looked at in an appropriate way or at sufficiently high energies, should really *be* identical. The structures of the electroweak and strong interactions are sufficiently similar to one another that it's reasonable to suppose that an accepted Grand Unified Theory (GUT) lies just beyond the horizon.

GUTs address a series of connected problems. The two families of fermions—quarks and leptons—are clearly related to each other, and yet one responds to the strong force, and the other only to the electroweak. Perhaps they are a subgroup of some larger symmetry. There is no consensus GUT, but we'd like to give a few tractable, falsifiable models to give a sense of what physics beyond the Standard Model might look like.

---

[4] Even to devotees of anthropic principles, the requirement of carbon-based life seems excessive. Most would accept any intelligent observer.

12.2.1 *SU(5)*

In 1974, Howard Georgi and Sheldon Glashow proposed arguably the first GUT in the sense that we now understand it: a symmetry group, SU(5), which is intended to bring together the electroweak and strong interactions [77].

SU(5) is special in a number of ways. We've seen that there are an infinite number of continuous groups of the form SU(N), SO(N), U(N), and their various combinations. However, the larger the symmetry group, the more fundamental particles in the theory, and the more mediators. For instance, we've seen that SU(N) has $N^2 - 1$ generators and thus $N^2 - 1$ mediators. To date, we've detected or inferred 12: the eight gluons predicted from SU(3), three weak mediators from SU(2), and the photon.

Out of a desire for simplicity, as well as to not aggravate the universe by predicting a bunch of new particles, we want the fundamental symmetry group to be as small as possible. SU(5) is the smallest group that contains SU(3), SU(2)$_L$, and U(1) as subgroups. But small as it is, SU(5) still has 24 mediators: twice the number we've detected experimentally.

Symmetry groups also provide a way of describing a fundamental particle in terms of a multiplet. For SU(5), one useful representation of a transformation and multiplet vector is

$$
\mathbf{M}\Psi = \left(\begin{array}{c|c} \text{SU}(2) & ? \\ \hline ? & \text{SU}(3) \end{array} \right) \begin{pmatrix} \nu \\ e^- \\ \bar{d}_r \\ \bar{d}_g \\ \bar{d}_b \end{pmatrix}_L , \tag{12.1}
$$

where the left-handed particles are enumerated explicitly, and block-diagonal components of the matrix representation of the group element illustrate that SU(2) and SU(3) are subgroups of SU(5).[5] The off-block-diagonal terms in principle allow a mixing between quarks and leptons.

If baryon and lepton number aren't conserved, then what is? Inspection of the multiplet vector for SU(5) reveals something interesting in that regard: both the color and the charge in the vector add to zero—a consequence of the fact that the generators of the group are traceless (required for "special" in SU(N)).

We can say a fair amount about the significance of the additional and off-diagonal elements in the transformation matrix. For one, they must contain 12 excess mediators, above and beyond those seen in the Standard Model. Since those terms can mediate interactions between antidown quarks and either electrons or neutrinos, conservation of electric charge dictates that the mediators must have charges of either +4/3 (X bosons, as they're known), or +1/3 (Y bosons).

---

[5] as is U(1), which simply introduces a phase over the whole matrix and which we will ignore for the time being.

But the abbreviated list of fundamental particles in equation (12.1) raises questions almost immediately. For instance, where are the two other generations of particles? The answer is that this, and most other GUTs, provide a framework for combining the quark and lepton sectors only within a single generation. Of deeper concern might be the conspicuous absence of up quarks. The Georgi-Glashow model deals with the missing particles by constructing them as antisymmetric superpositions of $\overline{\Psi}\overline{\Psi}$. There are 10 such particles that arise in this way:

	$\overline{\nu}$	e^+	d_r	d_g	d_b
$\overline{\nu}$	**0**	e^+	d_r	d_g	d_b
e^+	$-e^+$	**0**	u_r	u_g	u_b
d_r	$-d_r$	$-u_r$	**0**	$\overline{u}_b$	$\overline{u}_g$
d_g	$-d_g$	$-u_g$	$-\overline{u}_b$	**0**	$\overline{u}_r$
d_b	$-d_b$	$-u_b$	$-\overline{u}_g$	$-\overline{u}_r$	**0**

This representation, $\underline{10}$ (as the set of composite particles is colloquially known), contains all the downs, ups, and antiups, as well as the positron.

The SU(5) group generates 15 fundamental particles which are designated as left-handed—with their antiparticles designated as right-handed. A quick count leaves us with a bit of a problem: there is only a single left-handed neutrino and its antiparticle, the right-handed antineutrino. Neutrinos are, indeed, produced in a left-handed helicity state in the weak interaction, but they're only left-handed in *all* Lorentz frames if they are massless, and we know from neutrino oscillation measurements that at least two generations of neutrinos have mass. A working GUT should have room for a right-handed neutrino.

Even so, there is an enormous appeal to a theory like SU(5). Rather than three separate coupling constants, we anticipate only one, g_{GUT}, from which the other three would be expected to emerge from spontaneous symmetry breaking. We can even anticipate, based on our experience with the electroweak model, that the X and Y bosons should have masses similar to the GUT scale, around $10^{15}-10^{16}$ GeV.

This enormously high mass can help explain why it is that we've never produced X and Y bosons in the lab. But in principle, their presence could be inferred through the signature of **proton decay**,

$$p \rightarrow e^+ + \pi^0,$$

as illustrated in Figure 12.3. Proton decay is a fairly generic outcome of GUTs and is the final ingredient (along with CP violation) in the observed matter-antimatter asymmetry in the universe. It is a direct experimental test of conservation of baryon number.

We don't need to do a detailed analysis of the proton decay, as there are only two dominant effects which govern the amplitude: (1) the very massive mediator which contributes $\sim g^2/M_X^2$, and (2) the proton (the only other energy scale in the system) is

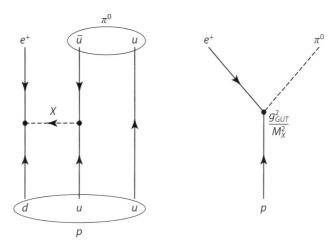

Figure 12.3. Left: A possible mechanism for proton decay in SU(5). Right: Simplified proton decay with the X-mediator treated as a point.

much, much more massive than either the outgoing pion or positron, and thus, by simple dimensional analysis[6]

$$\mathcal{A} \sim \frac{g_{GUT}^2 m_p^3}{M_X^2},$$

where we have ignored all dimensionless quantities. The decay rate may then be approximated as

$$\Gamma_p \sim \frac{\alpha_{GUT}^2 m_p^5}{m_X^4}. \tag{12.2}$$

That the decay rate is so sensitive to the mass of the X particle in some sense obscures dependence on other variables. Proton decay has not yet been observed, and the current experimental constraints are [116]

$$\tau_{\text{proton}} > 8 \times 10^{33} \, \text{yr}; \quad m_X \gtrsim 10^{16} \, \text{GeV},$$

which, unfortunately, is somewhat larger than the unification scale suggested by Figure 11.9—approximately 10^{15} GeV. A detailed analysis suggests that the proton lifetime in the Georgi-Glashow theory should be $\tau_{SU(5)} \lesssim 10^{32}$ yr, and thus the minimal model of SU(5) would seem to be ruled out [77, 116, 119].

Even if proton decay constraints didn't rule out SU(5), we'd still be left with a massless neutrino, unexplained mixing angles, a hierarchy problem, and a number of infinities to be explained. But at least a successful GUT will explain the baryon asymmetry in the universe, as well as the relative strengths of the fundamental forces.

---

[6] This is also extremely reminiscent of the Fermi model of neutron decay, wherein we substituted a very massive mediator for a dimensional coupling constant.

12.2.2 *Beyond SU(5)*

It's disappointing that the simplest possible unifying symmetry seems not to have been exploited by nature. Thus, choosing the next-best alternative becomes something of an aesthetic consideration. Georgi suggested SO(10), a supergroup of SU(5), as a candidate.[7] We have encountered SO(n) before, and in particular, SO(2) and SO(3). The groups represent rotations in n-dimensional real space. It may, at first, seem odd that real-valued rotations can be a supergroup of the complex transformations of SU(5), but bear in mind that groups like U(1) and SO(2) are simply isomorphisms of one another. They represent the same information—a rotation of a single angle—but presented in a different way and acting on a different type of object.

The SO(n) groups have $n(n-1)/2$ generators, which for SO(10) corresponds to 45 mediators, with the properties of the X and Y bosons we've seen previously. Most important, as SO(10) is a larger group than SU(5), there's room in the embedding of SO(10) for a right-handed neutrino, which is a shorthand for saying that SO(10) allows for neutrino masses.

A larger symmetry group also allows for a fair amount of freedom in the formulation of the various Higgs-like potentials. This, in turn, predicts significantly longer proton lifetimes, on the order of $10^{34}-10^{35}$ yr, depending on the choice of constants [93]. The good news is that there is a large parameter space yet to be explored by experiment.

The bad news is that the cost of detecting proton decay is proportional to the lifetime to be measured. The constraints on proton decay established by SO(10) and other simple groups like SU(8) and "flipped" SU(5), require a factor of 10 or more improvement in the current constraints.[8]

12.3 Supersymmetry

The Standard Model generally treats fermions as the "bosses," with bosons arising either primordially (Higgs) or via gauge symmetries (all the rest). There's no particular reason that particles should be segregated this way. **Supersymmetry**, or **SUSY** as it's frequently abbreviated, is predicated on unifying the particles of different *statistics*.[9] As early as the 1970s [43,58], a number of researchers began constructing supersymmetric theories, both for aesthetic reasons and, as we'll see, because SUSY addresses some very real shortcomings in the Standard Model.

The underlying premise of SUSY is that there is a whole host of particles beyond those we've already discovered, typically at higher energies than we've yet probed; for every known fermion, there is an undiscovered bosonic partner with a spin of 0, and for every

---

[7] Most accounts place the discovery in the same evening.

[8] Supersymmetric extensions of GUTs generally predict even longer lifetimes, making falsification tough but, in principle, doable.

[9] Though it may be clear from the context, "statistics" in this case refers to the distinction of fermions and bosons.

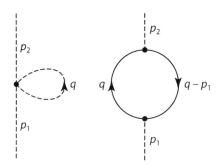

Figure 12.4. Two nearly identical diagrams which modify the self-interaction of a Higgs particle. Left: A second-order bosonic loop. Right: A second-order fermionic loop.

known boson, there's a new spin-1/2 fermion.[10] As to *why* we might suppose there's such an enormous number of unknown particles, we need to return to renormalization.

12.3.1 *The Higgs Mass*

The measured and bare mass of a particle are not the same, as we've seen. For the Higgs boson, however, the mass corrections have an interesting twist, depending on whether the virtual particle is a boson or a fermion (e.g., Figure 12.4).

Consider the interaction of a Higgs boson with a massive scalar particle with a dimensionless coupling constant λ_S. Application of the Feynman rules is straightforward and quickly yields a mass correction of

$$\Delta M_H^2 \simeq \frac{\lambda_S}{8\pi^2} \left[M^2 - m_B^2 \ln \left(\frac{M^2}{m_B^2} \right) \right], \tag{12.3}$$

where M is the high-energy mass cutoff, and m_B is the mass of the virtual boson. This simple loop provides some interesting insights:

1. Bosons give a positive mass perturbation to the Higgs. This is a fairly generic result. Whatever the spin of the boson (so long as it's an integer), the contribution to the amplitude will be positive.
2. The mass correction to the Higgs is of order the cutoff energy in the renormalization scheme—presumably the Planck mass.[11]

The fermionic contribution to the Higgs mass is a little more complicated. Assuming a coupling constant of $\lambda_F/\sqrt{2}$, we can quickly simplify the right-hand diagram in

---

[10] We give the forewarning that, to date, there's zero *direct* experimental evidence that supersymmetry manifests in nature. This is why it's physics *beyond* the Standard Model.

[11] Because we're solving for ΔM_H^2, this is known as a "quadratic divergence," but the contribution to the mass itself is linear.

Figure 12.4 to yield

$$\mathcal{A} = i \frac{\lambda_F^2}{2} \frac{1}{(2\pi)^4} \int d^4q \, \text{Tr} \left[\frac{(\slashed{q} + m_f)(\slashed{q} - \slashed{p} + m_f)}{(q^2 - m_f^2)((q - p)^2 - m_F^2)} \right],$$

which simplifies dramatically if we assume that the bare Higgs mass is small compared with m_F.[12] Doing so gives a mass correction to the Higgs:

$$\Delta M_H^2 = -\frac{\lambda_f^2}{4\pi^2} \left[M^2 - 3m_f^2 \ln \left(\frac{M^2}{m_f^2} \right) \right]. \tag{12.4}$$

Fermionic contributions *subtract* from bare Higgs mass, and twice as much in magnitude as the scalar correction.[13] Bosons and fermions would contribute potentially canceling linear mass corrections to the Higgs mass, with only logarithmic corrections remaining,[14] but only if there were an equal number of bosons and fermions in the Standard Model. There are not.

- Fermions: 90 degrees of freedom
  For each of three generations:

    – An up-type and a down-type quark, each with three possible colors and their antiparticles and two handedness states (24 total)
    – Charged lepton and its antiparticle in both spin states (4)
    – Left-handed neutrino and right-handed antineutrino (2)

- Bosons: 28 degrees of freedom

    – Photon in two spin states (2)
    – $W^\pm$ and Z^0, each in three spin states (9)
    – Scalar Higgs (1)
    – Eight massless gluons in two possible spin states (16)

There are *far* too many observed fermions.

For the experimental mass of the Higgs to be as small as it is, the bare mass would have to be on the order of the Planck scale. The differences of fermionic and boson loops would then combine to subtract off the bare mass from the renormalization correction to yield a final mass that is 1 part in 10^{17} times smaller than the original. That is some phenomenal fine-tuning!

We're not free to simply invent new particles to balance the contributions of fermions and bosons. Rather, if physics beyond the Standard Model is anything like physics within, new particles will come from new symmetries.

---

[12] It most certainly is, say, compared with the top quark. For low-mass terms, we get the same leading quadratic contribution, but the detailed calculation is more complicated.

[13] This is sensible, as fermions have twice as many spin states as do scalar particles, and thus twice as many degrees of freedom.

[14] Under those circumstances, the Higgs would be expected to be of the same order of mass as the heaviest fermions, but not the Planck scale.

12.3.2 What SUSY Is and What SUSY Does

The underlying prediction of all supersymmetric theories is straightforward: every Standard Model particle has a **superpartner** with the same color and charge and opposite statistics (fermion $\leftrightarrow$ boson). Naively, we'd expect the superpartners to also have the same mass as their "ordinary" counterparts, but this is manifestly not the case.[15]

Schematically, supersymmetric models introduce a new operator of the form

$$\hat{Q} \sim \sqrt{m}\hat{a}_b^\dagger \hat{c}_f; \quad \hat{Q}^\dagger \sim \sqrt{m}\hat{c}_f^\dagger \hat{a}_b,$$

where the subscript indicates whether the creation or annihilation operator is for a fermion or a boson.[16] There must be a supersymmetric operator per degree of freedom, but naturally, the masses and coupling constants will be very closely related to one another. Applying the commutation and anticommutation rules for bosons and fermions yields an anticommutator for the supersymmetric operator:

$$\{\hat{Q}, \hat{Q}^\dagger\} = m\left(\hat{a}_b^\dagger \hat{a}_b + \hat{c}_f^\dagger \hat{c}_f\right) = m(N_b + N_f). \tag{12.5}$$

This is simply the Hamiltonian for stationary particles!

If we introduce $\hat{Q}$ operators into the the fundamental Lagrangian, we immediately pave the way for a doubling of particles. Each of the known bosons acquires a fermionic spin-1/2 partner with names like the Zino, the photino, the higgsino, the gluino, and the Wino,[17] which is generally denoted with a tilde: $\tilde{Z}$. Likewise, fermions get a bosonic partner with identical charge but spin 0—the selectron, sneutrino, squark, and so on.

Under SUSY, conservation of spin, conservation of lepton number, and conservation of quark number clearly don't hold. Rather there is a new symmetry R,

$$(-1)^R = (-1)^{2s+3(B-L)},$$

which results in the three familiar conservation laws separately, only at low energy.

Supersymmetry must be broken, at least with regard to mass, as evidenced by the fact that we haven't produced a superpartner in a lab. This would be a simple feat if, for instance, selectrons and electrons had the same mass. Superpartners could act as virtual particles in scattering processes (Figure 12.5) and would be detected through high-precision measurements of cross sections.

However, superpartner masses can't be *too* high. If they were significantly larger than the Higgs, then the logarithmic terms in equations (12.3 and 12.4) wouldn't cancel nearly enough, and we'd be left with a *new* fine-tuning problem. Thus, the "natural" mass of superpartners should be in the range of a few hundred giga-electron-volts, the difference from ordinary particles a result of symmetry breaking.

---

[15] This is a very good thing for chemistry. If electrons could, via emission of a photino, transform into a selectron, then the most energetically allowed state would be all the bosonic selectrons in the ground state. Even if supersymmetry holds, the breaking of mass degeneracy helps us out by making electron to selectron transition energetically forbidden.

[16] This is a simplified model for particles at rest. Conceptually, it bears the same relationship to complete supersymmetric theories as did the quantum harmonic oscillator to bosonic fields.

[17] Pronounced, a bit stodgily, as the "weeno."

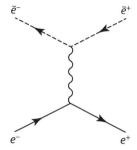

Figure 12.5. Supersymmetric pair production.

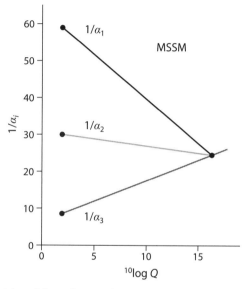

Figure 12.6. Adapted from the Particle Data Group [119], Fig. 16.1. Contra the result from the Standard Model, in the minimal supersymmetric model, all three coupling constants converge to within 1% at a GUT energy of approximately 10^{16} GeV.

Further, for the higgsino to mediate proton decay, it needs to have energies of order the GUT scale, some 13 orders of magnitude higher than the Higgs mass [135,157]. The "ordinary" Higgs typically becomes a doublet in supersymmetric models, while the higgsino is a color triplet. This unnatural difference between the two, known as the **doublet-triplet splitting problem** requires an extraordinary degree of fine-tuning—precisely the sort of thing supersymmetry was meant to address.

While not itself a Grand Unified Theory, supersymmetry was introduced because it solves a wide range of problems in the Standard Model and tends to complement GUTs. In addition to the mass corrections to the Higgs, SUSY adjusts the effective coupling constants for the Standard Model theories (e.g., equation 11.28). Recall that in nonsupersymmetric theories (Figure 11.9), the three coupling constants only *approximately* converged around 10^{15} GeV.

Under most SUSY models, the coupling constants meet at a point to within 1%, and the GUT scale increases by roughly an order of magnitude (Figure 12.6). This is aesthetically pleasing and also pushes the expected proton lifetime up by $\sim 10^{3-4}$, to timescales still allowed by experimental constraints.

Even more tantalizingly, superpartners might be potential candidates for dark matter particles. At first blush, this doesn't seem very promising. At masses of several hundred giga-electron-volts, it seems difficult to imagine that a supersymmetric particle will be stable. However, there are multiple species of **neutralinos** that are superpositions of the Zino, photino, and the higgsino. The lightest of these would have no natural decay channel and thus would be stable, abundant, and massive—precisely the desiderata for dark matter.

Supersymmetry solves a lot of problems, but is it correct?

The obvious smoking gun would be the experimental discovery of a superpartner. No such particles have been found in the energy ranges that have been explored. Phenomenological minimal supersymmetric models (pMSSM) seem to be ruled out by accelerator data at the LHC [2]. The high mass of the Higgs itself constrains the possible parameter space considerably. With 19 free parameters (including mass terms for all the partners), it is difficult to explore all possible models directly. But Monte Carlo analyses have suggested that a very small viable combination of parameters remains. The ATLAS collaboration has run analyses of more than 40,000 models, of which approximately 50 remain viable after the LHC Run 1. Supersymmetry can't be ruled out conclusively, but the prospects don't look good.

12.4 The Strong CP Problem

Neutralinos aren't the only dark matter candidate. Another possibility is introduced by a remaining mystery with the strong interaction. The strong force, so far as we know, is entirely CP invariant. But that need not be the case. An additional term may be added to the strong Lagrangian which satisfies the SU(3) and Lorentz symmetry requirements,

$$\mathcal{L}_{CP\ Strong} = \theta \frac{\alpha_S}{8\pi} F^{(i)}_{\mu\nu} \tilde{F}^{(i)\mu\nu}, \tag{12.6}$$

where the "tilded" version of the gluon fields are the permutation

$$\tilde{F}^{(i)}_{\mu\nu} = \frac{1}{2} \epsilon_{\mu\nu\sigma\rho} F^{(i)\sigma\rho},$$

and the ϵ object is a four-dimensional generalization of the Levi-Civita symbol. The strength and sense of the symmetry violation is governed by the value and sign of θ. This strong form may appear a bit confusing, so it's useful to consider what the electromagnetic analog would look like:[18]

$$\mathcal{L}_{CP\ EM} = \theta \frac{\alpha_e}{8\pi} \vec{E} \cdot \vec{B}. \tag{12.7}$$

---

[18] The electromagnetic version violates U(1), so there's no possibility of its existence. Nevertheless, we introduce it as a familiar form.

For any nonzero value of θ, this hypothetical interaction term is manifestly asymmetric under P transforms. The sign of $\vec{E}$ (as a vector) inverts under a P transform, but $\vec{B}$ (as an axial vector) doesn't. However, since both the electric and magnetic field yield a minus sign under a C transform, equation (12.7) is symmetric under C transforms and therefore asymmetric under CP (as is the corresponding term in the strong force).

An interaction Lagrangian of the form in equation (12.6) could, for instance, give rise to an electric dipole in the neutron, but so far no signal has been detected, yielding experimental estimates of θ that are unnaturally small [119]:

$$|\theta| < 10^{-10}.$$

Why should there be CP violation in the weak interaction but not the strong? In every gauge symmetry thus far introduced, every possible nonzero parameter which preserved the symmetry has been found to be nonzero.[19]

While it is possible that θ could be small simply by chance, a priori it seemed reasonable to suppose that it should lie with uniform probability between 0 and 2π. That the effective CP-violating charge should be as small as it is, is known as the **strong CP problem.**

In 1977 Roberto Peccei and Helen Quinn [125] suggested that the solution lay in introducing a *new* chiral symmetry for quarks, labeled $U(1)_{PQ}$:

$$\psi \rightarrow e^{-i\beta\gamma^5}\psi. \tag{12.8}$$

For a massless quark, the chiral symmetry is perfect. Indeed, if there was *any* massless quark, the chiral phase could be "rotated away" by symmetry, and $\theta = 0$ would simply be a gauge choice. However, all quarks have mass, which adds a decidedly non–gauge invariant term to the Dirac Lagrangian:

$$m\overline{\psi}\psi \rightarrow m\overline{\psi}\psi - 2i\beta\overline{\psi}\gamma^5\psi.$$

Peccei and Quinn introduced a new scalar field, the **axion**, to keep the Lagrangian invariant under the chiral symmetry. It may not be immediately obvious why the axion helps; if θ represents the phase of the axion (it does), then shouldn't the phase be uniformly likely around the unit circle?

The CP-violating term in the strong interaction allows for coupling between the axions and gluons, producing a "tilting" of a Higgs-like potential that yields a nonzero ground state which then *cancels* the value of θ selected by nature.

$$\mathcal{L}_{\text{Axion}} = \left(\overline{\theta} - \frac{\phi_A}{f_A}\right)\frac{\alpha_S}{8\pi}F^{(i)}_{\mu\nu}\tilde{F}^{(i)\mu\nu},$$

where ϕ_A is the axion field, f_A is the decay constant ($\sim 10^{10}$ GeV [119]), and $\overline{\theta}$ is the parameter to be canceled. This is quite a trick! The CP-violating term θ is randomly generated by some unknown process, but the Peccei-Quinn solution is constructed in just such a way to make it cancel identically.

---

[19] A possible exception is the CP-violating term in the neutrino PMNS matrix, for which no direct CP violation has been found, but that may speak to the weakness of neutrino interactions.

This is a lot of work to get rid of a term which experimentally seems to be zero, but in an important sense, it's testable. The axion itself is assumed to be a real particle, and just as the Higgs produces mass for the quarks and leptons, the coupling to the gluon field produces an axion mass of

$$m_A \simeq \frac{6 \times 10^{-7} \, \text{eV}}{f_A / 10^{10} \, \text{GeV}},$$

but current astrophysical and laboratory constraints put the possible axion mass at

$$m_A \lesssim 10^{-2} \, \text{eV} \quad \text{(astrophysical)}.$$

Even at those low masses, axions could have important cosmological implications: axions (if they exist) could be the elusive dark matter particle. They theoretically have many nice properties: they are numerous (produced, if they exist, at rough equilibrium with photons) and electrically neutral. Provided

$$m_A \gtrsim 2 \times 10^{-5} \, \text{eV} \quad \text{(dark matter)},$$

axions could contribute to some or all of the dark matter in the universe.

There is yet another problem. Particle detector constraints have steadily eaten away at the parameter space of possible axion masses and interaction cross sections [119]. Further, axions would have been produced early enough in the universe to produce **isocurvature** fluctuations in the density field—that is, a surplus of axions in one region of space would correspond to a deficit of other types of matter. However, large-scale observations of the distributions of galaxies have suggested that the universe was *not*, in fact, seeded by isocurvature fluctuations.

12.5 Some Open Questions

While a dark horse confirmation of SUSY or a successful Grand Unified Theory may address some of the holes in the Standard Model, those questions are just the tip of the iceberg. We don't know (and may never know) how finely tuned the Standard Model is, and for the most part, physicists have made their peace with the small knobs of nature. However, the Standard Model exists in a larger context of physics and cosmology, and some very important questions remain unanswered.

What Are Dark Matter and Dark Energy?

Most of the matter in the universe, approximately 85%, appears to be nonbaryonic. Constraints from galaxy rotation curves, gravitational lensing, and detailed analysis of the cosmic microwave background give every indication that dark matter is real, relatively cold (and thus relatively massive), and electrically neutral. But what is it? We've explored several possibilities, including axions,[20] neutrinos,[21] and the lightest supersymmetric partner.[22]

Dark energy, which accelerates the expansion of the universe, is even more troublesome. While we found that scalar fields can mimic dark energy, the true vacuum of the inflaton or Higgs field must be fine-tuned to enormous precision to account for the current dark energy density.

Likewise, the electroweak vacuum energy density behaves a lot like dark energy in terms of the equation of state but is a factor of 10^{120} times larger than the observed value. SUSY, naively, would seem to address this question, since a virtual photino field should partially counteract the electromagnetic field. But, alas, because the SUSY is a broken symmetry, even if it is correct, the SUSY Casimir effect would leave us with a vacuum energy "only" 10^{60} larger than the observed dark energy density. While technically an improvement, it seems like a Pyrrhic victory.

Strictly speaking, dark energy need not be reconciled by particle physics at all. General relativity allows for the possibility of a true cosmological constant embedded in the Einstein field equations. While this is unsatisfying, it leaves a question that at least may be answered by another branch of physics.

How Will Gravity Be Unified with the Other Forces, If At All?

The Casimir effect suggests that a nonzero vacuum energy is, in fact, real and large. And yet the smallness of dark energy suggests that gravity in some sense doesn't "see" all the source terms in the stress-energy tensor. What is the interplay between general relativity and the Standard Model?

While quantum field theory may be performed on a curved spacetime,[23] we've made significantly less progress addressing the converse question of how space is quantized. There are numerous approaches, from loop quantum gravity [19, 141] to superstring theory [164], both of which are beyond the scope of this book. We've seen that a quantized gravity is a nonrenormalizable theory, but it remains unclear whether there is a graviton and whether it can be experimentally detected.

Why Is There an Energy Desert?

Electromagnetic unification becomes important on a scale of mega-electron-volts. The electroweak force comes in at 100 GeV or so. But most GUT models tend to unify the three gauge forces at around 10^{16} GeV. There are 14 orders of magnitude in energy where a desert of new energy is expected (Figure 1.5). Why is this? It also seems strange that there are only three orders of magnitude between the GUT scales and the Planck scale.

---

[20] Which haven't been detected, and would produce primordial density perturbations inconsistent with observations.

[21] Which appear far too light to do the job. However, if neutrinos are Majorana particles, and the Majorana scale is low enough, sterile neutrinos could act as dark matter. But those are a lot of "ifs."

[22] Whose prospects are tightly bound with SUSY itself, and thus quite dim.

[23] This is the origin of the hypothetical Hawking radiation surrounding black holes [86].

We have frequently commented on the peculiarity of so much low-energy physics given that the most natural energy scale is the Planck, but the enormous gap between where we're able to probe experimentally and the Planck limit almost seems like a dare from nature.

Hierarchy, Parsimony, and All That

Why are the apparent strengths of the fundamental interactions in the Standard Model so different from one another under normal laboratory conditions? Why is the Higgs so much smaller than the Planck scale? We're beset by a number of hierarchy problems, but they all seem to arise from the fundamental issue that we are unable to set any absolute scales without inserting them manually.

It seems strange, to say the least, that the Higgs mass and other physical quantities are as finely tuned as they are without some governing physical principle at work, and yet we've seen that the extended Standard Model has 26 free parameters. Many physicists have a perhaps misguided belief in *parsimony*, what we might generally call Occam's razor.

This is something of an unfortunate state of affairs, since we're unable to say anything more than that we find it *distasteful* to have to set so many dials by hand. Science, of course, is supposed to be predictive, but as we have only universe to experiment on, we don't actually know at this point whether we'll ever be able to simplify a Theory of Everything to a point that won't offend our delicate aesthetic sensibilities.

Problems

12.1 Based on the construction of SU(5), give the charges and colors of the six elements in the upper right and lower left blocks of equation (12.1), and identify them as X, Y, $\overline{X}$, or $\overline{Y}$ bosons.

12.2 In our discussion of Grand Unified Theories, we found an approximate (and fairly general) relationship for proton decay:

$$\Gamma \sim \alpha_{GUT}^2 \frac{m_p^5}{M_X^4},$$

where the constraints on the rate are found by observing N protons for T years and in the absense of a decay, setting the lower bound to $\tau > NT$. (As we saw in problem 7.2, this isn't quite right, but we'll ignore the constant factor of a few.) Assume that $\alpha_{GUT} = 1$ and that the prefactors to the decay rate are also 1.

The Super-Kamiokande experiment comprises an underground pool of 50,000 tons of ultrapure water. Assume that all potential proton decays would be detectable via Cerenkov radiation.

(a) How many protons are in Super-K hydrogen?
(b) Suppose no proton decays were detected in 3 years. What is the minimum lifetime of the proton?
(c) What does that imply about the minimum mass of the X-boson?
(d) Suppose, instead, we observe for 100 years. How does that change your answer in part (c)?

12.3 Compute the anticommutator $\{\hat{Q}, \hat{Q}^\dagger\}$ from the supersymmetric operators, and verify the Hamiltonian (equation 12.5).

Further Readings

- Aitchison, Ian. *Supersymmetry in Particle Physics: An Elementary Introduction*. Cambridge: Cambridge University Press, 2007. This relatively slender volume provides a good overview of the minimal supersymmetric model (MSSM).
- Georgi, Howard, *Lie Algebras in Particle Physics: From Isospin to Unified Theories*, 2nd ed. Boulder, CO: Westview Press, 1999. Chapters 18 (on SU(5)) and 24 (on SO(10)) are especially relevant for discussions of physics beyond the Standard Model.
- Peacock, John A. *Cosmological Physics*. Cambridge: Cambridge University Press, 1999. While it's a little out of date at this remove, Chapter 8 has a nice discussion of the interplay of Grand Unified Theories and cosmology.
- Sikivie, P. The pooltable analogy to axion physics. *Phys. Today*, 49N12:22-27, 1996. While nontechnical in nature, this paper is a very endearing description of the beauty (and the seeming swindle) of the Peccei-Quinn mechanism for resolving the strong CP problem.
- Susskind, Leonard. (2010). *Supersymmetry and Grand Unification*. Downloaded from http://theoreticalminimum.com/courses/particle-physics-3-supersymmetry-and-grand-unification/2010/spring. Susskind has a delightful "order-of-magnitude" approach to supersymmetry and, in particular, in his discussions of renormalization and SU(5) (lecture 10), which connect to the material presented both in this and in the previous chapter.

Appendix A | Spinors and γ-Matrices

The Dirac equation for an isolated particle is written

$$\left(\pm \gamma^\mu p_\mu - m\right) \psi(p) = 0,$$

where the γ-matrices are written

$$\gamma^0 = \begin{pmatrix} 0 & \mathbf{I} \\ \mathbf{I} & 0 \end{pmatrix} \qquad \gamma^i = \begin{pmatrix} 0 & \sigma_i \\ -\sigma_i & 0 \end{pmatrix}, \tag{A.1}$$

where the Pauli matrices are defined by

$$\sigma_1 = \begin{pmatrix} 0 & 1 \\ 1 & 0 \end{pmatrix} \qquad \sigma_2 = \begin{pmatrix} 0 & -i \\ i & 0 \end{pmatrix} \qquad \sigma_3 = \begin{pmatrix} 1 & 0 \\ 0 & -1 \end{pmatrix} \tag{A.2}$$

subject to the commutation relation

$$\left[\sigma_i, \sigma_j\right] = 2i \sum_k \sigma_k \epsilon_{ijk}. \tag{A.3}$$

Likewise, the γ-matrices have the anticommutation relations

$$\{\gamma^\mu, \gamma^\nu\} = 2g^{\mu\nu}. \tag{A.4}$$

There is also the helicity gamma matrix γ^5

$$\gamma^5 = i\gamma^0 \gamma^1 \gamma^2 \gamma^3 = \begin{pmatrix} -\mathbf{I} & 0 \\ 0 & \mathbf{I} \end{pmatrix}, \tag{A.5}$$

which obeys the anticommutation relation

$$\{\gamma^5, \gamma^\mu\} = 0 \tag{A.6}$$

for all four of the γ-matrices.

4-vectors may also be contracted with the γ-matrices according to

$$\not{p} = \gamma^\mu p_\mu . \tag{A.7}$$

BISPINORS

For particles propagating along the z-axis, there are 4 orthogonal solutions to the Dirac equation:

$$u_{\pm}(p) = \frac{m}{\sqrt{E+p^z}}\begin{pmatrix} 1 \\ 0 \\ \frac{E+p^z}{m} \\ 0 \end{pmatrix}; \quad \frac{m}{\sqrt{E-p^z}}\begin{pmatrix} 0 \\ 1 \\ 0 \\ \frac{E-p^z}{m} \end{pmatrix} \tag{A.8}$$

$$v_{\pm}(p) = \frac{m}{\sqrt{E-p^z}}\begin{pmatrix} 0 \\ 1 \\ 0 \\ \frac{-E+p^z}{m} \end{pmatrix}; \quad \frac{m}{\sqrt{E+p^z}}\begin{pmatrix} 1 \\ 0 \\ -\frac{E+p^z}{m} \\ 0 \end{pmatrix}. \tag{A.9}$$

where the adjoint is defined by

$$\overline{\psi} \equiv \psi^{\dagger}\gamma^0. \tag{A.10}$$

The orthogonality condition is given by

$$\overline{u}_r(p)u_s(p) = 2m\delta_{rs}; \tag{A.11}$$

$$\overline{v}_r(p)v_s(p) = -2m\delta_{rs}, \tag{A.12}$$

and all contractions with u and v spinors (with the same 4-momentum) are zero.

CASIMIR'S TRICK

Many quantum field theory calculations produce amplitude terms of the form

$$\mathcal{A} \sim [\overline{u}(1)\gamma^{\mu}u(2)][\overline{u}(3)\gamma_{\mu}u(4)].$$

Casimir's trick involves computing the square of the amplitude over all combinations of spin states. While we do not prove the relationship, it can be shown that

$$\sum_{\text{All spins}} [\overline{u}(1)\gamma^{\mu}u(2)][\overline{u}(2)\gamma^{\nu}u(1)]^* = \text{Tr}[\gamma^{\mu}(\not{p}_2 + m_2)\gamma^{\nu}(\not{p}_1 + m_1)], \tag{A.13}$$

where typical scattering processes will involve two such traces. While we have written γ^{μ}, and γ^{ν} for concreteness, in reality, these may represent *any* 4×4 matrices constructed from products of γ-matrices, including combinations like $\gamma^{\mu}\gamma^5$.

For antiparticles, there is the similar relation

$$\sum_{\text{All spins}} [\overline{v}(1)\gamma^{\mu}v(2)][\overline{v}(2)\gamma^{\nu}v(1)]^* = \text{Tr}[\gamma^{\mu}(\not{p}_2 - m_2)\gamma^{\nu}(\not{p}_1 - m_1)]. \tag{A.14}$$

For cross terms (interference), things get a bit more complicated. For instance, it is not uncommon to have terms like

$$\sum_{\text{All spins}} [\overline{u}(1)\gamma^{\nu}u(3)][\overline{u}(3)\gamma^{\mu}v(4)][\overline{v}(4)\gamma_{\nu}v(2)][\overline{v}(2)\gamma_{\mu}u(1)]$$

$$= \text{Tr}\left[(\not{p}_1 + m_1)\gamma^{\nu}(\not{p}_3 + m_3)\gamma^{\mu}(\not{p}_4 - m_4)\gamma_{\nu}(\not{p}_2 - m_2)\gamma_{\mu}\right], \tag{A.15}$$

where we've assumed a cross term between fermions (1,3) and antifermions (2,4). This generalizes in a straightforward way.

Evaluation of the traces requires the relationships

$$\text{Tr}(\gamma^\mu) = 0$$

$$\text{Tr}(\gamma^\mu \gamma^\nu) = 4g^{\mu\nu}$$

$$\text{Tr}(\gamma^\mu \gamma^\nu \gamma^\lambda) = 0$$

$$\text{Tr}(\gamma^\mu \gamma^\nu \gamma^\lambda \gamma^\kappa) = 4(g^{\mu\nu}g^{\lambda\kappa} - g^{\mu\lambda}g^{\nu\kappa} + g^{\mu\kappa}g^{\nu\lambda})$$

$$\text{Tr}\left[\gamma^5 \gamma^\mu \gamma^\nu \gamma^\lambda \gamma^\sigma\right] = 4i\epsilon^{\mu\nu\lambda\sigma},$$

which can be used to eliminate all odd powers of γ-matrices. The permutation tensor $\epsilon^{\mu\nu\lambda\sigma}$ is the four-index analog of the Levi-Civita, but with $\epsilon^{0123} = 1$, and all elements with repeated indices equal to zero. Permutation of adjacent indices produces a reversal of sign.

A common trace term in Feynman diagrams involves variants of

$$\text{Tr}[\gamma_\mu \, \slashed{A} \gamma_\nu \, \slashed{B}] = 4[A_\mu B_\nu + A_\nu B_\mu - A \cdot B g_{\mu\nu}]. \tag{A.16}$$

In generality,

$$\sum_{\text{All spins}} [\bar{u}(1)\gamma^\mu u(2)][\bar{u}(2)\gamma^\nu u(1)]^* = 4\left[p_1^\mu p_2^\nu + p_1^\nu p_2^\mu + g^{\mu\nu}[m_1 m_2 - p_1 \cdot p_2]\right] \tag{A.17}$$

and similarly for antiparticles.

Casimir's trick simplifies dramatically in a couple of limits:

- In the limit of two stationary (or at least nonrelativistic) particles,

$$\vec{p}_1 = \vec{p}_2 = 0$$

yields

$$\sum_{\text{All spins}} [\bar{u}(1)\gamma^0 u(2)][\bar{u}(2)\gamma^0 u(1)]^* = 8m_1 m_2$$

for $\mu = \nu = 0$, and zero otherwise.
- For one massless particle:

$$\sum_{\text{All spins}} [\bar{u}(1)\gamma^\mu u(2)][\bar{u}(2)\gamma^\nu u(1)]^* = 4[p_1^\mu p_2^\nu + p_1^\nu p_2^\mu - g^{\mu\nu} p_1 \cdot p_2],$$

which for $\mu = \nu = 0$ yields

$$\sum_{\text{All spins}} [\bar{u}(1)\gamma^0 u(2)][\bar{u}(2)\gamma^0 u(1)]^* = 4[E_1 E_2 + \vec{p}_1 \cdot \vec{p}_2].$$

PARITY

The electroweak model does not respect P symmetry. As a result, it's useful to decompose spinors into left-handed and right-handed components using the γ^5 matrix

$$\psi_{L,R} = \frac{1}{2}\left(1 \mp \gamma^5\right)\psi, \tag{A.18}$$

where

$$\psi = \psi_L + \psi_R .$$

It is straightforward to show from the anticommutation relation (equation A.6) that

$$\overline{\psi}_L \psi_L = \frac{1}{4}\psi^\dagger \left[(1 - \gamma^5)\gamma^0(1 - \gamma^5)\right]\psi = 0, \tag{A.19}$$

with a similar relation for the right-handed terms. Thus,

$$\overline{\psi}\psi = \overline{\psi}_L \psi_R + \overline{\psi}_R \psi_L . \tag{A.20}$$

Appendix B | Decays and Cross Sections

DECAYS

An N-particle decay rate may be calculated via

$$\Gamma = S \int |\mathcal{A}|^2 (2\pi)^4 \delta(\Delta p) \frac{1}{2M} \prod_{\text{out states}} \left(\frac{d^3 p_i}{(2\pi)^3} \frac{1}{2E_i} \right) \tag{B.1}$$

in the center-of-mass frame. M is the mass of the decaying particle, S is the degeneracy factor for outgoing particles, and $\mathcal{A}$ is the amplitude of the process.

TWO-BODY DECAY

For a two-body decay, the rate may be dramatically simplified:

$$\Gamma = S|\mathcal{A}|^2 \frac{p_F}{8\pi M^2}, \tag{B.2}$$

where p_F is the magnitude of the uniquely determined outgoing momentum

$$p_F = \frac{1}{2M} \sqrt{(M - m_1 - m_2)(M - m_1 + m_2)(M + m_1 - m_2)(M + m_1 + m_2)}. \tag{B.3}$$

For larger numbers of outgoing particles, the result is not uniquely determined. Indeed, for N outgoing particles, there are $3N - 7$ free parameters for the decay.

THREE-BODY DECAY

For a three-body decay, the outgoing energies cannot be determined uniquely and depend instead on the directions of the particles (Figure B.1).

The decay rate may be simplified to a three-dimensional integral in full generality:

$$\Gamma = S \iiint \frac{|\mathcal{A}|^2}{(4\pi)^3 M} \frac{p_1^2 p_2}{E_1 E_3(E_1, E_2, \theta)} dp_1 d E_2 d(\cos\theta) \delta\left(M - E_1 - E_2 - E_3\right),$$

where the relationships between p_1 and E_1 are governed by the normal dispersion relation. Likewise, conservation of momentum demands

$$p_3^2 = p_1^2 + p_2^2 + 2 p_1 p_2 \cos\theta.$$

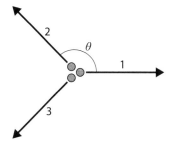

Figure B.1.

The decay rate doesn't simplify much more unless we place additional constraints on the masses of the particles and make a few assumptions. For many realistic systems, $m_2 \simeq 0$ (as might be the case for a weak decay involving a neutrino):

$$\Gamma = S \iint \frac{|\mathcal{A}|^2}{2(4\pi)^3 M} \frac{M^2 + m_1^2 - m_3^2 - 2ME_1}{E_1(M - E_1 + p_1\cos\theta)^2} p_1^2 dp_1 d(\cos\theta). \tag{B.4}$$

This expression, in turn, can't be simplified much further, since the scattering amplitude is likely to be a function of all three outgoing energies, which are in turn a function of the scattering angle

$$E_2 = \frac{M^2 + m_1^2 - m_3^2 - 2ME_1}{2p_1 \cos\theta + 2M - 2E_1}, \tag{B.5}$$

and

$$E_3 = M - E_1 - E_2.$$

However, we can consider the case for which one of the particles absorbs virtually all the energy of the decay, $m_3 \simeq M$, as is the case for neutron decay and other beta decay processes. Defining

$$\Delta E \equiv M - m_3,$$

and considering only systems wherein

$$m_1 < E_1 \ll M,$$

we may approximate the decay rate as

$$\Gamma = S \iint \frac{|\mathcal{A}|^2}{(4\pi)^3} \frac{\Delta E - E_1}{M^2} \sqrt{E_1^2 - m_1^2} \, dE_1 d(\cos\theta). \tag{B.6}$$

The angle θ between the two "light" particles in the decay does not play an explicit role in the rate, though it might enter the amplitude.

SCATTERING

The general form for a scattering cross section is

$$\sigma = S \int |\mathcal{A}|^2 (2\pi)^4 \delta\,(\Delta p)\, \frac{1}{4\sqrt{(p_1 \cdot p_2)^2 - (m_1 m_2)^2}} \prod \left(\frac{d^3 p_i}{(2\pi)^3 2\,E_i} \right), \tag{B.7}$$

where $\mathcal{A}$ is the amplitude of the scatter and is given by the Feynman rules.

Scattering is particularly simple in the case where there are two incoming particles of identical mass, and two outgoing particles, also of identical mass. In that case, the cross section may be written as

$$\frac{d\sigma}{d\Omega} = S \frac{|\mathcal{A}|^2}{(16\pi)^2\, E_0^2} \frac{|p_F|}{|p_I|}, \tag{B.8}$$

where E_0 is the energy of each particle in the system, and p_F and p_I are the magnitudes of the outgoing and incoming momenta, respectively, satisfying

$$E_0^2 = m_I^2 + p_I^2 = m_F^2 + p_F^2.$$

Appendix C | Feynman Rules

The following rules are developed in the text and are used to compute a scattering amplitude $\mathcal{A}$ for a Feynman diagram.

PRELIMINARIES
Regardless of the quantum field theory, a few initial steps will set the scene for QFT calculations.

1. Compute and simplify (as much as possible), the kinematic components of the decay rate (equation 7.12) or cross section (equation 7.19).
2. Compute the degeneracy term $1/n!$ for identical types of outgoing particles.
3. Draw *all* possible Feynman diagrams to as high an order as precision requirements demand. Each quantum field theory has a fundamental vertex which can be inferred directly from the classical interaction Lagrangian. A valid Feynman diagram can be twisted and rotated with abandon, so long as the appropriate number (and type) of particles go into and the appropriate number go out of each vertex.
4. Label each ingoing particle p_1, p_2, and spin, s_1, s_2, ... and each outgoing particle p_{n+1}, p_{n+2}, and similarly with spin. For each draw an arrow (or assume a convention) which indicates that the 4-momentum of the particles are going into or out of a vertex. Label each vertex with an index $\mu, \nu, \ldots$ to correspond to the 4-momentum component of the mediating boson.

THE AMPLITUDE
Take the product of all relevant terms from the diagram.

1 External Lines.

Particle	Contribution	Representation
Outgoing scalar	1	
Incoming scalar	1	

Particle	Contribution	Representation
Outgoing fermion	$\bar{u}_s(p)$	
Incoming fermion	$u_s(p)$	
Outgoing antifermion	$v_s(p)$	
Incoming antifermion	$\bar{v}_s(p)$	
Outgoing photon	$\epsilon^{(r)\mu*}$	
Incoming photon	$\epsilon^{(r)\mu}$	

MODIFICATIONS AND REMINDERS

i. The Greek index on the external bosons is the same as the index on the relevant vertex.

ii. In strong interactions, the color vector of ingoing quarks is labeled c_i, and of outgoing quarks, $c_j^{\dagger}$.

2 Vertex Factors.

Vertex	Contribution $\times \left[(2\pi)^4 \delta(\Delta p) \right]$	Representation
Toy scalar theory	$-i\lambda$	
Electromagnetism	$iq_e \gamma^{\mu}$	
Charged current (leptons)	$-i\dfrac{g_w}{2\sqrt{2}}\gamma^{\mu}(1-\gamma^5)$	ν_l, l^-, W^-
Charged current (quarks)	$-i\dfrac{g_w}{2\sqrt{2}}\gamma^{\mu}(1-\gamma^5)V_{ij}$	u_i, d_j, W^-

Vertex	Contribution$\times \left[(2\pi)^4 \delta(\Delta p) \right]$	Representation
Charged current (quarks)	$-i\dfrac{g_W}{2\sqrt{2}}\gamma^\mu(1-\gamma^5)V_{ij}^*$	
Neutral current	$-ig_W\gamma^\mu(c_V - c_A\gamma^5)$	
Higgs boson (charged)	$ig_W M_W g_{\mu\nu}$	
Higgs boson (neutral)	$ig_W\dfrac{M_Z}{\cos\theta_W}g_{\mu\nu}$	
2 Higgs boson (charged)	$i\dfrac{g_W^2}{2}g_{\mu\nu}$	
2 Higgs boson (neutral)	$i\dfrac{g_W^2}{2\cos^2\theta_W}g_{\mu\nu}$	
Higgs fermion	$-\dfrac{i}{2}\dfrac{m_f}{M_W}g_W$	
Strong (quark)	$-\displaystyle\sum_{k=1}^{8}\dfrac{ig_S}{2}\lambda_k\gamma^\mu$	
Strong (three-gluon)	$-g_S f^{jkl}\left[g_{\mu\nu}(q_1-q_2)_\lambda + g_{\nu\lambda}(q_2-q_3)_\mu + g_{\lambda\mu}(q_3-q_1)_\nu \right]$	
Strong (four-gluon)	$-ig_S^2\left[f^{kli}f^{mni}(g_{\mu\lambda}g_{\nu\rho} - g_{\mu\rho}g_{\nu\lambda}) + f^{kni}f^{lmi}(g_{\mu\nu}g_{\lambda\rho} - g_{\mu\lambda}g_{\nu\rho}) + f^{kmi}f^{nli}(g_{\mu\rho}g_{\nu\lambda} - g_{\mu\nu}g_{\lambda\rho}) \right]$	

MODIFICATIONS AND REMINDERS

i. The conservation of 4-momentum term $\delta(\Delta p)$ represents the difference of the incoming and outgoing particles. For internal lines, especially, the "direction" of the 4-momentum (into or out of the vertex), while arbitrary, must be clearly specified.

ii. For vertices involving vector bosons, the Greek index refers to the outgoing boson.

iii. Follow each fermionic line backward and write the terms from left to right. Doing so will produce scalar combinations of the form

$$[\bar{u}(2)\gamma^\mu u(1)].$$

iv. For weak interactions, u is an up-type quark (up, charmed, top), d is a down-type quark, and V_{ud} is the element of the CKM matrix.

v. For neutral current interactions,

$$c_V = T_{3,:} - Q\sin^2\theta_W \qquad c_A = T_{3,L},$$

where $T_{3,L}$ is the weak isospin for a left-handed particle, and Q is the electric charge.

vi. For each quark-gluon vertex, follow the quark lines backward to collect terms like

$$[c_2^\dagger \lambda_k c_1].$$

vii. For three-boson vertices, the momenta are assumed to point inward. Outgoing momenta should get a minus sign in the amplitude contribution.

3 *Propagators.* For each internal line, label the momentum (and include the direction with an arrow) q_i, and integrate over all possible values.

Propagator	Contribution	Representation
Scalar	$\int \dfrac{d^4 q_i}{(2\pi)^4} \dfrac{i}{q_i^2 - m^2}$	$q_i \longrightarrow$
Fermions	$i\int \dfrac{d^4 q_i}{(2\pi)^4} \dfrac{(\not{q}_i + m_e \mathbb{I})}{q_i^2 - m_e^2}$	$q_i \longrightarrow$
Photons	$-i\int \dfrac{d^4 q_i}{(2\pi)^4} \dfrac{g_{\mu\nu}}{q_i^2}$	$q_i \longrightarrow$
$W^\pm$	$-i\int \dfrac{d^4 q_i}{(2\pi)^4} \dfrac{\left(g_{\mu\nu} - \frac{q_{i,\mu} q_{i,\nu}}{M_W^2}\right)}{q_i^2 - M_W^2}$	$q_i \longrightarrow$
Z^0	$-i\int \dfrac{d^4 q_i}{(2\pi)^4} \dfrac{\left(g_{\mu\nu} - \frac{q_{i,\mu} q_{i,\nu}}{M_Z^2}\right)}{q_i^2 - M_Z^2}$	$q_i \longrightarrow$
Gluons	$-i\int \dfrac{d^4 q_i}{(2\pi)^4} \dfrac{g_{\mu\nu}\delta^{jk}}{q_i^2}$	$q_i \longrightarrow$

MODIFICATIONS AND REMINDERS

i. For each internal fermionic loop, multiply by -1, and take the trace of the resulting term.

ii. For bosonic loops use the "upstairs" version of the metric $g^{\mu\nu}$ to contract the indices.

4 *Cancel the Delta Function.* Multiply the particle, vertex, and propagator contributions together, and integrate over virtual particles $\{q_i\}$. The result yields

$$-i(2\pi)^4 \mathcal{A}\delta(\Delta p)$$

for the overall process, from which you may read out the amplitude.

5 *Antisymmetrization.* Include a minus sign between diagrams that differ only in the interchange of two incoming or outgoing fermions.

Appendix D | Groups

All groups, continuous and discrete, have the following requirements.

1. *Closure.* If g_i and g_j are part of the group, then

$$g_i \circ g_j = g_k \tag{D.1}$$

   must be as well.

2. *Identity.* There must exist an element $\mathbf{I}$ such that

$$g_i \circ \mathbf{I} = \mathbf{I} \circ g_i = g_i \tag{D.2}$$

   for all members of the group.

3. *Inverse.* For every element g_i there exists an inverse $g_j = (g_i)^{-1}$ such that

$$g_j \circ g_i = g_i \circ g_j = \mathbf{I}$$

4. *Associativity.*

$$g_i \circ (g_j \circ g_k) = (g_i \circ g_j) \circ g_k.$$

Most of the groups encountered in the Standard Model are continuous Lie groups whose elements may be written in the form

$$\mathbf{M} = e^{-i\vec{\theta}\cdot\vec{\mathbf{X}}}, \tag{D.3}$$

where the generators are $\mathbf{X}_i$. The structure constants of a group may be written as

$$[\mathbf{X}_i, \mathbf{X}_j] = \sum_k 2i f_{ijk} \mathbf{X}_k, \tag{D.4}$$

where for an Abelian group, all the structure constants are necessarily zero.

The commutation relations are extremely important. Supposing X and Y are two noncommuting operators or functions, the Baker-Campbell-Hausdorff relation yields

$$e^X e^Y = e^{X+Y+\frac{1}{2}[X,Y]}, \tag{D.5}$$

and, importantly in non-Abelian gauge theories:

$$e^X Y e^{-X} = Y + [X, Y]. \tag{D.6}$$

SOME IMPORTANT LIE GROUPS IN THE STANDARD MODEL

- **Special (S)**: The determinant of the matrix is +1.
- **Orthogonal (O)**: The inverse is equal to the transpose, $\mathbf{M}\mathbf{M}^T = \mathbf{I}$
- **Unitary (U)**: The inverse of a complex-valued $\mathbf{M}$ is equal to the Hermitian transpose, $\mathbf{M}\mathbf{M}^\dagger = \mathbf{I}$

So:

- U(1) has 1 generator, $\mathbf{X} = 1$, representing a phase

  $\mathbf{M} = e^{-i\theta}$.

- SU(N) is reducible to an $N \times N$ matrix with $N^2 - 1$ generators:

  - SU(2): The three generators are the Pauli spin matrices.
  - SU(3): The eight generators are the Gell-Mann matrices.

- SO(N) is reducible to a real-valued $N \times N$ matrix with $N(N-1)/2$ generators and describes rotations in N dimensions.

Bibliography

[1] G. Aad and the ATLAS Collaboration. Observation of a new particle in the search for the standard model higgs boson with the ATLAS detector at the LHC. *Phys. Lett. B*, 716(1): 1–29, 2012.

[2] G. Aad and the ATLAS Collaboration. Summary of the ATLAS experiment's sensitivity to supersymmetry after LHC Run 1 – interpreted in the phenomenological MSSM. 2015.

[3] G. Aartsen, M. M. Ackermann, J. Adams, et al. Atmospheric and astrophysical neutrinos above 1 TeV interacting in IceCube. *Phys. Rev. D*, 91: 022001, January 2015.

[4] S. Abachi, B. Abbott, M. Abolins, et al. Search for high mass top quark production in pp collisions at s = 1.8 TeV. *Phys. Rev. Lett.*, 74: 2422–2426, March 1995.

[5] F. Abe, H. Akimoto, A. Akopian, et al. Observation of top quark production in $\overline{p}p$ collisions with the Collider Detector at Fermilab. *Phys. Rev. Lett.*, 74: 2626–2631, April 1995.

[6] K. Abe, K. Abe, R. Abe, et al. Observation of large CP violation in the neutral B meson system. *Phys. Rev. Lett.*, 87: 091802, August 2001.

[7] K. Abe, J. Adam, H. Aihara, et al. Measurements of neutrino oscillation in appearance and disappearance channels by the T2K experiment with 6.6×10^{20} protons on target. *Phys. Rev. D*, 91: 072010, April 2015.

[8] K. Abe, Y. Hayato, T. Iida, et al. Evidence for the appearance of atmospheric tau neutrinos in Super-Kamiokande. *Phys. Rev. Lett.*, 110: 181802, May 2013.

[9] S. Abe, T. Ebihara, S. Enomoto, et al. Precision measurement of neutrino oscillation parameters with KamLAND. *Phys. Rev. Lett.*, 100: 221803, June 2008.

[10] C. V. Achar, M.G.K. Menon, V. S. Narasimham, et al. Detection of muons produced by cosmic ray neutrinos deep underground. *Phys. Lett.*, 18: 196–199, August 1965.

[11] P. Adamson, C. Andreopoulos, K. E. Arms, et al. Study of muon neutrino disappearance using the Fermilab Main Injector neutrino beam. *Phys. Rev. D*, 77: 072002, April 2008.

[12] N. Agafonova, A. Aleksandrov, O. Altinok, et al. Observation of a first candidate event in the OPERA experiment in the CNGS beam. *Phys. Lett. B*, 691(3): 138–145, 2010.

[13] S. Alekhin, A. Djouadi, and S. Moch. The top quark and Higgs boson masses and the stability of the electroweak vacuum. *Phys. Lett. B*, 716(1): 214–219, 2012; revised October 2013, arXiv:1207.0980v3.

[14] C. D. Anderson. The positive electron. *Phys. Rev.*, 43: 491–494, March 1933.

[15] L. Anderson, É. Aubourg, S. Bailey, et al. The clustering of galaxies in the SDSS-III Baryon Oscillation Spectroscopic Survey: Baryon acoustic oscillations in the Data Releases 10 and 11 galaxy samples. *MNRAS*, 441: 24–62, June 2014.

[16] Philip W. Anderson. Plasmons, gauge invariance, and mass. *Phys. Rev.*, 130: 439, 1963.

[17] G. Arnison, A. Astbury, B. Aubert, et al. Experimental observation of isolated large transverse energy electrons with associated missing energy at s = 540 GeV. *Phys. Lett. B*, 122(1): 103–116, 1983.

[18] Y. Ashie, J. Hosaka, K. Ishihara, et al. Evidence for an oscillatory signature in atmospheric neutrino oscillations. *Phys. Rev. Lett.*, 93: 101801, September 2004.

[19] Abhay Ashtekar. New variables for classical and quantum gravity. *Phys. Rev. Lett.*, 57: 2244–2247, November 1986.

[20] J. J. Aubert, U. Becker, P. J. Biggs, et al. Experimental observation of a heavy particle. *J. Phys. Rev. Lett.*, 33: 1404–1406, December 1974.

[21] M. Auger, D. J. Auty, P. S. Barbeau, et al. Search for neutrinoless double-beta decay in ^{136}Xe with EXO-200. *Phys. Rev. Lett.*, 109(3): 032505, July 2012.

[22] J.-E. Augustin, A. M. Boyarski, M. Breidenbach, et al. Discovery of a narrow resonance in e^+e^- annihilation. *Phys. Rev. Lett.*, 33: 1406–1408, December 1974.

[23] John N. Bahcall, Neta A. Bahcall, and Giora Shaviv. Present status of the theoretical predictions for the ^{37}Cl solar-neutrino experiment. *Phys. Rev. Lett.*, 20: 1209–1212, May 1968.

[24] John N. Bahcall and Raymond Davis Jr. The evolution of neutrino astronomy. *PASP*, 112: 429, 2000.

[25] M. Banner, R. Battiston, P. Bloch, et al. Observation of single isolated electrons of high transverse momentum in events with missing transverse energy at the CERN p̄p collider. *Phys. Lett. B*, 122(56): 476–485, 1983.

[26] D. P. Barber, U. Becker, H. Benda, et al. Discovery of three-jet events and a test of quantum chromodynamics at PETRA. *Phys. Rev. Lett.*, 43: 830–833, September 1979.

[27] John D. Barrow and Frank J. Tipler. *The Anthropic Cosmological Principle*. Oxford: Oxford University Press, 1988.

[28] J. Beringer et al. Review of particle physics (RPP). *Phys. Rev. D*, 86: 010001, 2012.

[29] Hans Bethe. Energy production in stars. *Phys. Rev.*, 55: 434, March 1939.

[30] E. D. Bloom, D. H. Coward, H. DeStaebler, et al. High-energy inelastic $e - p$ scattering at 6° and 10°. *Phys. Rev. Lett.*, 23: 930–934, October 1969.

[31] Satyendra Nath Bose. Plancks Gesetz und Lichtquantenhypothese. *Z. Physik*, 26: 178–181, 1924.

[32] M. Breidenbach, J. I. Friedman, H. W. Kendall, et al. Observed behavior of highly inelastic electron-proton scattering. *Phys. Rev. Lett.*, 23: 935–939, October 1969.

[33] Nicola Cabibbo. Unitary symmetry and leptonic decays. *Phys. Rev. Lett.*, 10: 531, 1963.

[34] Nigel Calder. *Key to the Universe: Report on the New Physics*. New York: Penguin, 1977.

[35] H.B.G. Casimir. On the attraction between two perfectly conducting plates. *Proc. Kon. Nederland. Akad. Wetensch.*, B51: 793–795, 1948.

[36] J. Chadwick. Possible existence of a neutron. *Nature*, 129: 312, February 1932.

[37] S. Chatrchyan and the CMS Collaboration. Observation of a new boson at a mass of 125 GeV with the CMS experiment at the LHC. *Phys. Lett. B*, 716(1): 30–61, 2012.

[38] J. H. Christenson, J. W. Cronin, V. L. Fitch, and R. Turlay. Evidence for the 2π decay of the K_2^0 meson. *Phys. Rev. Lett.*, 13: 138–140, July 1964.

[39] Rudolf Clausius. On the moving force of heat and the laws regarding the nature of heat itself which are deducible therefrom. *Ann. Phys.*, 79: 368–397, 500–524, 1850. English translation by *Philosophical Magazine* and *Journal of Science*, July 1851.

[40]   C. L. Cowan Jr., F. Reines, F. B. Harrison, H. W. Kruse, and A. D. McGuire. Detection of the free neutrino: A confirmation. *Science*, 124: 103–104, July 1956.

[41]   G. Danby, J-M. Gaillard, K. Goulianos, et al. Observation of high-energy neutrino reactions and the existence of two kinds of neutrinos. *Phys. Rev. Lett.*, 9: 36–44, July 1962.

[42]   Raymond Davis, Don S. Harmer, and Kenneth C. Hoffman. Search for neutrinos from the sun. *Phys. Rev. Lett.*, 20: 1205–1209, 1968.

[43]   Savas Dimopoulos and Howard Georgi. Softly broken supersymmetry and SU(5). *Nucl. Phys. B*, 193(1): 150–162, 1981.

[44]   P.A.M. Dirac. On the theory of quantum mechanics. *Proc. R. Soc. London, Ser. A*, 112: 661–677, October 1926.

[45]   P.A.M. Dirac. The quantum theory of emission and absorption of radiation. *Proc. R. Soc. London, Ser. A*, 114: 243–265, March 1927.

[46]   P.A.M. Dirac. The quantum theory of the electron. *Proc. R. Soc. London, Ser. A*, 117: 610–624, February 1928.

[47]   P.A.M. Dirac. Quantised singularities in the quantum field. *Proc. R. Soc. London, Ser. A*, 133: 60–72, September 1931.

[48]   P.A.M. Dirac. A new notation for quantum mechanics. *Math. Proc. Cambridge Philos. Soc.*, 35: 416–418, 1939.

[49]   DONUT Collaboration, K. Kodama, N. Ushida, et al. Observation of tau neutrino interactions. *Phys. Lett. B*, 504: 218–224, April 2001.

[50]   F. J. Dyson. The radiation theories of Tomonaga, Schwinger, and Feynman. *Phys. Rev.*, 75: 486–502, February 1949.

[51]   Freeman J. Dyson. Energy in the universe. *Scientific American*, 225: 50–59, 1971.

[52]   A. Einstein. On a heuristic point of view about the creation and conversion of light. *Ann. Phys.*, 322: 132–148, 1905.

[53]   Albert Einstein. On the electrodynamics of moving bodies. *Ann. Phys.*, 17(10): 891–921, 1905.

[54]   Albert Einstein. Die Feldgleichungen der Gravitation. *Sitz. d. Preuss. Akad. d. Wiss. zu Berlin*, pp. 844–847, 1915.

[55]   F. Englert and R. Brout. Broken symmetry and the mass of gauge vector mesons. *Phys. Rev. Lett.*, 13: 321–323, August 1964.

[56]   G. F. Hinshaw et al. Nine-year wilkinson microwave anisotropy probe (wmap) observations: Cosmological parameter results. *Astrophys. J. Suppl.*, 208: 19H, 2013.

[57]   L. D. Faddeev and V. N. Popov. Feynman diagrams for the Yang-Mills field. *Phys. Lett. B*, 25: 29, 1967.

[58]   P. Fayet. Spontaneously broken supersymmetric theories of weak, electromagnetic and strong interactions. *Phys. Lett. B*, 69: 489–494, 1977.

[59]   Enrico Fermi. On the quantization of the monatomic ideal gas. *Rend. Lincei*, 3: 145–149, 1926. English translation by Aberto Zannoni, arXiv:cond-mat/9912229.

[60]   Enrico Fermi. *Nuclear Physics*. Chicago: University of Chicago Press, 1950.

[61]   Richard Feynman. Space-time approach to quantum electrodynamics. *Phys. Rev.*, 76: 769–789, 1949.

[62]   Richard Feynman. The development of the space-time view of quantum electrodynamics. In *Nobel Lecture*, 1965. Cited at http://www.nobelprize.org/nobel_prizes/physics/laureates/1965/feynman-lecture.html.

[63]   Richard P. Feynman. *QED: The Strange Theory of Light and Matter*. Princeton, NJ: Princeton University Press, 1985.

[64]   Richard P. Feynman. The reason for antimatter. In *The 1986 Dirac Memorial Lectures*. Cambridge: Cambridge University Press, 1987.

[65] Richard P. Feynman, Robert B. Leighton, and Matthew L. Sands. *The Feynman Lectures on Physics*, vols. 1,2,3. Boston: Addison-Wesley, 1970.

[66] George Francis FitzGerald. The aether and the earth's atmosphere. *Science*, 13(328): 390, 1889.

[67] Benjamin Franklin. From Benjamin Franklin to Peter Collinson, 25 May 1747. http://founders.archives.gov/documents/Franklin/01-03-02-0059.

[68] Craig G. Fraser. Théorie des fonctions analytiques. Chapter 19 in *Landmark Writings in Western Mathematics, 1640–1940*, ed. I. Grattan-Guiness. Amsterdam: Elsevier, 2005.

[69] A. Friedmann. Über die Krümmung des Raumes. *Z. Phys.*, 10: 377–386, 1922.

[70] Y. Fukuda, T. Hayakawa, E. Ichihara, et al. Evidence for oscillation of atmospheric neutrinos. *Phys. Rev. Lett.*, 81: 1562–1567, August 1998.

[71] Galileo Galilei. *Dialogue Concerning the Two Chief World Systems* (1635). Trans. S. Drake. Oakland: University of California Press, 1953.

[72] Galileo Galilei. *[Discourses and Mathematical Demonstrations one] Two New Sciences* (1638). Trans. and ed. S. Drake. Madison: University of Wisconsin Press, 1974.

[73] Évariste Galois. Manuscrits de Évariste Galois. [Papiers et écrits mathématiques.] Publiés par Jules Tannery. http://name.umdl.umich.edu/AAN9280.0001.001. Accessed: November 26, 2014.

[74] Murray Gell-Mann. Symmetries of baryons and mesons. *Phys. Rev.*, 125: 1067–1084, 1962.

[75] Murray Gell-Mann. A schematic model of baryons and mesons. *Phys. Rev. Lett.*, 8: 214–215, 1964.

[76] Murray Gell-Mann. *The Quark and the Jaguar: Adventures in the Simple and the Complex.* New York: Henry Holt, 1995.

[77] Howard Georgi and Sheldon Glashow. Unity of all elementary-particle forces. *Phys. Rev. Lett.*, 32: 438, 1974.

[78] Sheldon Glashow. Partial symmetries of weak interactions. *Nucl. Phys.*, 22, 1961.

[79] Jeffrey Goldstone. Field theories with superconductor solutions. *Nuovo Cimento*, 19: 154–164, 1960.

[80] David Griffiths. *Introduction to Elementary Particles,* 2nd ed. New York: Wiley-VCH, 2008.

[81] D. J. Gross. Twenty five years of asymptotic freedom. *Nucl. Phys. Proc. Suppl.*, 74: 426–446, March 1999.

[82] David J. Gross and Frank Wilczek. Ultraviolet behavior of non-Abelian gauge theories. *Phys. Rev. Lett.*, 30: 1343–1346, June 1973.

[83] Franz Gross. *Relativistic Quantum Mechanics and Field Theory.* New York: Wiley-VCH, 1993.

[84] G. S. Guralnik, C. R. Hagen, and T. W. Kibble. Global conservation laws and massless particles. *Phys. Rev. Lett.*, 13: 585–587, November 1964.

[85] A. H. Guth. Inflationary universe: A possible solution to the horizon and flatness problems. *Phys. Rev. D*, 23: 347–356, January 1981.

[86] S. W. Hawking. Black hole explosions? *Nature*, pp. 30–31, 1974.

[87] W. Heisenberg. Über den anschaulichen Inhalt der quantentheoretischen Kinematik und Mechanik. *Z. Phys.*, 43: 172–198, March 1927.

[88] P. W. Higgs. Broken symmetries, massless particles and gauge fields. *Phys. Lett.*, 12: 132–133, September 1964.

[89] D. Hilbert, J. von Neumann, and L. Nordheim. Über die Grundlagen der Quantenmechanik. *Math. Ann.*, 98(1): 1–30, 1928.

[90] E. Hubble. A relation between distance and radial velocity among extra-galactic nebulae. *Proc. Natl. Acad. Sci. USA*, 15: 168–173, March 1929.

[91] Johannes Kepler. *De Cometis Libelli Tres.* Avgvstae Vindelicorvm, A. Apergeri, 1619.

[92] Makoto Kobayashi and Toshihide Maskawa. *CP*-violation in the renormalizable theory of weak interaction. *Prog. Theor. Phys.*, 49: 652–657, 1973.

[93]   Helena Kolešová and Michal Malinský. Proton lifetime in the minimal $SO(10)$ GUT and its implications for the LHC. *Phys. Rev. D*, 90: 115001, December 2014.

[94]   Joseph Louis Lagrange. *Théorie des Fonctions Analytiques.* L'Imprimerie de La République, 1799.

[95]   S. K. Lamoreaux. Demonstration of the Casimir force in the 0.6 to 6 μm range. *Phys. Rev. Lett.*, 78: 5, 1997.

[96]   Lev Landau. On the quantum theory of fields. In *Niels Bohr and the Development of Physics.* ed. Wolfgang Pauli. Oxford: Pergamon Press, 1955.

[97]   C.M.G. Lattes, H. Muirhead, G.P.S. Occhialini, and C. F. Powell. Processes involving charged mesons. *Nature*, 159: 694–697, May 1947.

[98]   Leon Lederman. The upsilon particle. *Scientific American*, 239: 72, 1978.

[99]   Tsung-Dao Lee and Chen-Ning Yang. Question of parity conservation in weak interactions. *Phys. Rev.*, 106: 1371, 1956.

[100] J. P. Lees, V. Poireau, V. Tisserand, et al. Observation of time-reversal violation in the B^0 meson system. *Phys. Rev. Lett.*, 109: 211801, November 2012.

[101] G. Lemaître. Un Univers homogène de masse constante et de rayon croissant rendant compte de la vitesse radiale des nébuleuses extra-galactiques. *Annal. Soc. Sci. de Bruxelles*, 47: 49–59, 1927.

[102] Marius Sophus Lie and Friedrich Engel. *Theorie der Transformationsgruppen (Theory of Transformation Groups)*, 3 vols. Leipzig: B. G. Teubner, 1888–1893.

[103] A. D. Linde. Chaotic inflation. *Phys. Lett. B*, 129: 177–181, September 1983.

[104] Hendrik Antoon Lorentz. The relative motion of the earth and the aether. *Zittingsverlag Akad. v. Wet.*, 1: 74–79, 1892.

[105] Gerhart Lüders. On the equivalence of invariance under time reversal and under particle-antiparticle conjugation for relativistic field theories. *R. Dan. Acad. Sci. Lett.*, 5: 1–17, 1954.

[106] Michael Sean Mahoney. *The Mathematical Career of Pierre de Fermat, 1601–1665.* Princeton, NJ: Princeton University Press, 2nd ed., 1994.

[107] Ettore Majorana. Teoria simmetrica dell'elettrone e del positrone (A symmetric theory of electrons and positrons). *Il Nuovo Cimento*, 14: 171–184, 1937.

[108] Z. Maki, M. Nakagawa, and S. Sakata. Remarks on the unified model of elementary particles. *Prog. Theor. Phys.*, 28: 870–880, November 1962.

[109] Albert A. Michelson and Edward W. Morley. On the relative motion of the earth and the luminiferous ether. *Am. J. Sci.*, 34: 333–345, 1887.

[110] Stanislav P. Mikheyev and Alexei Yu. Smirnov. Resonance oscillations of neutrinos in matter. *Sov. Phys. Usp.*, 30: 759–790, 1987.

[111] Hermann Minkowski. The basic equations for the electromagnetic processes in moving bodies. *Nachr. v. d. Ges. d. Wiss. zu Göttingen, Math.-Phys. Klasse*, pp. 53–111, 1908.

[112] Hermann Minkowski. Space and time. *Phys. Z.*, 10: 75–88, 1908.

[113] Yoichiro Nambu. Quasi-particles and gauge invariance in the theories of superconductivity. *Phys. Rev.*, 117: 648–663, February 1960.

[114] Seth H. Neddermeyer and Carl D. Anderson. Note on the nature of cosmic-ray particles. *Phys. Rev.*, 51: 884–886, May 1937.

[115] Dwight E. Neuenschwander. *Emmy Noether's Wonderful Theorem.* Battimore, MD: Johns Hopkins University Press, 2010.

[116] H. Nishino, S. Clark, K. Abe, et al. Search for proton decay via $p \to e^+\pi^0$ and $p \to \mu^+\pi^0$ in a large water Cherenkov detector. *Phys. Rev. Lett.*, 102: 141801, April 2009.

[117] Emmy Noether. Invariante variationsprobleme. *Nachr. v. d. Ges. d. Wiss. zu Göttingen, Math.-Phys. Klasse*, pp. 235–257, 1918. Trans. M. A. Tavel, http://arxiv.org/abs/physics/0503066v1.

[118] B. Odom, D. Hanneke, B. D'Urso, and G. Gabrielse. New measurements of the electron magnetic moment using a one-electron quantum cyclotron. *Phys. Rev. Lett.*, 97: 030801, July 2006.

[119] K. A. Olive and Particle Data Group. Review of particle physics. *Chin. Phys. C*, 38(9): 090001, August 2014.

[120] Wolfgang Pauli. Uber den Zusammenhang des Abschlusses der Elektronengruppen im Atom mit der Komplexstruktur der Spektren. *Z. Phys.*, 31: 765, 1925. Translated as On the connection between the completion of electron groups in an atom with the complex structure of spectra.

[121] Wolfgang Pauli. Letter to Tübingen conference participants. December 1930. English translation in The idea of the neutrino, by Laurie M. Brown, Phys. Today, p. 27, September 1978.

[122] Wolfgang Pauli. Relativistic field theories of elementary particles. *Rev. Mod. Phys.*, 13: 203–32, 1941.

[123] Wolfgang Pauli. *Niels Bohr and the Development of Physics; Essays Dedicated to Niels Bohr on the Occasion of His Seventieth Birthday*. Oxford: Pergamon Press, 1955.

[124] John A. Peacock. *Cosmological Physics*. Cambridge: Cambridge University Press, 1998.

[125] Roberto D. Peccei and Helen R. Quinn. CP conservation in the presence of pseudoparticles. *Phys. Rev. Lett.*, 38: 1440, 1977.

[126] Olaf Pedersen. *Early Physics and Astronomy: A Historical Introduction*. Cambridge: Cambridge University Press, 1974.

[127] M. L. Perl, G. S. Abrams, A. M. Boyarski, et al. Evidence for anomalous lepton production in e^+e^- annihilation. *Phys. Rev. Lett.*, 35: 1489–1492, December 1975.

[128] Michael E. Peskin and Dan V. Schroeder. *An Introduction to Field Theory*. Boulder, CO: Westview Press, 1995.

[129] Max Planck. On the law of distribution of energy in the normal spectrum. *Ann. Phys.*, 309: 553, 1901.

[130] Planck Collaboration, P.A.R. Ade, N. Aghanim, et al. Planck intermediate results. XVI. Profile likelihoods for cosmological parameters. *Astron. Astrophys.*, 566: A54, June 2014.

[131] Henri Poincaré. Toward the dynamics of the electron. *Rend. Circ. Mat. Palermo*, 21: 129–176, 1906.

[132] H. D. Politzer. Reliable perturbative results for strong interactions? *Phys. Rev. Lett.*, 30: 1346–1349, June 1973.

[133] B. Pontecorvo. Mesonium and anti-mesonium. *J. Exp. Theor. Phys. (JETP)*, 33: 549–551, 1957.

[134] B. Pontecorvo. Neutrino experiments and the problem of conservation of leptonic charge. *Sov. J. Exp. Theor. Phys.*, 26: 984, May 1968.

[135] L. Randall and C. Csa'ki. The doublet-triplet splitting problem and Higgses as pseudogoldstone bosons. ArXiv:hep-ph/9508208, August 1995.

[136] M. J. Rees. Numerical coincidences and 'tuning' in cosmology. *Astrophys. Space Sci.*, 285(2): 375–388, 2003.

[137] F. Reines, M. F. Crouch, T. L. Jenkins, et al. Evidence for high-energy cosmic-ray neutrino interactions. *Phys. Rev. Lett.*, 15: 429–433, August 1965.

[138] Michael Riordan. *The Hunting of the Quark: A True Story of Modern Physics*. New York: Simon & Schuster, 1987.

[139] G. D. Rochester and C. C. Butler. Evidence for the existence of new unstable elementary particles. *Nature*, 160: 855–857, December 1947.

[140] B. Rossi and D. B. Hall. Variation of the rate of decay of mesotrons with momentum. *Phys. Rev.*, 59: 223–228, February 1941.

[141] Carlo Rovelli and Lee Smolin. Loop space representation of quantum general relativity. *Nucl. Phys. B*, 331(1): 80–152, 1990.

[142] Ernest Rutherford. The scattering of α and β particles by matter and the structure of the atom. *Philosoph. Mag.*, 21: 669–688, 1911.

[143] Ernest Rutherford. Collision of α particles with light atoms. IV. An anomalous effect in nitrogen. *Philosoph. Mag.*, 37: 581, 1919.

[144] A. D. Sakharov. Special Issue: Violation of CP invariance, C asymmetry, and baryon asymmetry of the universe. *Sov. Phys. Usp.*, 34: 392–393, May 1991.

[145] J. J. Sakurai. *Modern Quantum Mechanics.* Boston: Addison-Wesley, 1967.

[146] Abdus Salam. Elementary particle theory. *Proc. 8th Nobel Symposium*, ed. N. Svartholm. Stockholm: Almqvist and Wiksell, 1968.

[147] Erhard Schmidt. Über die Auflösung linearer Gleichungen mit unendlich vielen Unbekannten. *Rend. Circ. Mat. Palermo*, 25(1): 53–77, 1908.

[148] Julian Schwinger. Quantum electrodynamics. I. A covariant formulation. *Phys. Rev.*, 74: 1439, 1948.

[149] Julian Schwinger. Gauge invariance and mass. *Phys. Rev.*, 125: 397, 1961.

[150] F. A. Scott. Energy spectrum of the beta-rays of radium E. *Phys. Rev.*, 48: 391–395, 1935.

[151] Dmitri Skobeltsyn. Über eine neue Art sehr schneller β-Strahlen. *Z. Phys.*, 54: 686, 1929.

[152] M. J. Sparnaay. Attractive forces between flat plates. *Nature*, 180: 334–335, August 1957.

[153] Max Tegmark. On the dimensionality of spacetime. *Class. Quan. Grav.*, 14: L69–L75, 1997.

[154] J. J. Thomson. Cathode rays. *Philosoph. Mag.*, 44: 293, 1897.

[155] Sun-Itiro Tomonaga. On a relativistically invariant formulation of the quantum theory of waves. *Prog. Theor. Phys. (Kyoto)*, 27, 1946.

[156] Steven Weinberg. A model of leptons. *Phys. Rev. Lett.*, 19: 1264–1266, 1967.

[157] Steven Weinberg. Supersymmetry at ordinary energies. Masses and conservation laws. *Phys. Rev. D*, 26: 287–302, July 1982.

[158] Herman Weyl. Eine neue Erweiterung der Relativitätstheorie. *Ann. Phys.*, 59: 101–133, 1919.

[159] Hermann Weyl. *Symmetry.* Princeton, NJ: Princeton University Press, 1952.

[160] Hermann Weyl and H. P. Robertson (trans.). *Theory of Groups and Quantum Mechanics.* Mineola, NY: Dover, 1928.

[161] G. C. Wick. Properties of Bethe-Salpeter wave functions. *Phys. Rev.*, 96: 1124–1134, November 1954.

[162] Frank Wilczek. The future of particle physics. In *Physics in the 21st Century*, 71–96. Singapore: World Scientific, 1997.

[163] Fred L. Wilson. Fermi's theory of beta decay. *Am. J. Phys.*, 36(12): 1150–1160, 1968.

[164] Edward Witten. String theory dynamics in various dimensions. *Nucl. Phys. B*, 443(12): 85–126, 1995.

[165] L. Wolfenstein. Neutrino oscillations in matter. *Phys. Rev. D*, 17: 2369–2374, May 1978.

[166] C. Wu, E. Ambler, R. Hayward, D. Hoppes, and R. Hudson. Experimental test of parity conservation in beta decay. *Phys. Rev.*, 105: 1413–1415, February 1957.

[167] C. N. Yang and R. L. Mills. Conservation of isotopic spin and isotopic gauge invariance. *Phys. Rev.*, 96: 191–195, October 1954.

[168] Hideki Yukawa. On the interaction of elementary particles. *Proc. Math. Soc. Japan.*, 17: 48, 1935.

[169] Pieter Zeeman. The effect of magnetisation on the nature of light emitted by a substance. *Nature*, 55: 347, February 1897.

[170] George Zweig. An SU(3) model for strong interaction symmetry and its breaking. Version 1. *CERN Report*, 8182/TH.401, January 1964.

[171] George Zweig. An SU(3) model for strong interaction symmetry and its breaking. Version 2. *CERN Report*, 8419/TH.412, February 1964.

Index

Note: Page numbers in bold font indicate where a particular term is defined.